# Synthese Library

Studies in Epistemology, Logic, Methodology,
and Philosophy of Science

Volume 424

**Editor-in-Chief**
Otávio Bueno, Department of Philosophy, University of Miami, USA

**Editors**
Berit Brogaard, University of Miami, USA
Anjan Chakravartty, University of Notre Dame, USA
Steven French, University of Leeds, UK
Catarina Dutilh Novaes, VU Amsterdam, The Netherlands

The aim of *Synthese Library* is to provide a forum for the best current work in the methodology and philosophy of science and in epistemology. A wide variety of different approaches have traditionally been represented in the Library, and every effort is made to maintain this variety, not for its own sake, but because we believe that there are many fruitful and illuminating approaches to the philosophy of science and related disciplines.

Special attention is paid to methodological studies which illustrate the interplay of empirical and philosophical viewpoints and to contributions to the formal (logical, set-theoretical, mathematical, information-theoretical, decision-theoretical, etc.) methodology of empirical sciences. Likewise, the applications of logical methods to epistemology as well as philosophically and methodologically relevant studies in logic are strongly encouraged. The emphasis on logic will be tempered by interest in the psychological, historical, and sociological aspects of science.

Besides monographs *Synthese Library* publishes thematically unified anthologies and edited volumes with a well-defined topical focus inside the aim and scope of the book series. The contributions in the volumes are expected to be focused and structurally organized in accordance with the central theme(s), and should be tied together by an extensive editorial introduction or set of introductions if the volume is divided into parts. An extensive bibliography and index are mandatory.

More information about this series at http://www.springer.com/series/6607

Amihud Gilead

# The Panenmentalist Philosophy of Science

From the Recognition of Individual Pure
Possibilities to Actual Discoveries

 Springer

Amihud Gilead
Department of Philosophy
University of Haifa
Mount Carmel, Haifa, Israel

Synthese Library
ISBN 978-3-030-41126-8          ISBN 978-3-030-41124-4    (eBook)
https://doi.org/10.1007/978-3-030-41124-4

This Springer imprint is published by the registered company Springer Nature Switzerland AG.
The registered company address is: Gewerbestrasse 11, 6330 Cham, Switzerland

*To Alina*

# Acknowledgments

I owe a great deal to Otávio Bueno, the editor of Springer's Synthese Library. From the very beginning of my manuscript proposal to Springer Nature, he was dedicated to this book.

Istvan Hargittai immediately recognized my contribution in interpreting the philosophical significance of the discovery of quasicrystals and other scientific discoveries. For years, he has encouraged me to continue in developing my panenmentalist philosophy of science. Without his encouragement, I would not have continued in my efforts in this direction.

The anonymous reviewers of Springer Nature contributed to the improvement of this book with their most helpful comments.

Marion Lupu faithfully edited the English.

I associate the book with the memory of my late parents. After the death of my father, Mordechai (Istvan) Siegler-Gilead, I found among his belongings his matriculation certificate, in which his marks were recorded. To my great surprise, his achievements in philosophy and mathematics were awarded the highest marks. He never told me about this. Instead, his deep curiosity and interest in any scientific discovery and in mathematics have always inspired me. I was captured by his deep intellectual interest. As I was confined to bed for more than a year of my childhood, due to a severe attack of rheumatic heart disease, and to my fascination, my father used to bring me any book I wanted and thus enabled me to spend all my time in reading. Sometimes, he just guessed about my interests and found the necessary books for me, even if this cost him much effort and money. I loved my father most deeply and owe him so much.

My late mother, Bracha Gilead (né Krasny), gave birth to me twice: my biological birth on 13 January 1947 and my second birth which thankfully brought to an end my disastrous hospitalization due to the aforementioned rheumatic fever. Against the advice of the hospital doctors, her maternal instincts instructed her to take me back home, where her endless love together with the devoted care of one of the most eminent doctors in Jerusalem at that time, the late Professor Eli Davis, actually saved my life (though the doctors had reckoned that the probability of my survival was very low). Can the reader imagine how great is the love of a son who was born twice to the same mother?

Ruthie, my late wife, to whom I dedicated most of my previous books, gave me 36 years of a very happy marriage. She followed me devotedly in every important step of my life. Her love and dedication helped me to overcome all the obstacles in my life. She deserved to see the day in which this book appears. Alas, I have remained without her to see this day.

I dedicate this book to Alina, the dearest friend of my heart. Alina has helped me greatly over the most difficult period of my life. Her love, wisdom, devotion, and determination have been a blessing. With her love and support, I succeeded in finishing three books on which I worked for many years, including the present volume. Thanks to her, "after I am waxed old, I have pleasure" (following Genesis 18:12).

# Contents

1   What Are Individual Pure Possibilities and for What
    Are They Good? .......................................... 1

2   How Many Pure Possibilities Are There? Or *Contra* Actualism..... 11

3   A Panenmentalist Reconsideration of the Identity
    of Indiscernibles .......................................... 27

4   Two Kinds of Discovery: An Ontological Account................. 37
    4.1   Some Examples of Two Kinds of Discoveries.............. 37
    4.2   Creation or Invention ................................. 41
    4.3   Conjectures, Hypotheses, or Predictions ................... 42
    4.4   Fictions and Thought-Experiments ...................... 43
    4.5   Conventions ......................................... 43
    4.6   Stipulation ......................................... 44
    4.7   Epistemic Aids and the Discovered Existents ............... 44
    4.8   Calculation and Measurement .......................... 45
    4.9   Abstract or Ideal Entities ............................. 46
    4.10  Possible Existents as Individual Pure Possibilities .......... 48
    4.11  How Is *a priori* Accessibility to Pure Possibilities Possible? ... 51
    4.12  Panenmentalism and Its Uniqueness and Originality.......... 52
    4.13  A Metaphysical Platform .............................. 55
    4.14  Two Kinds of Discovery............................... 56

5   Mathematical Possibilities and Their Discovery ................ 59
    5.1   The Questions........................................ 59
    5.2   Pure Mathematical Entities Are Not Actualities ............. 59
    5.3   Mathematical Entities Are Individual Pure Possibilities ....... 61
    5.4   Individual Pure Possibilities and What Mathematics Really Is .. 67
          5.4.1   Hersh and a Failure in Proving That There
                  Are Square Circles........................... 72
    5.5   An Alternative Indispensability Argument.................. 80

5.6     The *a priori* Epistemic Accessibility to Mathematical
        Pure Possibilities. . . . . . . . . . . . . . . . . . . . . . . . . . . . . . . . . .   83
5.7     Pure Possibilities Are Discoverable, Not Inventible . . . . . . . . . .   84
5.8     Some Examples of Discoveries of Mathematical Pure
        Possibilities . . . . . . . . . . . . . . . . . . . . . . . . . . . . . . . . . . . . .   87
5.9     Mathematical Language, Conceptual Possibilities,
        and the Applicability of Mathematics . . . . . . . . . . . . . . . . . . . .   88
5.10    Is It Valid to Quantify over Mathematical Pure Possibilities? . . .   89
5.11    Proofs and the Reality of Mathematical Pure Possibilities . . . . .   90
5.12    Mathematical Fictions, Creativity, and Discoverable Truths. . . .   91
5.13    Mathematical Discoveries Are of Pure Possibilities,
        Not of Actualities . . . . . . . . . . . . . . . . . . . . . . . . . . . . . . . . .   93
5.14    How Does Panenmentalism Challenge Parsons's Recent
        View on Mathematical Objects? . . . . . . . . . . . . . . . . . . . . . . . .   95
5.15    Conclusion: The Answers to the Abovementioned
        Three Questions. . . . . . . . . . . . . . . . . . . . . . . . . . . . . . . . . . .   97

6    **A Panenmenalist Approach to Molyneux's Problem
     and Some Empirical Findings** . . . . . . . . . . . . . . . . . . . . . . . . . . .   99
6.1     Molyneux's Problem . . . . . . . . . . . . . . . . . . . . . . . . . . . . . . . .   99
6.2     Purely Geometrical Objects and a Philosophically
        Modal Approach to Molyneux's Problem. . . . . . . . . . . . . . . . . .  100
6.3     Empirical Answers to Molyneux's Question . . . . . . . . . . . . . . .  109
6.4     Conclusions . . . . . . . . . . . . . . . . . . . . . . . . . . . . . . . . . . . . .  111

7    **Pure Possibilities and Some Striking Scientific Discoveries** . . . . . . .  113
7.1     Introduction . . . . . . . . . . . . . . . . . . . . . . . . . . . . . . . . . . . . .  113
7.2     Linus Pauling's Discovery of the Alpha Helix. . . . . . . . . . . . . .  119
7.3     Árpád Furka's Discovery of a New Field—Combinatorial
        Chemistry . . . . . . . . . . . . . . . . . . . . . . . . . . . . . . . . . . . . . . .  122
7.4     F. Sherwood Rowland and Mario J. Molina's Discovery—The
        Destruction of the Atmospheric Ozone Layer. . . . . . . . . . . . . . .  123
7.5     Kary B. Mullis's Polymerase Chain Reaction . . . . . . . . . . . . . .  125
7.6     Neil Bartlett's Discovery of "Noble" Gas Compounds . . . . . . . .  126
7.7     James Watson and Francis Crick's Discovery
        of the DNA Structure. . . . . . . . . . . . . . . . . . . . . . . . . . . . . . .  130
7.8     Leo Szilard's Idea of a Nuclear Chain Reaction . . . . . . . . . . . .  131
7.9     George Gamow's Big Bang Model . . . . . . . . . . . . . . . . . . . . . .  133
7.10    Frances H. Arnold and a Novel Way to Open up Chemical
        and Evolutional Possibilities. . . . . . . . . . . . . . . . . . . . . . . . . . .  134
7.11    Conclusion . . . . . . . . . . . . . . . . . . . . . . . . . . . . . . . . . . . . . .  138

8    **The Philosophical Significance of Alan Mackay's Theoretical
     Discovery of Quasicrystals**. . . . . . . . . . . . . . . . . . . . . . . . . . . . .  139

9    **Shechtman's Three Question Marks: Possibility, Impossibility,
     and Quasicrystals** . . . . . . . . . . . . . . . . . . . . . . . . . . . . . . . . . . . .  161

**10  Eka-Elements as Chemical Pure Possibilities.** . . . . . . . . . . . . . . . . . . .  173

**11  Quantum Pure Possibilities and Macroscopic Actual Reality.** . . . . . .  185
    11.1    Introduction . . . . . . . . . . . . . . . . . . . . . . . . . . . . . . . . . . . . . . . .  186
    11.2    Quantum Possibilities and Reality . . . . . . . . . . . . . . . . . . . . . . .  189
    11.3    The Relationship Between Quantum Pure Possibilities
             and Classical Actualities . . . . . . . . . . . . . . . . . . . . . . . . . . . . . . .  199
    11.4    The Two-Slit Experiment, Interference, and Entanglement
             in the Light of Panenmentalism. . . . . . . . . . . . . . . . . . . . . . . . .  202
    11.5    Yakir Ahronov: Free Will, Time, Teleology, and Kant. . . . . . . . .  204
             11.5.1   Aharonov . . . . . . . . . . . . . . . . . . . . . . . . . . . . . . . . . . .  204
             11.5.2   Kant's Duality Problem: Natural Causality
                       and Causality of Freedom . . . . . . . . . . . . . . . . . . . . .  206
             11.5.3   Another Metaphysical Contribution Concerning
                       Alternative Possibilities for Free Choice . . . . . . . . . . .  211
             11.5.4   Some Philosophical and Scientific Lessons . . . . . . . . . .  214
    11.6    Conclusions . . . . . . . . . . . . . . . . . . . . . . . . . . . . . . . . . . . . . . . . .  216

**12  Brain Imaging and the Human Mind** . . . . . . . . . . . . . . . . . . . . . . . .  217
    12.1    A Current Ambition Concerning Brain Imaging
             Technologies . . . . . . . . . . . . . . . . . . . . . . . . . . . . . . . . . . . . . . . .  217
    12.2    Some Preliminary Doubts . . . . . . . . . . . . . . . . . . . . . . . . . . . . . .  219
    12.3    What Does It Really Mean that We Know What Other
             People Think? . . . . . . . . . . . . . . . . . . . . . . . . . . . . . . . . . . . . . . .  221
    12.4    The Irreducibility of Individual or Personal Differences . . . . . . .  224
    12.5    Psychical Possibility and Its Physical Actualization;
             Psychophysical Inseparability or Unity. . . . . . . . . . . . . . . . . . . .  228
    12.6    Why Is There No Psychical Inspection Either from Within
             or from Without? . . . . . . . . . . . . . . . . . . . . . . . . . . . . . . . . . . . .  230
    12.7    Why Our Body and Behavior Do Not Allow Any Epistemic
             Access to the Mind . . . . . . . . . . . . . . . . . . . . . . . . . . . . . . . . . . .  231
    12.8    Psychical Reflection and Clinical Picture as Opposed
             to Imaging . . . . . . . . . . . . . . . . . . . . . . . . . . . . . . . . . . . . . . . . .  233
    12.9    The False Dualistic Picture of the Psychophysical Unity . . . . . .  236
    12.10  Real Physicalism and Brain Imaging . . . . . . . . . . . . . . . . . . . . .  237
    12.11  Conclusions . . . . . . . . . . . . . . . . . . . . . . . . . . . . . . . . . . . . . . . . .  238

**13  Neoteny and the Playground of Pure Possibilities** . . . . . . . . . . . . . . .  239
    13.1    The Role of Neoteny in Human Evolution and Life . . . . . . . . . .  239
    13.2    Winnicott: Playing and Reality . . . . . . . . . . . . . . . . . . . . . . . . . .  243
    13.3    Panenmentalism, Saving Pure Possibilities, and Neoteny . . . . . .  245

**14  Stanley Milgram's Experiments and the Saving
    of the Possibility of Disobedience.** . . . . . . . . . . . . . . . . . . . . . . . . . . .  249

**15  Singularity and Uniqueness: Why Is Our Immune System
      Subject to Psychological and Cognitive Traits?** ................  263
    15.1    My Psychophysical Assumptions. ......................  263
    15.2    Applying Psychological Terms to Immunology:
              Breznits's Contribution .............................  267
    15.3    A Comment on the "Cognition" and "Individuation"
              of Bacteria, Social Amoeba, and Social Insects ............  269
    15.4    Irun Cohen's Novel Cognitive Paradigm for Immunology .....  270
    15.5    Tauber's Criticism of Cohen and of Other Immunological
              Cognitivists .......................................  274
    15.6    Biological Agency, Metaphors and Biological Individuality:
              Comments on the Views of Wilson and Dennett............  276
    15.7    The Application of My Psychophysical Assumptions
              to Immunology ....................................  280
    15.8    Immunology and Panenmentalism: Why Does Our Immune
              System Actualize Psychical and Cognitive Possibilities?......  287
    15.9    Conclusions .......................................  288

**References** ....................................................  289

**Author Index** .................................................  305

**Subject Index** ................................................  311

# Chapter 1
# What Are Individual Pure Possibilities and for What Are They Good?

**Abstract** This introduction clarifies the nature of panenmentalism, which is an original modal metaphysics that the author introduced in 1999 and has elaborated on since then. Unlike phenomenology, which endorsed individual pure (i.e., non-actual) possibilities that are mind-dependent, panenmentalism is realist about mind-independent individual pure possibilities, while rejecting the notion of possible worlds. The introduction also specifies the reasons for rejecting this problematic notion.

The notion of possible worlds has become very popular among philosophers in various fields, and only few philosophers have voiced serious doubts about this notion and its uses (Lowe 2006; Jacobs 2010; Vetter 2011; Fine 2012; Millgram 2015; Marmodoro and Mayr 2019). It should be noted that there are many disagreements and debates about this notion and there is no general consent among philosophers about it. Possible-world semantics is one of the fields in which this notion has been proven to be useful, and one can hardly conceive modal logic without it. Nevertheless, considering ontology and metaphysics, there should be some serious doubts about this very popular notion.

For which reasons do I consider the notion of possible worlds dispensable or even wrong and much prefer the notion of individual pure possibility instead?

But first let me say something about pure possibilities. Phenomenologists have used the concept of pure possibilities, which are clearly mind-dependent, whereas in this concept I understand mind-independent possibilities that are discoverable and not invented. There are also some mind-dependent individual pure possibilities, those that play roles in our creativity and invention, especially in fiction. Many of these possibilities are a means to discover individual pure possibilities that are mind-independent (especially in mathematics and producing theories). Whenever I mention or use the concept of pure possibilities in this book, I mean individual pure possibilities and not worlds or generalities. General concepts have to do with the relations between individual pure possibilities and their properties as well. I use "pure" in a sense analogous to "pure mathematics" to be distinguished from "applied mathematics." Instead of "applied," I use "actual." One should distinguish between actual possibilities and pure, non-actual ones.

© Springer Nature Switzerland AG 2020
A. Gilead, *The Panenmentalist Philosophy of Science*, Synthese Library 424, https://doi.org/10.1007/978-3-030-41124-4_1

If it is argued that modal logic cannot dispense with possible worlds as truth-makers, why should we not prefer individual possibilities instead? Instead of quantifying over possible worlds, why should we not prefer modal quantifiers of individual possibilities? Individual possibilities can serve adequately as truth-makers, because such possibilities and their relationality, namely the general or universal ways in which they relate to each other, are the truth-makers of the propositions about them. Individual possibilities are simpler and clearer entities than possible worlds which are much more complicated entities that need many more assumptions. If any possible world is a total or maximally comprehensive state of affairs, or a complete world history, what can be more complicated? Who is the human being capable of comprehending such totalities? Individual possibilities are thus much preferable at least in this respect. Nevertheless, a challenger of such an idea may argue that the number of individual pure possibilities is so huge that the violence of the reasonable principle of Occam's razor is an inevitable consequence of relying upon them. This is not a sound argument, however, for the number of possible worlds is not less huge than that of individual possibilities; in fact, it is much greater. According to the notion of possible worlds, any deviation from or change of actual reality for any individual entity requires reliance upon a possible world. This renders the number of possible worlds as infinite or much greater than the number of actual particulars, as the counterpart of each such particular may be different in each possible world. Thus the number of the individual pure possibilities of all actual entities, which is precisely the number of these actualities, is smaller than that of the possible worlds concerning these actualities and their "counterparts." Hence, the idea of individual pure possibilities is much more parsimonious than that of possible worlds.

Moreover, the notion of possible worlds raises at least four apparently unsolvable epistemic problems, the first of which is well known. First, how can we have epistemic access to any world that is different from our actual world and entirely separate from it? Such possible worlds take no part in our actual world and have no connections with it, causal, spatiotemporal, or otherwise. If, on the contrary, they are parts of our actual world, they are not worlds at all but only non-actual parts of this world; as a result, there are no possible worlds but only one, actual world. And, alternatively, if the possible worlds and the actual one overlap, at least to some extent, the distinction between them must become a serious problem. Second, if a necessary truth is true in (or valid for) all possible worlds, how can we be familiar with *all* of the possible worlds in order to decide whether a truth is necessary? Third, if possible worlds are total, how can we conceive such totalities which are not formal or mathematical but replete with concrete entities down to their last detail? Are not such totalities beyond our cognitive capability? Fourth, we do not discover possible worlds, we only can stipulate them but, if such is the case, how can we get any solid *truth* about them? Are not all such truths simply stipulated and subject to our arbitrary decisions or whims? What is the value or significance of such truths? In contrast, individual pure possibilities are discoverable by means of our intellect and imagination. We can discover the individual possibilities of entities, whether these entities are actual or merely possible. We can make such discoveries simply by considering entities regardless or independently of any spatiotemporal or causal conditions or restrictions. In this way, we can discover the individual possibility of any entity that we encounter or even imagine.

Let me explain or demonstrate this by an example. Suppose that on an island, completely isolated and far from the mainland, the inhabitants had never encountered bats, nor heard anything about them. Nevertheless, these people are educated, intelligent, and imaginative enough to surmise the pure ("mere") possibilities of flying mammals. Without stipulating any possible world, they form quite a few *true* propositions about these merely possible creatures, each of which is an individual possibility. Forming these propositions, they use their cognitive capability of transcending actual, empirical data, simply by relying upon their intellect and imagination (which are not confined to empirical or actual data as well as to any spatiotemporal and causal restrictions). Such propositions are: (1) "(merely possible) flying mammals are not birds"; (2) "they are not reptiles"; (3) "they may feed on fruits growing on tall trees"; (4) "some of them may feed on flying insects"; (5) "they have wings suitable for flying fairly fast and high"; (6) "their offspring feed on the milk of female flying mammals"; (7) "like other mammals, they are also intelligent creatures," and many more true propositions such as these. Most of these truths are necessary (1, 2, 5, 6, and 7). None of these truths relies upon the notion of possible worlds.

Hence, philosophers are too generous and careless about their usage of the concept of possible worlds and they should be much more doubtful or circumspect about it. I sincerely advise them to dispense with such usage entirely. Philosophers have argued that without such a concept we cannot have a modal semantic, not to mention a modal philosophy. Of course, there have been philosophers, such as Quine, who have attempted to dispense with modality, but it is not reasonable to dispense with modal sentences and propositions that play decisive roles in our discourse and ways of thinking. As I will argue below, individual pure possibilities are much more suitable to satisfy each demand that modality requires.

To make justice with a very different, quite radical, objection to the existence of individual pure possibilities instead of possible worlds, I should refer to Ori Simchen's view.

Ori Simchen argues:

> … if there really is no $\varphi$, and if nothing could [be] $\varphi$ as a matter of mere possibility without already $\varphi$-ing, then it turns out that there it is impossible that there be a $\varphi$ after all. … A proper appreciation of what it means to deny that there are merely possible things requires us to deny (2) [namely, that it is possible that there be a $\varphi$] in cases where the property in question is uninstantiated and nothing could serve as a possible instance without already being an instance. (Simchen 2006, pp. 10–11)

This is obviously wrong. Take, for example, the first impressionist painting, before the advent of the term "impressionism," before the emergence of this movement, and just before the particular picture—Claude Monet's 1872 *Impression, soleil levant (Impression, Sunrise)*— was actually created (namely, before its actual existence). Monet grasped the pure possibility of such a painting and then, or at the same time, he created it as an actuality. This *individual pure possibility* was, atemporally but *in fact* at that time already, *the pure possibility of the first impressionist painting*! Simchen could not argue that such was not the case, for, in fact, this is the individual pure possibility of the first impressionist painting, although the stipulations that Simchen demands are not satisfied.

Or, take an example from the history of science. On the morning of April 8, 1982, Dan Shechtman observed through an electron-microscope a crystal that he produced and was considered then, on the basis of classical crystallography, as simply impossible. At that time, Shechtman knew nothing about the purely geometrical possibility of such a crystal. On orthodox theoretical basis, he considered it as impossible. Nevertheless, he had to acknowledge that what he certainly observed was an actual crystal. In fact, some time before that discovery, a *theoretical* crystallographer, Alan Mackay, described concretely such a *geometrical pure possibility*. Nowadays, such crystals are called "quasicrystals" but, at the time, there was no such name. Thus, the aforementioned individual pure possibility, without any reliance upon the concept of possible worlds, was certainly there even though at the time no crystallographer, including Mackay, knew about the actual existence of such a crystal. Mackay simply did not deny the pure possibility of a quasicrystal. Notwithstanding, in fact, Simchen denies the individual pure possibilities whose actualities are unknown or yet-unknown. There are many counterexamples of his stance in his paper. The counterexamples are taken from everyday life, the history of art, and that of science.

Now let me take the paragraph cited above (Simchen 2006, pp. 10–11) and "translate" it in the light of my possibilist terms:

> … if actually there were no impressionist paintings, and if nothing could be an actual example of such a painting, an impressionist painting as a pure (mere, non-actual) possibility without (before) the actual existence of impressionism is simply impossible. Thus, it turns out that it is impossible that there be an impressionist purely possible painting after all. …A proper appreciation of what it means to acknowledge that there are purely possible things requires us to admit cases where the property in question is uninstantiated and something could serve as an instance without already being an actual instance.

For Simchen, there are individual actualities but not individual possibilia! With Simchen, only general possibilities exist.

Simchen begins the paragraph cited above as follows: "if there really is no …", whereas it should be: "if there *actually* is no ..." Conflating "actually" for "really" means to assume the begged conclusion from the outset! Why should "real" be equivalent to "actual"? Why may purely possible objects not be real for us, no less than actualities? The fear of a future possible earthquake can be much stronger than one's confidence in the actual solid construction of one's house. Merely possible illness can be more robust for one than one's actual sound body, even in the cases in which one knows about no actual case of such an illness. Think of the first impressionist picture, before the term "impressionism" existed and before any other actual impressionist picture was painted, before the actual creation of such a picture, what was the *real, concrete object* in the mind of the artist who was its producer? Certainly, an individual pure possibility of that picture. Before an actual story written by Kafka, such a story, while Kafka surmised its pure possibility, had existed as an individual pure possibility, as an object that Kafka discovered.

Simchen gives an example of the eka-element of atomic number 117 (without mentioning the term "eka-element")—ununseptium (Uus)—and considers it as a natural kind. He also mentions "the kind proton" (Simchen 2006, p. 12, footnote 16). Nevertheless, any eka-element or sub-atomic particle, such as proton, are *names* (though not private names) and they are not kinds. Even if "Ori Simchen", like any other proper name, may be the name of several persons (so far, unknown to me), each of these persons is not a member of a kind. Similarly, proton or ununseptium is not a kind but a *name*. To begin with, it is a name of an individual pure possibility, which belongs to a "family" of individual pure possibilities bearing the same name. As I see them, sub-atomic particles and eka-elements are, first of all, individual pure possibilities whose kinds or general properties depend upon these possibilities and not vice versa, very much unlike Simchen's consideration of "cases." Individual pure possibilities are entirely different from kinds or types, which are general.

At the beginning of the abovementioned footnote, Simchen refers to the formula $\Box x \varphi x$. This logical formula, pertaining to the predicate calculus (namely, first-order predicate calculus or predicate logic), is too meager and not sensitive enough to capture individual pure possibilities, which are not defined by any general property. In contrast, the use of ordinary, non-artificial, non-formal language captures, represents, and communicates with individuals directly without relying upon their general properties. For instance, "Someone is called Socrates," "Amihud Gilead was born on 13 January 1947," "Someone exists who was born ..." without using any predicate. The fact that existential quantifiers and universal quantifiers belong to the predicate calculus has no metaphysical or ontological significance. The language of this calculus cannot be adequate for my metaphysical or ontological discussions.

I am convinced that: (a) modal propositions are meaningful, some of them must be true and others—false; (b) these propositions are either about entities existing in our actual world or about possible individuals, and, thus, no possible worlds are needed for making such propositions possible; and (c) individual pure possibilities are ontologically and epistemologically necessary for a solid knowledge and understanding of anything actual and, thus, they are indispensable for scientific progress.

In other words, nothing actual can exist and can be known and understood without assuming the mind-independent existence of its individual pure possibility.

To ascribe existence to pure ("mere", non-actual) possibilities may sound outlandish. Nevertheless, consider the following: (1) Impossible entities do not and could not exist; (2) To exist, any entity has first *to be* purely possible, to be a pure possibility; and (3) This prior necessary condition has, indispensably, an ontological aspect, too, and not only logical, epistemological and modal ones: If an individual entity fails to exist as an individual pure possibility first, it cannot exist at all. Each existent, whether actual or not, has first to satisfy this ontological condition. Having this primary ontological condition satisfied and only then, the secondary ontological condition as to what are the contingent spatiotemporal and causal circumstances under which this entity can or cannot actually exist, may or can be satisfied. Hence, the existence of any individual entity, whether actual or not, depends primarily on the existence of its pure possibility.

To understand the last paragraph better, the reader should know what individual pure possibilities are according to my panenmentalist metaphysics, as this will be demonstrated in the following chapters and which I have introduced and elaborated on in several books (Gilead 1999, 2003, 2009, 2011) and papers (Gilead 2005a, 2010, 2013, 2014a, b, c, 2015a, b, c, d, e, 2016, 2017).

What does the term "panenmentalism" mean? Following the term "panentheism," which means that everything ("all") that exists is comprised, as a part, in the whole that is God, the term "panenmentalism" means that everything that actually exists is part of the total whole of the mental, and the mental (to be distinguished from the psychical, which has to do with subjectivity and with singularity, as we shall see) is the purely possible. The physical is simply the spatiotemporal and causal restricted part of the mental, of the purely possible. The term "panenmentalism" is replaceable by the term "panenpossibilism." Indeed, panenmentalism is a special kind of possibilism.

Note that the mental and the psychical are by no means one and the same. Only the psychical depends on minds and subjects, whereas the mental or the nonphysical is absolutely independent of them. By "mental" I mean the conceptual, or something like that which Spinoza entitled the Attribute of Thought, with the clear difference that, while in Spinoza's philosophy Thought and Extension are parallel, and the one is not extensive or "wider" than the other, in panenmentalism, in contrast, the mental is wider and more comprehensive than the physical-actual, which is included in the former (hence the "en" in "panen" or "all-in").

Each existent, each individual existing entity, must be possible otherwise it could not have existed. This necessity, this "must," is mind-independent otherwise it would have no ontological meaning. As a metaphysics, panenmentalism has also an ontological commitment and not only an epistemological one. If modal propositions can be true, they must also be about existents that are mind-independent. After all, existence, which needs an ontological commitment, is under discussion, namely, the necessary existence of individual pure possibilities.

But what are individual *pure* possibilities? With panenmentalism, "pure" is similar to "pure mathematics" (to be distinguished from applied mathematics), namely, mathematics that is independent of anything actual or any application. The objects of pure mathematics do not exist in space and time and are not causally functional. Likewise, individual pure possibilities, such as numbers and geometrical figures, do not spatiotemporally and causally exist. They exist in a different manner from actual entities.

Individual pure possibilities are individual existents (namely, they exist mind-independently, unlike phenomenological pure possibilities), for they are not possible worlds; they are, instead, the existents without which individual actual things (in short, actualities, entities existing in space and time and are causally subject) cannot exist and cannot be identified, known, and understood.

When, in April 1983, Dan Shechtman using an electron microscope, watched a very strange crystal that he had himself created, he could not, at first, believe his own eyes, because such a strange crystal, demonstrating a tenfold symmetry, was considered simply *impossible* according to both applied crystallography (using X-ray tests in empirical crystallography) and theoretical-mathematical crystallography known to him at that time. Both kinds of crystallography, the pure one and the applied-actual one, referred then only to 230, no more and no less, crystallographic space groups, whereas the crystal that Shechtman examined clearly did not belong to any of these structures. It appeared that such a crystal was simply *impossible*. Hence, Shechtman wrote in his notebook, in Hebrew: "Ten-fold symmetry??? There is no such an animal!" ["אין חיה כזו"]).

Had Shechtman been familiar with the theoretical-geometrical pure considerations by Penrose and Mackay, he would not have questioned such a *possibility*, theoretically based, in the first observation of the very strange crystal. While Shechtman, in the first moments, questioned the pure, the very possibility, of such a crystal, Linus Pauling, until his death, in fact excluded it. Such an exclusion could prevent the dramatic revolution in crystallography in the twentieth and twentieth-first centuries. I am referring to the real, mind-independent existence of crystalline individual pure possibilities (the acknowledged existence of quasicrystals has changed the definition of crystals and crystallography).

Similarly, there have been missing elements in the Periodic Table of Elements. These missing elements have been called "eka-elements." Mendeleev had already used this term and he predicted the existence of several elements that were not known as actual existents at the time. Mendeleev described the chemical and physical properties of those eka elements quite accurately. Several years after he made these predictions, the actual elements were discovered, and they fitted his descriptions of them. At present, there are many more eka-elements in the present Periodic Table, and even more trans-uranium elements may be found in nature as actual or be produced as actual in laboratories (see Chap. 10 below).

As I see it, the Periodic Table is a summary of all the pure and actual possibilities of the chemical elements as well as of their possible and actual combinations. Thus, the table aspires to comprise all the individual pure possibilities of the chemical elements. Without assuming the mind-independent existence of eka-elements, chemists would be incapable of identifying the relevant actual elements while encountering them in actual reality!

Another example of a mind-independent pure possibility is the Higgs boson. On 8 October 2013, the Royal Swedish Academy of Sciences announced its decision to award the Nobel Prize in Physics for 2013 to François Englert and Peter W. Higgs "for the *theoretical discovery* of a mechanism that contributes to our understanding of the origin of mass of subatomic particles, and which recently was confirmed through the discovery of the predicted fundamental particle, by the ATLAS and CMS experiments at CERN's Large Hadron Collider" (my italics). Note that the discovery under discussion is a theoretical one, whereas the discovery of the actual Higgs boson was made some years later, following the actual discovery of a Higgs boson at CERN's Large Hadron Collider. Translating this into panenmentalist terms, the theoretical discovery is of the pure possibility of the boson, while the actual discovery is of the actual boson. Without the first discovery, the second discovery could not have taken place and no actual Higgs boson would have been identified and recognized.

The most plentiful examples of individual pure possibilities are the existents and objects of pure mathematics. As we shall see, in pure mathematics, to mathematically exist is, in fact, to be mathematically possible. To purely mathematical exist is to exist as an individual pure possibility.

There are many more fine examples of the usefulness and vitality of individual pure possibilities in the sciences. In this book, I will discuss some of them as well as the aforementioned ones.

In sum, individual pure possibilities are all the possibilities that are mind-independent and which are not spatiotemporally and causally existent.

Individual pure possibilities are preferable to possible worlds. Paradoxically, the existence of individual pure possibilities is more parsimonious relative to that of possible worlds. Assuming the existence of an individual pure possibility does not necessarily imply the existence of any other possibilities, whereas if assuming the existence, say, of a possible talking donkey, implies a possible world, such an assumption implies more pure possibilities, namely, all those that the existence of a talking donkey needs to imply. Hence, the existence of individual pure possibilities is much more parsimonious than that of possible worlds. We simply refer to such donkeys as individual pure possibilities that do not require any possible world for them to exist.

As individual pure possibilities are necessary, both ontologically and epistemologically (their existence is indispensable for scientific progress), assuming that, their existence does not breach the imperative of Occam's razor. Assuming it, we do not multiply existence beyond the necessary.

These few words of introduction are by no means enough to describe panenmentalism or its principles. Nevertheless, in each of the following chapters all that is needed by panenmentalist principles to understand the contents will be explicated in greater detail.

As I have mentioned panenmentalism above, let me say something more about this original modal metaphysics, which is realist about individual pure possibilities.

Let us begin with a simple and basic metaphysical question: What does exist and what can we know about it? There are two kinds of existents: Possible entities, which are individual pure possibilities, and actual entities, which are named actualities. Each actuality is an actualization of an individual pure possibility. Each actuality is subject to spatiotemporal and causal conditions, whereas individual pure possibilities are exempt from such conditions. Actualities are only empirically known, that is, they are cognized by means of empirical observations, experiments, or experience, whereas individual pure possibilities are cognized by means of our intellect and imagination.

As two different kinds of existents, individual pure possibilities and their actualities cannot be reduced to each other; instead, they are united. Each actuality is united with its individual pure possibility.

Panenmentalism is devoted to a special view of the psychophysical unity. According to it, our body is an actuality of our mind, which is a special individual pure possibility. The only way to have an epistemic contact with individual pure possibilities is, as I have argued above, by means of our intellect or imagination. That is, only minds can refer to individual pure possibilities, which change nothing in the realm of actualities. There is no causal relationship or connection between individual pure possibilities or between them and any actuality. Only because our mind is an individual pure possibility of a special kind (singularity), can it relate to other individual pure possibilities. Our body is the spatiotemporal and causal part of our mind. This means that only some of the possibilities that are open to us are actualized in our body). In general, the whole realm of actualities is but the confined and conditioned part of the whole, total realm of all individual pure possibilities.

Minds, especially conscious minds, make a special kind of individual pure possibilities. No two individual pure possibilities are identical. If there were two identical pure possibilities, they would be one and the same possibility, as there is nothing spatiotemporal and causal to make any difference between such "two" possibilities. While two individual pure possibilities cannot be identical, two psychical individual pure possibilities cannot be even similar. That is, each psychical individual pure possibility, such as a mind, is a singular possibility. Singularity is the mark of the psychical or of the conscious.

As each individual pure possibility is different from all the others, each individual pure possibility relates to all the others. This relationality includes also singular individual pure possibilities. As wholly different from each other and from all the other individual pure possibilities, the psychical possibilities necessarily relate to one another.

No individual pure possibility can be general, whereas generality or universality stems from the relations between individual pure possibilities.

# Chapter 2
# How Many Pure Possibilities Are There?
# Or *Contra* Actualism

**Abstract** Independently or regardless of any actualization, possibilities are pure. Are such possibilities real? Actualism excludes the existence of individual pure possibilities, altogether or, at least, as existing independently of actual reality. In contrast, possibilism emphatically acknowledges the existence of mind-independent, non-actual (pure) possibilities, which are entirely independent of actual reality. Possibilism, thus, is realist about the existence of such possibilities. Challenging the views of Nicholas Rescher and other actualists, beginning with Quine, I attempt to defend a realism of individual pure possibilities. For this purpose, I suggest some counterexamples that appear to render such views groundless. Indeed, no answer can be given to the question: How many pure possibilities are there? Yet, notwithstanding Rescher's critique, such non-answerability does not endanger or challenge realism of pure possibilities or any possibilist realism. Such non-answerability is also valid for genuine literary works of art, in which only what makes a difference is necessarily there and subject to our questions. Such works of art maintain a sort of necessity, exclusively pertaining to individual pure possibilities and their relations, which are necessarily within these works. Individual pure possibilities are as real as actualities, albeit in a different sense.

Independently or regardless of any actualization or actuality, possibilities are pure. Suppose that pure possibilities or *possibilia* are not possible worlds but individual, concrete possibilities. How many pure possibilities are there? As I would like to show in this Chapter, although no answer can be given to such a question, it does not mean that this non-answerability endangers or challenges realism of pure possibilities or any possibilist realism, notwithstanding Nicholas Rescher's critique (Rescher 1999, 2003).

To show that "the currently fashionable *realism* of possible worlds is deeply problematic and needs to be replaced by a suitable—and ontologically more modest—version of *conceptualism*" (Rescher 1999, p. 403), Rescher raises the question of how many possible worlds are to be identified, individuated, and counted (ibid.). Since no answer can be given to such a question, Rescher suggests replacing a

---

A first version of this chapter was published in *Metaphysica* 5:2 (2004), pp. 85–103.

possibilism that is substantively oriented (*de re*) by one that is proportionately oriented (*de dicto*).

In spite of Rescher's report, apart from possible worlds realism (such as David Lewis's), possibilism is not at all currently fashionable and actualism is in vogue instead. Moreover, as the representative selections of the views taking part in the debate over actualism and possibilism clearly show,[1] although many actualists adopt the idea of possible worlds, all of them explicitly reject the existence of purely possible individuals or particulars, namely, individual pure possibilities. Hence, at the moment, possibilism needs a strong defense against various attacks, actualist and otherwise. Actualism states that everything that exists are actualities, whereas possibilism argues that possibilities that are entirely independent of actualities or actual reality in general, also exist. In the following chapters, I will refer to some *actualist fallacies* in which, on the basis of actual findings alone, scientists dogmatically excluded some possibilities that at the time were not considered actual but which, later, turned to be actual, too (for instance, concerning eka-elements, that transuranium elements are in fact impossible, or that it must be impossible to include "noble elements" within chemical reactions and thus they should be considered inert elements).

Rescher argues that ostensive confrontation as regards to *possibilia* is lost and that the purely descriptive individuation of nonexistent (that is, nonactual) individuals is an "altogether impractical project" (Rescher 1999, pp. 403 and 411). In what follows, I will show that individuation and reference can be independent of description. If this indeed is the case, is the individuation of *possibilia* altogether impractical project?

What I termed "eka-fallacy" (Gilead 2003, pp. 65–70) is sufficient to indicate overwhelming counterexamples, which would make Rescher's argument against possibilism groundless. The phenomenon of predictable, yet nonactual, chemical elements enabled Mendeleev and others to fully identify and to exhaustively describe possible, though actually missing, chemical elements. The places of such eka-elements in the periodic table could or can, yet must not, be occupied by actual elements. Even today, chemists predict the existence of many possible chemical elements that so far we have no evidence of their actual existence. The list of

---

[1] See Loux 1979; Fitch 1996; Tomberlin 1998; and Gendler and Hawthorne 2002. In a more recent paper, Rescher states that "the metaphysics of possibility has been a growth industry in recent years" (Rescher 2003, p. 363). Christopher Menzel defines actualism and possibilism as follows: "Actualists ... deny that there are any non-actual individuals. Actualism is the philosophical position that everything there is—everything that can in any sense be said to be—*exists*, or is *actual*. Put another way, actualism denies that there is any kind of being beyond actual existence; to be is to exist, and to exist is to be actual. Actualism therefore stands in stark contrast to possibilism, which ... takes the things there are to include possible but non-actual objects" (Menzel 2011/2016). Contrary to actualism, my possibilist approach distinguishes between two kinds of existents—that of individual pure possibilities and that of actualities. In my view, individual pure possibilities *exist* independently of actualities. Even though many actualists adopt the idea of possible worlds, all of them explicitly reject the existence of merely, purely possible individuals or particulars, which are entirely independent of the actual.

eka-elements is not exhausted and it is still open, yet the identification and the description of any eka-element are practical, possibly useful, heuristic, and fully satisfy all we need from identification and description. The description under consideration is by no means schematic, and no person is entitled to describe it as a "mere scenario," for it provides all the needed chemical details. This kind of possibilism thus obviously gains a scientific standing and yet is entirely incompatible with Rescher's critique as above.

Even when the predictability of any eka-element is rendered actual, the identity, reference, and description of such an element are entirely independent of any such actualization. Having been found actual, the chemical properties of the element do not change; the only change lies in the name of the element. All eka-elements thus meet the requirement needed for possibilities to be pure. Indeed, eka-elements are pure chemical particular possibilities. Each of which has its particular place in the periodic table, however open and expandable; owing to that openness or expandability, radioactive and transuranic elements, unknown at Mendeleev's time, are arranged in rows later added to the table. This open nature of the table is entirely compatible with that of the realm of pure possibilities. By contrast, the particular, individual status of any eka-element as a pure possibility is clearly incompatible with Rescher's view about the ontological furniture of the world, possible or actual (ibid., p. 408). Eka-elements, particular fictions, or pure possibilities in general must not be abstract objects, mere schemata for possible individuals, or mere thought-instruments (to borrow from ibid.). They can supply some ontological furniture, for instance, in chemistry as a realist scientific theory. What they cannot provide is the actual ontological furniture, which only experience and observation can provide. In other words, the actual ontological furniture is empirically acquirable alone. Yet other, no less real, ontological furniture exists, consisting of individual pure possibilities. As I see it, each eka-element satisfies the condition that Rescher puts to particularity, namely, particularity demands identification (ibid., p. 409). Any eka-element qualifies as an identified particular, and not a general schema for an element. Hence, arguing that "hypotheses enable individuals to be discussed in the abstract but not to be identified in the concrete" (ibid.), Rescher commits what I have entitled "the eka-fallacy."

Suppose that we accept Rescher's stance according to which "only a description that is saturated and complete could possibly manage to specify or individuate a merely possible particular individual. For any genuinely particular individual must be property-decisive, and a nonexistent possible individual can obtain this decisiveness only through the route of descriptive saturation" (Rescher 2003, p. 378). Eka-elements precisely meet even such a demand for their decisive physical and chemical properties are entirely sufficient to secure identification as well as endlessly recurrent re-identification. If the demand from any possible individual "cannot be vague or schematic but must issue a committal yea or nay with respect to every property whatsoever" (ibid.), each eka-element has perfectly met that demand. We have all the descriptive saturation we need from the periodic table to secure perfect identification of an element as purely possible or actual. From the epistemological point of view, at least, such identification should not raise real problems. After all, identification and re-identification are epistemological issues.

As for the ontological-metaphysical background, chemical elements, as participating in the periodic table, are chemical pure possibilities, like notes on a musical scale, which are independent of actualization. If you assume that the periodic table is merely a picture or representation of the actual chemical reality, you are missing the whole point, especially as far as eka-elements are concerned. Like any natural science, chemistry has its own theoretical basis, which consists of pure possibilities and their relationality. As much as the mathematical foundations of any natural science are pure possibilities and not actualities, so are the chemical possibilities arranged in the periodic order. The possibilities-identities and their relationality are there, completely, in the table. As such, they are clearly existents, they are obviously real, by no means Rescher's nonexistents. They are not just "verbally or mentally intended referents" but real referents. They are not merely "*de dicto*" possibilia, but possibilia *de re*. Our thought and language do not invent or create them, but rather capture them as discoveries of chemical pure possibilities. No eka-element has been invented or created; it has been merely discovered as a pure possibility.

We are not entitled to compare any eka-element to, for instance, the philosopher's stone, which is a "putative item" or a "suppositional being ... the [linguistically engendered] artifact of an interpersonally projected supposition or assumption," "a pseudo-object that is no object at all" (ibid., p. 379). The metaphysical-ontological status of any chemical element as a particular, concrete pure possibility is well established. These pure possibilities constitute a part of the ontological furniture of the world of chemistry as a scientific theory. Whether the chemical elements in the periodic table are merely eka-elements or actual elements, their well-established identification is beyond any doubt. At least for the time being, their chemical and physical description is complete or saturated enough, quite sufficient for all the theoretical and practical needs of chemistry as a scientific theory and absolutely sufficient for any chemical identification. The qualifications "enough" and "for the time being" are needed because the future of chemistry, like that of any other science, is beyond our present knowledge.

Yet at this point a serious, one might say unsolvable, problem arises for a possibilist view that excludes multiple actualization of any pure possibility, for each chemical element (eka or not) has innumerable actualities or "tokens." Is an eka-element a pure possibility singly actualizable? The case appears to be just the contrary, and, if so, we cannot meet, as it were, Rescher's demand of uniqueness: "where only a single unique realization is possible" (1999, p. 413; I would prefer "unique actualization"). To solve this problem, we should distinguish between the particular chemical possibility-identity of an element (eka or not) and the chemical *name*, as a part of the *language* or *terms* of chemistry. Name and identity are by no means identical. Any name, as taking part in a language, is general, as no language is private. As much as the proper name "James" is general, serving as a common name for all persons named "James," so Germanium is the common name for all existing atoms or pieces of Germanium, each of which has a single, unique possibility-identity in a metaphysical, panenmentalist respect. Under such a nominalist view, every genus, species, kind, or type is merely a name, which is general. Thus, the element of Germanium in the periodic table serves as a double meaning or significance: as a name and as an assemblage of possibilities-identities sharing an intrinsic similarity that all the atoms of Germanium have. Each actual atom of Germanium shares the same name with any other atom of Germanium, each of which has an exclusive pure possibility-identity. The periodic table secures for each of these pure possibilities-identities the common locus, serving as a general name, in the table. To recognize a piece of matter as Germanium is to entitle it with a general name, shared by all the Germanium atoms, but the identification, like any identification, is particular: *This* piece of matter, *here and now*, is a piece *of* Germanium.

As for eka-elements, the distinction between name and possibility-identity is even simpler or more manifest. As long as chemists use eka-elements in the periodic table, no evidence appears to the actual existence of any of these elements, and hence only a single representative possibility-identity in each case of those eka-elements has to be referred to (or mentioned in the table), whereas the name in each case is general (even though no single actual case is known yet). Thus, prior to the actual discovery of Germanium, its pure possibility-identity was named as eka-silicium. This name, like any other, was general, yet the possibility-identity mentioned was individual, indicating the locus of each identity-possibility of each atom of Germanium, all of which are intrinsically similar. And this locus has been secured since the advent of Mendeleev's periodic table. In this way, each pure possibility-identity has only "a single unique" actualization ("realization" in Rescher's actualist term). The pure possibility-identity of Germanium, known before its actual discovery by the name "eka-Silicium," satisfies all of what Rescher demands of identification or individuation although, under that name or independently of actualization, it is merely a pure possibility! In any event, no eka-element can be considered as abstractly general, for it could not be abstracted from anything actual.

Finally, no eka-element can be considered *ens rationis*, a mere thought-object or thought-entity, such as the equator or the north pole (which Rescher mentions on p. 414). Since the reality of each eka-element is necessitated by the periodic law, which excludes possible gaps or vacancies in the periodic order or system, no eka-element is treated as *ens rationis*, which is the ontological standing of conventions, none of which is treated as real, let alone as necessary. The reality of any eka-element as a pure possibility is necessitated, whereas no *ens rationis* is ontologically necessitated.

Arguing that "the actual identification and introduction of … *possibilia* is effectively impossible" (ibid., p. 403), Rescher appears to commit a fallacy, especially concerning the introduction of pure particular possibilities. I would like to name that fallacy after Jules Verne—the Verne fallacy. In *Paris in the Twentieth Century* (written in 1863 yet published in 1994), many years before the advent of any kind of Internet, Verne introduces a possible Internet without relying on anything actual (except for electricity and electric conductivity). He thus most effectively introduces, identifies, and describes the pure possibility of such a device in full detail, without relying upon any actual device, for no such invention, in fact, existed at that time. Such a counterexample (and many more), I believe, are quite sufficient to render Rescher's arguments against possibilism *de re* invalid or groundless. Many similar examples exist of the introduction and identification of pure possibilities, *possibilia* in Rescher's term, which are quite practical and effective for various purposes.

Could Rescher argue that such counterexamples are fictions? First, if not misleading, fictions can do great service for us in searching for new discoveries, many of which are strictly scientific. Second, no matter how we discover novel possibilities, what is decisive at this point is that prior to their actual existence and quite independently of it or of anything actual, as pure possibilities alone, they were discovered by scientists, thinkers, writers, and the like. Thus, Verne introduced, identified, and described an Internet many years before its actual appearance. He referred then to such objects as pure possibilities; he substantively oriented toward them. To characterize such a reference (or "orientation") adequately, we certainly need a possibilism *de re*, very much contrary to Rescher's view.

Another counterexample of Rescher's view is the numerical series. Allegedly following Plato's *Republic* VII, Rescher mistakenly considers numbers abstract things (ibid., p. 404, note 1; cf. 2003, p. 376: "*abstracta* such as numbers"). First, Plato does not consider Ideas, mathematical or metaphysical-dialectical, as abstract entities, although it is quite true that they are exempt from processuality and, hence, from dispositional character. On the contrary, for Plato, sensible entities are *abstracta*, which are copies or mere reflections-participants of more real, substantial, concrete beings—the Platonic Ideas. Sensible or actual entities thus depend upon their Ideas, and not the other way round, which is the case of anything abstract. In Plato's philosophy, numbers clearly belong to the realm of the mathematical Ideas, which manifestly makes them non-abstract. Second, altogether independently of Plato's philosophy, as pure possibilities, numbers are not *abstracta* at all. Instead, they are concrete beings. To argue that numbers are *abstracta*, as if numbers were abstracted out of actual things, Rescher takes an actualist stance, despite his manifest efforts not to do so. If Rescher means by "abstract" simply an entity that does not exist spatiotemporally or causally, I do not follow such a concept of abstract, for although individual pure possibilities are not spatiotemporal and causal existents, they are, nevertheless, *concrete* and not general or universal. Only the relations of such possibilities (their relationality) are general or universal.

Since we regard numbers independently of any actualization or actual entities, and since their existence is exempt from any spatiotemporal and causal conditions, we should consider numbers as pure possibilities and not as actual entities. Yet the identity, reference, and description of numbers is undoubtedly altogether practical. Though no end exists to the number of numbers, no philosopher is entitled to argue on the basis of this indisputable truth that numbers are not real enough. No realism about numbers is endangered by the argument that the question—How many numbers are there?—is unanswerable. Numbers can be considered quite real, although they are not actual beings but merely pure possibilities and there is no end to their number. As pure possibilities, numbers are substantively oriented, practically referable, fully individuated, satisfactorily describable, and subsumable to ostensive confrontation. Contrary to Rescher's view, possibilism concerning numbers is both quite meaningful and committed to substantively oriented (*de re*) pure possibilities. No need exists to replace it with any "more modest" version of conceptualism.

As for the more recent version of Rescher's view (2003), the crucial problem that he confronts is: what does fix the identity of an individual? Is this an actual factor or not? What appears to be Rescher's answer is that it must be an actual factor that fixes the identity of an individual (ibid., p. 368). For instance, "the Hubert Humphrey we know and love" is an actual individual, whose identity has been fixed or settled "irrespective of what worlds or what descriptions may be involved" (ibid., p. 367). Indeed, such is the case: Humphrey's identity is independent of all these. The problem remains: what does determine his identity? What secures its persistence or survival of various contingencies and changes in his life? As I see it, the identity of an individual has not to do with possible worlds, transworld identity, or actual reality. We have to face the same problem whether we identify a pure possibility, say, an eka-element, or an actuality: what alterations or modifications can x, purely possible or actual, undergo and still retain its (or his or her) identity? To identify a member in a symmetrical mathematical group, which is altogether purely possible and not actual, or to identify the actuality of the subatomic particle omega-minus, requires no recourse to anything actual. Rather the contrary: in both cases, the purely mathematical and the physical-actual, we rely upon *theoretical* criteria, which are *purely mathematical* or *purely physical possibilities* and not actualities.

We have to face the same problem: the problem of reference and identification, while referring to a pure possibility and identifying it or to an actuality and identifying it. An ostensive identification is equally applicable to a pure possibility ("this member of this mathematical group") or to actuality ("this is the trajectory of omega-minus" or "this is the mark of omega-minus"). Contrary to Rescher's view (ibid., p. 374), spatiotemporal positioning is not a necessary condition for ostensive identification. Hence, we can ostensively identify pure possibilities, although they are exempt from any spatiotemporality and causality and are not actualities subject to experience, experiment, and observation. We can point to them as much as we can point to actualities. To identify or to refer to something, we can do without reference to actualities or to the actual world, just as we do while identifying numbers, members of mathematical groups, eka-elements, and so on.

Similarly with numbers, Rescher treats fictional objects as mere *abstracta*: "Fictional 'objects' are abstractions and not concrete *possibilia*" (Rescher 1999, p. 408). Again, we are not entitled to consider Verne's fictions as *abstracta*, for they are quite independent of actual reality and by no means abstracted from it. Second, they are not schematic but quite concrete *possibilia* within Verne's texts. As for literary fictional characters, Rescher is also wrong. Hamlet, Madame Bovary, Anna Karenina, Swann, and many other fictional figures in fine literature are by no means abstract objects, schemata, pseudo-individuals, and the like. Such are the mark of literary failures or bad literature. Although we do not normally treat any of these characters as actual, we certainly relate to them as concrete, as individuals bearing the mark of singularity and genuineness. Their ontological status is not in short of that of actual individuals, although it is quite different. Fictional characters may affect us no less than actual beings, sometimes even more. They can be very real, especially for us, and by no means as abstract but, rather the contrary, as concrete and particular as much as possible. There is a necessity about real fictional figures in literary works of art, which no actual being can have. Aristotle points out such a necessity in artistic tragedies, contrary to an actual history that may be contingent.[2] Supposedly, Rescher would not agree with such an Aristotelian idea, which enables us to realize what is the special nature of great works of literary art and especially what is meaningful and significant about them. Rescher does not ignore meaningful discussions and reasoning of "merely possible states of affairs and scenarios" or stories (2003, p. 380). Yet he leaves them to "abstract generality" alone (ibid.). Such is not the case as I see it. Literary masterpieces deal with concrete, particular pure possibilities as well as with the necessity for them. I will discuss below the necessity of pure possibilities in literary works of art.

But worse is yet to come. Rescher leaves "merely possible individuals and worlds viewed as particulars" without the "disposal of our latter-day modal realists" (ibid.). Instead, the infinite depth of the requisite details of such possibilities "confines them to the province of God alone" (ibid., p. 381). Thus, "only God can realize the idea of nonexistent particularism" (ibid.). Such is not the case at all. Over the years, literary artists, theoretical scientists, mathematicians, and the like have discovered particular pure possibilities, because no infinite depth of a complete description has been needed at all for this purpose. All they have needed has been their capability of discovering new particular pure possibilities, which are within the reach of human beings who are imaginative enough, who are not enslaved to the actual. However confined or limited, the freedom from the actual is in our nature and at our disposal. Equally, the capability of relating to pure possibilities as existents, although obviously nonactual, is very much in our nature as conscious psychical beings. Possibilia are undoubtedly within the reach of our psychical and intersubjective or interpersonal life, moreover, such life consists of them. Unless we confine all there is to the actual alone, but then nothing would be left of psychical or intersubjective reality.

As I see it, Rescher appears to miss the point of the identification of fictional characters. He asks whether the mysterious stranger in the first chapter of a novel is the same person whose corpse is mentioned in the fifth chapter. They are one and the same person, he answers, "only if the author says so—there are no facts of the matter apart from those our novelist specifies. In the absence of such specification all that can be said about the issue of identity is—absolutely nothing" (ibid., p. 370). Indeed, no actualities, no facts of the matter exist to provide us with an answer. Nevertheless, and this is the point that has been missed, if the novel has been written in a masterly way, everything relevant is necessarily there and the relations between the specific details are as necessary as they are. Thus, even if the narrator says absolutely nothing about such identification, the reader, following the inner necessity of the novel, may find the answer by herself. Nothing is arbitrary about such identification, and no recourse to contingent actualities is needed to realize it. Readers can supply the missing parts for themselves.

---

[2] *Poetics* 1451a25–b11, b34–35, and 1454a34–38.

Contrary to Rescher's view, we do not arbitrarily assume, postulate, or suppose pure possibilities as "objects that are projected in discussion" (ibid.). Although no facts of the matter determine the existence or reality of pure possibilities, they are not arbitrarily postulated or assumed. Just as in pure mathematics, in logic or in fine literature, nothing is arbitrary about pure possibilities. Contrary to Rescher's view (ibid.), they have independent characteristics that we have to discover, as much as eka-elements have had them. The modal metaphysics—panenmentalism—that I have introduced attempts to show precisely this. It is an actualist fallacy to assume that only given facts are discoverable and that "nonexistent [i.e., non-actual] possible … individuals are never given to us" (ibid., p. 376); pure possibilities are as discoverable and as given as actualities. They are given in a different way from the way that actualities or facts of the matter are given. For we discover actualities by empirical means; and these cannot capture pure possibilities.

In the final account, Rescher relies on the prominent manifesto of actualism, namely, Quine's "On What There Is" (Rescher 1999, p. 413, note 9; cf. Rescher 2003, p. 376). Undoubtedly, the following is an actualist view: "Thought and language move off in their way, and existence and reality go off on their way, and only where there is actual *adequatio ad rem* do they come together" (Rescher 2003, p. 379). Actual reality or existence does not exhaust reality as a whole. Thought and language have realities of their own and they exist as much as actual reality exists, although in different senses. Thought exists psychically, subjectively, or privately; language exists intersubjectively or interpersonally; and actual reality exists objectively or publicly. To ascribe reality only to the latter is what actualism is all about. To consider pure possibilities as nonexistents or to state that there is "no way to identify and individuate nonexistent [nonactual] possible individuals" (ibid., p. 376) is a manifest actualism.

Rescher leaves us one choice: "all or nothing: either a (distinctly problematic) metaphysical realism of self-subsistent possibilities or else a (somewhat unappealing) nominalism of mere verbal possibility talk, of possibility not as a matter of genuine fact but merely the product of an imaginative fictionalizing by linguistic manipulations" (ibid., p. 381). I entirely accept the first alternative, yet my view is a nominalist realism of pure particular possibilities in the following sense: what is general about pure possibilities is only their relationality. Furthermore, our imagination is capable of utilizing various illuminating fictions ("real fictions") to *discover* new pure possibilities, which are as real as actualities, although in a different sense. Real fictions thus do for us what no telescope can (to allude to Kripke's metaphor that is mentioned ibid., p. 377 and mistakenly ascribed to David Lewis). Verne's literary fictions provide us with one kind of example; eka-elements—with another. Let us leave linguistic manipulations to rhetoricians, copywriters, propagandists, preachers, and the like. Owing to an insightful metaphysics, philosophers can be realistically possibilists without being linguistically manipulated.

To limit or reduce possibilism to conceptualism or conceivability is to limit and confine the realm of pure possibilities unnecessarily. Possibilities, such as a round circle and $\sqrt{2}$ that is not a fraction, may exist beyond our current conceivability. To the extent that our current conceivability is concerned, they are deemed "impossibilities." Yet although we cannot conceive, at least at the moment, such possibilities, which are incompatible with our current logico-mathematical knowledge, we can nevertheless relate to them. We should not accept any restriction of the realm of pure possibilities to the limits of our current conceivability or to those of our current logico-mathematical knowledge. For this reason, I do not accept the idea that meta-physical possibility is "less expansive than narrow logical possibility" (Gendler and Hawthorne 2002, p. 5). Nor can I accept the view that conceivability or conception and possibility are coextensive or congruent. As I see it, conceivability, conception, imagination, employing fictions, and the like are the ways in which we discover pure possibilities, which are new for us. These possibilities are ontologically or metaphysically independent of the ways in which we discover them. Hence, the conceivable (or the like) and the possible are not identical.

Rescher is quite right in arguing that the description of any real thing is in principle inexhaustible (ibid., p. 405), but this is all the more valid for pure possibilities. Dispositional characterization aside, the infinitude and inexhaustibility of the relationality of any pure possibility to all the others must be beyond any doubt. Each pure possibility is different from any other pure possibility, for no two identical pure possibilities can exist. The law of the identity of the indiscernibles is especially valid for pure possibilities. Since each pure possibility is different from all the others, each pure possibility necessarily relates to all the others. Hence, its relationality is infinite and inexhaustible. This holds particularly for numbers. The open nature of the realm of pure possibilities as a whole is strictly compatible with infinitude and inexhaustibility concerning such possibilities. As Rescher states, "Endlessly many true descriptive remarks can be made about any actual physical object" (ibid.), but this also holds for Verne's Internet. The readers of Verne's works of fiction in the nineteenth century, much before the advent of actual Internet, could inexhaustibly imagine, describe, and refer to such a purely possible object. Moreover, each of Verne's readers could imagine this purely possible object under different conditions and circumstances, in the same way as the observers of the actual objects mentioned by Rescher. In both cases, of pure possibilities and of actual things, no end exists to "the perspectives of consideration that we can bring to bear on things" (ibid.).

The trouble is that stating this, Rescher has only actual things in mind. Yet an endless variety of cognitive viewpoints equally holds for individual pure possibilities and actual things. Pure possibilities enjoy descriptive perspectives as much as actual things do. Hence, Rescher's assumption that "*fictional* particulars ... are of finite cognitive depth" (ibid., p. 407) is simply groundless. Rescher's pre-commitment to description-transcending features, essential to our conception of any real, concrete object (ibid., p. 406), is certainly valid not only for actual objects but equally for individual pure possibilities. Owing to the infinite relationality of any individual pure possibility to all the others, its description is never exhaustive. However fictional a figure in a novel may be, there is an infinity of ways of relationality to it, and, hence, an infinity of possible descriptions. The more artistically rich and profound a novel, the more classic its nature, and we can realize more clearly that it is subject to more various interpretations or descriptions, the number of which has no end. Any fictional figure means or signifies different things for different readers, the numbers of which is indefinite. Novelties always wait for interpreting and describing the fictional as much as for the actual.

Rescher rests identification on the basis of description, and, given that no complete description of any particular is possible—the descriptive incompleteness or inexhaustibility—he concludes that we cannot distinguish any individual from all other possible or imaginable individuals (ibid., p. 410). As I see it, this is not the case at all. In principle, we can distinguish any individual, as a pure possibility, from all the others independently of description or relationality. No matter how we conceive them, no two individual pure possibilities can be identical, which means that we, as a matter of course, distinguish between any individual pure possibility and all the others. We do not need any description or relationality to distinguish any individual pure possibility, but the other way round. Distinguishing individual pure possibilities one from the other is the most primary or primitive act of the mind. Such is the mind's accessibility to any individual pure possibility. The reference to individual pure possibilities is direct as much as the reference to actualities is direct, and both kinds of direct reference are independent of description (Gilead 2003, pp. 56–58). Finally, since the identification of individual pure possibilities can be independent of any world, possible or actual, I do not accept Rescher's postulate that the "only feasible way to identify a possible individual would be with reference to the world to which it belongs" (ibid., p. 412). We can do without the dispensable idea of possible worlds. Instead, we are entitled to postulate the open realm of all individual pure possibilities, in which no two possibilities can be identical. As a prior or primary mental act, identification of, or reference to, individual pure possibilities is independent even of this realm too.

Direct reference—reference independent of description, interpretation, or narrative—is possible not only for actual referents but also, and even primarily, for purely possible referents, each of which is an individual, whether particular or singular (each psychical possibility is singular). Ostension to individual pure possibilities is possible and practical like ostension to actualities, given that individual pure possibilities are discoverable as are actualities. As necessarily atemporal, individual pure possibilities are discoverable and, in the last account, cannot be created, contrived, or invented (contrary to Rescher 2003, p. 364). Each individual pure possibility exists independently of its discovery, descriptions, narratives, interpretations, or significance, but obviously not the other way round. We can point out individual pure possibilities, as much as actual referents, independently of any description. Literary works of fiction may begin with direct reference to, or with introduction of, individual pure possibilities that the reader can easily follow.

At the very beginning of *Anna Karenina*, Tolstoy writes, "Everything had gone wrong in the Oblonsky household. The wife had found out about her husband's relationship with their former French governess and had announced that she could not go on living in the same house with him" (Tolstoy 1969, p. 13). In these two opening sentences, three direct referents are introduced and pointed out for the first time, all of which belong or relate to the same household: Oblonsky, his wife, and the governess. Given that the relations existing between the referents must not be confounded with descriptions of any sort, no description whatsoever is needed to refer to those fictional referents, which are not actualities but merely pure possibilities, entirely independent of their description and of any actualization as well.

Equally direct or independent is the reference or the ostension at the very beginning of Kafka's "Before the Law": "Before the Law stands a doorkeeper on guard. To this doorkeeper there comes a man from the country who begs for admittance to the Law" (Kafka 1961, p. 61). You can easily think of many other examples, not necessarily literary or fictional, including mathematical or theoretical examples. Writers can introduce, directly refer to, or point out fictional persons or objects and fix their names, independently of description whatsoever. Nothing should be schematic or hypothetical about these fictional figures; they can be particular or concrete. Nothing of contingency is left about them in a genuinely literary piece of art. All we have to know about them is necessarily there. All other questions that have nothing to do with such a necessity should not be asked about them. They are quite different from actualities, the basis for answering questions about which is necessarily empirical.

We can introduce or directly refer to fictional characters or objects, not only independently of any description but also of any narrative. Narrative may be the means to capture or discover these possibilities. Literary fiction serves us well in touring the land of individual pure possibilities, existing independently of our discovering them by narratives or by other means. Narrative, like description, may help us discover, capture, or find out individual pure possibilities, to which we may directly refer, on the ground that each of them is an individual possibility, different from any other possibility in the entire realm of individual pure possibilities. Furthermore, you can directly refer to or point out any of your personal, private, subjective possibilities, with or without naming them. While naming them, you intersubjectively refer to your personal pure possibilities. In this case, you utilize language and other means of communication, which does not render this reference indirect, given that it remains strictly independent of any description, interpretation, or narrative and directly accessible to you.

Asking with Rescher, how many lumps of coal lay in Sherlock Holmes's grate, we appear to have no fact-of-the-matter answer (Rescher 1999, p. 407). Indeed, relying on the text alone, the reader cannot answer such a question, for these lumps are not subject to his or her observation or experience. But this fact of uncountability does not render their reality less real, although they are real in a non-actual sense. As fiction, they are as real as actual things, otherwise they are senseless, meaningless, or insignificant for the readers. If Sherlock Holmes lights his pipe by means of a lump of coal, at least one such lump exists in his grate. If he says, "Not even a lump of coal remains in my grate, how can I light my pipe then?"—this would mean or signify something different for the reader, yet it would make sense as regards this text. As for property-decisiveness (*ibid.*, p. 408), it depends on the significance or meanings that the particular item has in the text, in the interpretation, or under the description that the reader has in mind. Second, do no actual, concrete, or particular things exist that are not property-decisive? For instance, the spatiotemporal properties of photons are clearly indecisive, and yet the existence of electrons is beyond any doubt.

But the most significant flaw in this argument by Rescher is of not distinguishing between two kinds of description: that of actualities and that of pure possibilities. Description or interpretation of actualities decisively meets such questions that Rescher suggests, owing to the contingent nature of actualities. Because of this nature, we must rely upon experience and observation to answer such questions. The case of individual pure possibilities is quite different. Describing or interpreting them, we should relate to the necessity about them. In a good "piece" of pure possibilities, for instance, in a literary work of art, in a scientific system such as the periodic table of elements, or in a mathematical system, a necessity determines each detail that makes a difference. If the question about the number of the lumps of coal lying in Sherlock Holmes's grate makes any difference as regards the text, if it has meaning and significance in the context of the story, we are entitled to ask it, and a decisive answer should be found in the text, if and only if it is artistically well made. If not, the question in this context is about an "external" contingent fact that is entirely irrelevant as regards this text, since it does not make a difference or bear significance in it, since no necessity about it can be found within this text.

Consider Kafka's "Before the Law" again. This concise fable is entirely free from any superfluous detail and it does not give rise to any distinction that does not make a difference. Suppose that the reader may ask, nevertheless, for distinctions and details that the fable does not mention at all. For instance, it mentions the fleas in the doorkeeper's fur collar, which the man from the country asked for helping him persuade the doorkeeper to change his mind and to allow him admittance to the Law (Kafka 1961, p. 63). Does it make sense to ask how many fleas are there? Or, how many fleas the man has asked for help? The answer to these questions must be negative, for such questions do not make any sense, insofar as such a literary piece of art is concerned. To raise such questions means to ask for a distinction that makes no difference, at least insofar as the fable is concerned. That is, such a distinction is not necessary at all and, insofar as the fable goes, this distinction or detail is merely contingent, playing no role or bearing no meaning and significance within it. The number of the fleas makes no difference to the fable's significance and meanings. The reality that the fable depicts is not actual, whereas such questions make sense and are valid or legitimate only when we address them to actual reality, in which contingencies naturally occur.

Necessity about individual pure possibilities is what I have entitled "determinism of pure possibilities" (Gilead 2003, pp. 137–141, 146–147), which means that nothing about such possibilities remains undeterminable or contingent, provided that we deal with their significantly relevant relationality. Hence, within the context of a literary work of art or within a psychical reality each pure possibility and its relationality are necessarily determined. As a result, contrary to Rescher, no "ontology of schematically fuzzy, descriptively undetermined possible worlds and individuals" (Rescher 1999, p. 417) should have any room within such contexts. Within them, each pure possibility, which is a real, concrete individual, is necessary, determined, and descriptively decisive.[3] My view of fine stories or illuminating fictions is quite different from Rescher's or other actualists' views of fictions and stories. Questions about actualities are quite different from questions about individual pure possibilities, for the first deal with contingencies and the second with necessities.

As for practical innumerablity, Rescher himself mentions meaningful discussion concerning unanswerable questions about, for instance, the number of individuals who lived thousands of years ago (ibid., p. 415). Though unanswerable in practice, such questions concern significant facts in the history of human evolution (ibid.). By contrast, the number of the lumps of coal as above makes no significance, sense, or meaning at all, for it makes no difference at least insofar as our knowledge or understanding is concerned. Such questions are unresolved as well as meaningless.[4]

Like many other actualists (to begin with Quine), declared or in fact, Rescher has one kind of existence in mind—actual existence. Against this background, he wrongly employs the distinction between possibility *de dicto* ("it is possible for individuals") and possibility *de re* ("there are possible individuals"). Discussing the proposition, "it is possible for spiders to weigh 80 lbs.," Rescher writes, "this does not mean that there is somewhere—in the 'realm of possibility'—some there-actual spider that has achieved this weight" (ibid., 417). Of course, in the realm of possibilities no such *actual* spider exists, but there certainly is a pure possibility of such a spider, since it is not identical with any other possibility, which is all we need to individuate it practicably. Unlike Rescher's view, the concept of reality bears two senses—actual and possible. Equally, *de re* too bears two different senses—actual and possible. Nevertheless, Rescher, like any actualist, does not make such differences at all, on the contrary—he reduces them to the actualist alternative.[5]

As an individual pure possibility, such a spider exists *de re*, although obviously not in the actual sense. *Possibilia* are as real as actualities, and certainly, contrary to Rescher (ibid., p. 418), we have practical ways of identifying or individuating particular pure possibilities, as long as they are not identical one with the other. Contrary to Rescher (ibid., p. 417), possible individuals are not "just like" actual individuals "in nature but merely different in content," for individual pure possibilities are ontologically and epistemologically independent of actualities. The case of eka-elements clearly demonstrates this.

We are absolutely entitled to commit ourselves to ontological realism of the individual pure possibilities, *possibilia*, or possible beings, to possibilism *de re*, which Rescher explicitly excludes (ibid., p. 420). Such possibilism is not conceptualism, which reduces possibility to conceivability. Pure possibilities are independent even of our conceivability of them. We discover them; we do not invent them. Insofar as

---

[3] Contrary to Rescher's "pseudo-individuals," putative individuals, or fictional particulars.

[4] Contrary to Rescher's idea of perfectly meaning unresolved, questions (Rescher 1999, p. 415).

[5] The same holds for the distinction between the possible/contingent and the purely possible/necessary and for that between realization/actualization as well.

pure possibilities are concerned, "invention" is indeed a personal discovery. As a result, I do not accept Rescher's *de dicto* possibilism or conceptualism—the "ontology" of conceptualizable possibilities (ibid.)—for it reduces or limits possibilism to mere conceptualism.

In sum, Rescher's ironic question—How many *possibilia* are there?—is as senseless as the question: How many numbers are there? This inescapable uncountability of numbers by no means renders them unreal or lacking individuality or identification, and the same holds for other *possibilia* or individual pure possibilities. Indeed, what we cannot individuate we cannot count (and Rescher is right on this point), but not everything that we can individuate can we count.

# Chapter 3
# A Panenmentalist Reconsideration of the Identity of Indiscernibles

**Abstract** If we consider any two entities (such as the two spheres in Max Black's well-known thought-experiment) as individual possibilities, pure or actual, they cannot be considered indiscernible at all. Since allegedly indiscernible possibilities are necessarily one and the same possibility, any numerically distinct (at least two) possibilities must be discernible, independently of their properties, "monadic" or relational. Hence, any distinct possibility is also discernible. Metaphysically-ontologically, the identity of indiscernibles as possibilities is thus necessary, even though epistemic discernibility is still lacking or does not exist. Because any actuality is of an individual pure possibility, the identity also holds for actual indiscernibles. The metaphysical or ontological necessity of the identity of indiscernibles renders, I believe, any opposition to it entirely groundless.

The principle of the identity of indiscernibles has been supported and also strongly attacked.[1] Max Black's attack (1952) on it deserves special attention.[2] As I will show below, the identity of indiscernibles can be secured on a panenmentalist basis regardless of any form of the principle of sufficient reason or any other Leibnizian consideration.

Black suggests the following counter-example to the identity of indiscernibles:

> Isn't it logically possible that the universe should have contained nothing but two exactly similar spheres? ... every quality and relational characteristic of the one would also be a property of the other. Now if what I am describing is logically possible, it is not impossible

---

A first version of this chapter was published in *Metaphysica* 6:2 (October 2005), pp. 25–51.

---

[1] Leibniz, Russell, Whitehead, F. H. Bradley, and McTaggart supported it, whereas Wittgenstein (the *locus classicus* is *Tractatus* 5.5302, criticizing Russell and arguing that two distinct objects may have all their properties in common), C. S. Peirce, G. E. Moore, C. D. Broad, and Max Black are among its strong opponents. The support may adopt an idealistic stance, while the opposition is clearly anti-idealistic or empiricist.

[2] Black's arguments have been discussed by Hacking (1975), Adams (1979), Casullo (1982), Denkel (1991), Landini and Foster (1991), French (1995), Cross (1995), O'Leary-Hawthorne (1995), Vallicella (1997), Zimmerman (1998), and Rodriguez-Pereyra (2004). Nevertheless, there is still room enough for alternative treatments of it on quite different grounds (especially different from those of fictionalism, the bundle theory, or haecceitism).

© Springer Nature Switzerland AG 2020

A. Gilead, *The Panenmentalist Philosophy of Science*, Synthese Library 424,
https://doi.org/10.1007/978-3-030-41124-4_3

for two things to have all their properties in common. This seems to me to *refute* the
Principle. (ibid., p. 156)

This counter-example consists of a possible world ("universe") in which no observer
is present and exact duplicates, exactly similar objects, identical twins, and the like,
all of which are indiscernible but not identical, may exist (ibid., pp. 160–62). I will
show why on some panenmentalist grounds no such possible world could exist.
Thus, independently of the question of common properties, relational or not, of
bundles of properties as universals, or of "predicative functions" (the term that
Russell and Whitehead's theory of types employs), I will show why indiscernibles
(or indistinguishables) that are not identical are metaphysically impossible. Even if
Black's aforementioned possible world is logically possible, it is nonetheless metaphysically or ontologically impossible.

Let us begin with the definitions of some terms that I will use in this Chapter.
Regardless or independently of any actuality or actualization, all possibilities are
pure. By "possibilities" I have no possible worlds in mind but individual possibilities (or possible individuals) instead. My possibilist stance is entirely independent
of any conception or semantics of possible worlds. Possibilism is an ontological or
metaphysical view according to which pure possibilities do exist. In contrast, actualism is the view that only actualities exist, and possibilities are merely the ways in
which such actualities might have existed. Possible worlds have been considered
among such ways. Hence, actualism is compatible with some conceptions of possible worlds but not with any ontological standing of pure possibilities (possibilities
*de re*). When we apply "existence" to pure possibilities, the term serves us in a non-
actualist sense. Since pure possibilities are individuals and not universals or bundles
of universals, there are no instances of them. Against many current views (such as
Rescher's 1999, 2003), we are capable of identifying and quantifying or enumerat-
ing individual pure possibilities (Williamson 1998, 1999, 2000, discussing individual "mere" or "bare" possibilities; Gilead 2004b, which is Chap. 2 above).
Furthermore, we can rely upon individual pure possibilities as the identities of actualities. If each actuality is an actualization of an individual pure possibility and of
no other possibility, the pure possibility serves as the identity of the actuality in
question. As pure, such possibility-identity is not spatiotemporally or causally conditioned, whereas any actuality is inescapably so conditioned. Actualities are accessible by empirical means, whereas pure possibilities—logical, mathematical,
metaphysical, or otherwise—are accessible to our thinking and imagination. As thus
accessible, pure possibilities are discoverable as much as actualities are (think of the
discoveries of purely mathematical or logical possibilities, which are not empirical
at all), but this must remain beyond the present Chapter (see Gilead 2004b or Chap.
2 above). As I will argue below, when it comes to individual possibilities, any distinction makes a qualitative difference.

To return to Black's thought-experiment, first we need a criterion of identification to denote or name something. To defend the principle of the identity of indiscernibles, I assume a criterion of identification of pure possibilities that does not
rely upon relational properties and spatiotemporal distinctions. Were such properties

and distinctions inescapably required to establish the principle, Black's view would have appeared to be more sound. Is Black right in stating that mere thinking is not enough to identify or name a thing (ibid., p. 157)? Black assumes that to identify or name anything we need a denotation of an actual object or a unique description of it (ibid.). Such needs not be the case at all. Think, for instance, of eka-elements in the periodic table. Each such element is not actual but is a predicted pure possibility (Gilead 2003, pp. 65–70). Many mathematical theories, let alone all the pure possibilities which they comprise, were discovered only by creative thinking or imagination, while identifying, naming, and describing any of these possibilities have been quite practical with no recourse to actualities. Indeed, to discover, refer to, identify, or name pure possibilities, thinking or imagination is more than enough. We are certainly capable of denoting pure possibilities, each of which is uniquely describable, for, as I will argue below, no two individual pure possibilities can be indiscernible. Second, pure possibilities-identities are necessary for identifying, denoting, searching for, detecting, and describing the relevant actualities, although we also need empirical means to do so.

There are two ways to interpret Black's thought-experiment, which is a counter-example to the principle of the identity of indiscernibles. First, the two spheres are merely two individual pure possibilities.[3] Second, the two spheres are actualities. In the second case, they must be subject to spatiotemporal and causal conditions, as no actuality is exempt from them. In the first case, they are exempt from such conditions altogether, for no pure possibility can be subjected to them. In both cases, the spheres are possible, for any actual thing is possible, too. This means that in both cases we have two possible spheres with the following difference: in the first case, the possibilities in question are pure, whereas in the second—they are actual.

What is precisely the distinction between $b$ as an individual pure possibility and $b$ as an individual actual possibility? The pure possibility in question comprises all the pure possibilities that are open to $b$ under one and the same identity, whereas $b$ as an actual possibility comprises only some of them, namely, only those that have been actualized. The actualization of any of these possibilities does not change the individual pure possibility-identity of $b$, which is one and the same possibility despite any change that $b$ as an actuality may undergo. For instance, James Joyce could have not written *Finnegans Wake* and yet he would have been the same James Joyce under one and the same individual pure possibility-identity (namely, the only possible author of *Dubliners, Ulysses, Finnegan Wake*, or other masterpieces). Note that $b$ as a pure possibility and $b$ as an actuality are one and the same $b$, both comprised in one and the same individual pure possibility-identity. All these distinctions

---

[3] Individual pure possibilities are exempt from any spatiotemporality. Can a sphere as a pure possibility be exempt from space? Yes, it can. Think of any figure, such as sphere, in the analytical geometry, which transforms any spatial distinction to algebraic properties. Although in Kantian terms, even algebraic properties are subject to temporality, since the arithmetic series is subject to it, yet my view is by no means Kantian, especially concerning spatiotemporality and the identity of indiscernibles. As a result, as individual pure possibilities, the two spheres are entirely exempt from spatiotemporality.

are within one and the same pure possibility-identity, which does not render it into separate individuals. In other words, *b* as actual and changeable or *b* as a pure possibility, which is neither changeable nor spatiotemporally and causally conditioned, takes part in one and the same pure possibility-identity. As actual, *b* is the spatiotemporally and causally conditioned part of *b* as an individual pure possibility. No actual individual exhausts all the possibilities that are open to it; it might always have been actually different and yet necessarily remaining one and the same individual under ("comprised in") one and the same individual pure identity-possibility. This possibilism *de re* requires no transworld identity, possible worlds, possible counterparts, or any haecceity (qualitative or nonqualitative "thisness," such as Adams's), each of which appears to give rise to further problems and vagueness instead of providing us with some clear answers.

No two individual pure possibilities might be indiscernible and yet not identical. Independently of any properties, "monadic" or relational, any allegedly "two" indiscernible pure possibilities, discoverable by means of our imagination or thinking, are indeed one and the same possibility. To think about or to imagine two individual pure possibilities necessarily means to distinguish between them, to discern the one from the other, with no recourse to spatiotemporal distinctions at all. Any individual pure possibility is exempt from any spatiotemporal or causal conditions. Hence, no individual pure possibility is spatiotemporally located. If, nevertheless, there are really two of them, they are distinct because they are qualitatively different, not because they are in different places at the same time. They relate one *to* the other because they are different one *from* the other, not the other way round. Since any actuality is of a single pure possibility-identity, necessarily, according to such metaphysics, no indiscernible yet non-identical pure or actual possibilities exist.

Could any actualist counter-argue that s/he had not the slightest idea of how could one have any access to the individual pure possibilities-identities of the two exactly similar spheres in one of the above possible interpretations of Black's thought-experiment? No, for all we need is something like such a thought-experiment to have access to the individual pure possibilities-identities of these two spheres. Indeed, Black unknowingly "provides" these possibilities in his imaginary experiment or logically possible universe, which is not confined to the actual one. All we need is our imagination, within the domain of logical possibilities (as Black assumes on p. 156) or without it, to be acquainted with pure possibilities such as these two. Even if no such spheres existed in our actual universe, Black could suggest his aforementioned thought-experiment because he, like any person who is endowed with imagination, has access to the realm of the purely possible. What makes such an experiment possible is simply our accessibility to that realm by means of our imagination, logic, mathematics, metaphysics, and other ways of thinking, all of which should not be confined to the actual. My interpretation that the two spheres can be either individual pure possibilities or actualities that actualized these pure possibilities holds true for Black's thought-experiment. Black would certainly agree that no two individual possibilities whatever can be identical, for "two" identical individual possibilities are really one and the same possibility. Note that, in this book, whenever I mention "pure possibilities," I have individual pure possibilities in mind and by no means possible worlds or any general or universal possibilities. Generality or universality has to do with the relationality of pure possibilities, not with these possibilities themselves.

The question is: are these two spheres, as pure or actual possibilities alike, not only non-identical but also indiscernible? Like "two" identical possibilities, "two" indiscernible possibilities are simply one and the same. There are not two of them at all. It is easier to realize that in the case of individual pure possibilities discernibility must be obvious. For in that case we have no recourse to actualities or to any of their conditions or terms. Can you think of, or imagine, two individual pure possibilities without discerning one from the other? No, since there are no two indiscernible pure possibilities. Indiscernibility of pure possibilities, if possible at all, would necessarily imply that there were no pure possibilities but only one. As far as pure possibilities are concerned, indiscernibility implies identity. If the two aforementioned spheres are pure possibilities, they must be discernible as well as not identical.

As we shall realize, the same holds true for the two spheres as actual possibilities. As far as actual possibilities are concerned, they too are necessarily discernible as well as not identical. Otherwise, the two spheres, as actual possibilities, would not have been considered two actual possibilities but only one.

Yet Black could answer back on another basis. He would restate his claim that there is no way of telling the spheres apart (ibid., p. 156), which implies, to return to my view, that even if we have access enough to the pure possibilities-identities of the spheres, how can we ascribe possibility $b$, for instance, to one of the spheres, given that we are entirely incapable of telling the spheres apart? In other words, how can I identify one of the spheres as an actuality of possibility $b$ rather than of possibility $c$? In this case, my accessibility to the pure possibilities-identities of the spheres appears not to be helping me to identify any of the actual spheres. Which is which if there is no difference to tell? Yet this would not help Black at all. For the problem of identification or recognition of actualities is epistemological and empirical, not ontological-metaphysical. We have to distinguish between identity, which is ontic, and identification, which is epistemic. We have also to distinguish between identification of pure possibilities, which requires no empirical means, and that of actualities, which requires such means in addition to the identification of the relevant pure possibilities-identities. Suppose that I cannot know which actual sphere is which, I still know for sure that either sphere must be ontologically-metaphysically discernible, for each is an actuality of a different possibility-identity (which is necessarily an individual pure possibility), whether I can tell the difference between the actual spheres or not.

If the spheres in question are actual, they must be different one from the other, for no two actualities can be of one and the same pure possibility-identity. Elsewhere I have shown that multiple actualization or "realization" of any pure possibility should be excluded (Gilead 1999, pp. 10, 28, 2003, p. 94). Apart from this, since any actuality is also a possibility (but not the other way round), and since any indiscernible or non-distinct possibilities are identical, and are one and the same possibility, any two—namely, at least numerically distinct—possibilities cannot be identical and are discernible on ontological-metaphysical grounds. The epistemological discernibility must follow the ontological-metaphysical discernibility of possibilities, pure or actual, not the other way round.

On the grounds of possibilities alone, the identity of indiscernibles is metaphysically secured beyond any possible doubt. Even regardless of their properties, "predicative functions," and relationality, absolutely, no two possibilities can be metaphysically indiscernible, otherwise they would have been merely one and the same possibility. Hence, with possibilities, pure or actual, numerical distinctness and qualitative difference are entirely compatible. No spatiotemporality, any other possible principle of individuation, or property is needed for the discernibility of any possibility. No two possibilities can be indiscernible, let alone identical, whatever are their properties, relational or not. The identity of each actuality is necessarily determined by its individual pure possibility-identity alone. No two actualities can share one and the same possibility-identity.

Note that my panenmentalist view does not acknowledge any spatiotemporal principle of individuation. All those classical empiricists or Kant (according to whom space and time are the forms of intuition or the only factors of individuation), who endorse spatiotemporal principle of individuation (*principium individuationis*) challenge the principle of the identity of indiscernibles in general or Leibniz's principle in particular. For they all assume the irreducibility of spatiotemporal differences to more fundamental or "primitive" factors of individuation. In this respect, Kant challenges that principle. According to him, like Locke, indiscernibles all of whose properties are common are not identical, for they exist in different places at the same time. Hence, this is sufficient to make indiscernibles numerically distinct. In contrast, my view, like Leibniz's, is that numerical distinctness of actualities indicates qualitative difference. Since actualities differ qualitatively; they are numerically different, not the other way round.

Black's possible world in which indiscernibles—duplicated particulars or worlds—are not identical is a narcissistic nightmare: "A kind of cosmic mirror producing real images … except that there wouldn't be any mirror" (ibid., p. 160). For a possible world in which "everything that happened at any place would be exactly duplicated at a place an equal distance on the opposite side of the center of symmetry" (ibid., p. 161) is a world in which no difference exists between an object and its mirror image. Suppose now that on epistemic grounds we cannot distinguish between two poles of a gravitational or magnetic field, two electrons, and the like (Black's examples on p. 162). If Black's possible world is a cosmic mirror, it is inferior to any world in which mirrors exist and in which we can distinguish between any object and its mirror image. Only due to some brain damage do adults become incapable of distinguishing between themselves and their mirror images or of recognizing such images as theirs. Notwithstanding, suppose that we know for sure that two things (two poles, two electrons, an object and its mirror image, and the like) exist in Black's possible world although there is no way to realize any difference between them, such indiscernibility, then, carries no ontological commitment whatever. All we can say is that we do not detect any difference, which is an epistemological question, but we are absolutely not entitled to conclude that no such difference *exists* at all. Unlike Black's examples, in which the presence of an observer changes the possible universe (ibid., which follows some interpretations of quantum mechanics), pure possibilities-identities are discoverable by us yet their existence and the differences they "make" or bear are entirely independent of our knowledge. Think again of eka-elements, mathematical pure possibilities, and the like; these were all discovered, not invented.

The two exactly similar or duplicated spheres that "exist" in Black's possible world are not identical only because, contrary to his argument, they are discernible. For, first, if they are merely pure possibilities, they are necessarily discernible, as no two ("numerically distinct") pure possibilities can be indiscernible. And, secondly, if the spheres are actual, either must be an actuality of a different pure possibility-identity, no matter what relations, spatiotemporal or otherwise, exist between the spheres or between any of them and any possible observer. Thus, contrary to Black's view (ibid., p. 163), there is always a way in which any thing, purely possible or actual, is different from any other. On these grounds, Black's arguments should not convince the readers at all, contrary to the ending of the article (ibid., p. 163), in which interlocutor A in Black's imaginary dialogue declares himself not convinced by B (Black)'s argument, while B responds, "Well, then, you ought to be" (ibid.). This is an excellent example for an "overwhelming" argument, which A is unable to refute and which, yet, is entirely blind to an illuminating insight about the ontological-metaphysical necessity or indispensability of the identity of indiscernibles.[4] I strongly recommend following that insight, which may open one's eyes to realize why that identity is a metaphysical necessity. In this Chapter, I have attempted to support this insight with a panenmentalist argument.

But suppose that Black rejects any possibilist view. Suppose that he argues against me that pure possibilities are merely nonsense (or that they are only *de dicto*, never *de re*), that only actual things can exist, and that his possible world or thought-experiment is not about pure possibilities but about actualities in the very actual world in which we live. Nevertheless, I could answer him again that since any actual thing is possible too, and since two possibilities that no difference exists between them are merely one possibility, the identity of indiscernibles is well secured. In other words, merely on modal grounds, actualist or otherwise, Black's view against the identity of indiscernibles holds no water. On the other hand, if he will not take modality seriously, and if the possible, pure or actual, implied no ontological commitment whatever, Black could defend his view at some unbearable cost, that is, rendering modality and especially possibility ontologically insignificant.

To attempt to persuade the actualist who does not accept any possibilist assumption or principle, the argument that the two spheres are actual possibilities should be good enough. If the term "pure possibilities-identities" do not make sense for actualists, they, nevertheless, must consider the two spheres either as actual possibilities or as the possible modes ("ways") in which the actual spheres might have existed.

---

[4] For some other instructive examples of blind arguments versus illuminating insights see Gilead 2004a.

In either case, those spheres are possibilities too, and no two indiscernible *possibilities* that are not identical can make sense for actualist or possibilist metaphysicians alike.

Let us reconsider the case of two actual "indiscernible" spheres from the aspect of spatiotemporality. In Euclidean space the case appears to be to some opponents of the identity of discernibles, from Kant on, that indiscernibles are not identical, for, sharing all their qualities, they are still "spatially dispersed, spatially distant from one another" (Adams 1979, p. 14), which makes them numerically distinct. Surely, as far as the space in Black's possible world is Euclidean, there are two spheres although no difference between them is discerned. Consider now these two actual spheres as actually possible, namely, as two actual *possibilities*. As possibilities, they are not spatially or temporally dispersed (at most they are spatially or temporally dispers*ible*), for no possibility, pure or actual, is spatially or temporally locatable. As actually possible, the spheres are two, not because they are spatially or temporally dispersed but rather because they are two qualitatively different possibilities and, *hence*, numerically distinct. Temporally dispersed actualities (namely, events) must be first and foremost qualitatively different because their ontological grounds or "primitives"—their possibilities—are qualitatively different. The *possibility* of being spatially or temporally dispersed, which is not spatiotemporally conditioned, is metaphysically prior to any actual spatial or temporal dispersal. In the final account, the pure possibilities-identities, which are absolutely exempt from any spatiotemporality, are the metaphysical-ontological grounds of the qualitative difference as well as the numerical distinctness of any individual actuality. In any case, were the two spheres not actually possible in the first place, they could not be two actual spheres spatially distant from one another. They would have been then one and the same sphere, namely, identical to itself. In this way, too, the identity of indiscernibles is *necessarily* maintained. Individual distinctness, such as numerical distinctness, is intelligible only dependently of qualitative difference (contrary to Adams 1979, p. 17). Black's counterexample to the identity of indiscernibles is thus refuted even when actual spheres in Euclidean space are concerned.

As for a non-Euclidean space or curved time, it has already been shown that on the grounds of spatial or temporal dispersal two indiscernible actualities can be identical.[5] In such space or time, one and the same object may be spatially or temporally distant from itself. Yet, the point is not to show that the identity of indiscernibles is possible but rather that on metaphysical grounds it is necessary, to show that there is no possible single example in which indiscernibles are not identical. Bearing in mind my arguments so far, I have shown that there is no such example and that no such example can be found. As a result, the identity of indiscernibles is necessary, not only possible.

The apparent advantage of my panenmentalist treatment of the question of the identity of indiscernibles is, I think, that it equally holds for pure possibilities and

---

[5] Consult Adams (1979, pp. 13–17), following Black (1952, p. 161) and Hacking (1975). Cf., however, Denkel (1991, pp. 214–15, footnote 3), Landini and Foster (1991, pp. 55–60), and French (1995, pp. 461–466).

actualities and, hence, clearly demonstrates that it is impossible for indiscernibles not to be identical. Both Leibniz's illustration of the discernibility of each leaf of an actual tree and, considering all the differences, C. S. Peirce's "no doubt, all things differ; but there is no logical necessity for it"[6] are aimed at actual things. What I have shown above is that there is a metaphysical or ontological necessity for the identity of indiscernibles, which, I believe, renders any opposition to it entirely groundless. For those who oppose this identity and who also assume that metaphysical and logical necessity are one and the same, the case appears that I have also proven that the identity of indiscernibles is logically necessary. In sum, my arguments, possibilist or otherwise, clearly show that the non-identity of indiscernibles is merely impossible, logically, ontologically, and metaphysically alike.

Finally, it is because any pure possibility is discernible from any other that the possibilities in question do not share all their properties, relational or otherwise, and not the other way round. Because any two pure possibilities are discernible, they must differ in their properties, too. Because any two individual pure possibilities are necessarily distinct and different one *from* the other, they necessarily relate one *to* the other, not the other way round. Hence, Leibniz's principle of the identity of indiscernibles should be modified on that possibilist basis. Every thing must be distinct and different from any other thing, not just because they do not share all their properties, but primarily because their pure possibilities-identities necessarily differ one from the other. Because of this difference, they cannot also share all their properties.

---

[6]As quoted in Black (1952, p. 163); cf. Casullo (1982, p. 595–596), Landini and Foster (1991, pp. 54–55, 58–60).

# Chapter 4
# Two Kinds of Discovery: An Ontological Account

**Abstract** What can we discover? As the discussion in this Chapter is limited to ontological considerations, it does not deal with the discovery of new concepts. It raises the following question: What are the entities or existents that we can discover? There are two kinds of such entities: (1) actual entities and (2) possible entities, which are individual pure possibilities. The Chapter explains why the first kind of discovery depends primarily on the second kind. The Chapter illustrates the discoveries of individual pure possibilities by presenting examples such as the Higgs particle, Dirac's positron, and Pauli-Fermi's neutrino.

## 4.1 Some Examples of Two Kinds of Discoveries

The Standard Model, which is the model that physicists assume to describe correctly all the sub-atomic particles working in nature, is strongly associated with some predictions and fascinating discoveries. Of particular interest are the sub-atomic particles W and Z, which are the weak force carriers, and the Higgs boson, the sub-atomic particle that endows sub-atomic particles with mass. From now on, I will simply mention "particles" instead of "sub-atomic particles."

The 2013 Wolf Prize in physics was awarded to Robert Brout, François Englert, and Peter Higgs "for pioneering work that has led to the insight of mass generation … in the world of sub-atomic particles."[1] This has been considered as an outstanding

---

[1] See the site of the Wolf Foundation (2004). For clear and precise representations of the theory concerning the Higgs boson, consult Gross (2009) and Shears and Heinemann (2006). On 8 October 2013, the Royal Swedish Academy of Sciences announced its decision to award the Nobel Prize in Physics for 2013 to François Englert and Peter W. Higgs "for the theoretical discovery of a mechanism that contributes to our understanding of the origin of mass of subatomic particles, and which recently was confirmed through the discovery of the predicted fundamental particle, by the ATLAS and CMS experiments at CERN's Large Hadron Collider" (italics added). As an anonymous reviewer to Springer Nature commented to me, Brout died in 2011, and there is no posthumous Nobel. Moreover, the Israeli late physicist, Yuval Ne'eman, predicted correctly the mass of the theoretical Higgs boson. See Hargittai and Hargittai 2004, p. 51. In this conversation, Ne'eman mentioned Mendeleev and Linnaeus for their theoretical discoveries thus: "Returning to SU(3), this work is often compared to the work of Mendeleev, who classified the chemical elements. Here in Sweden, the one-hundred-crown bill carries the picture of Linnaeus, who

© Springer Nature Switzerland AG 2020

A. Gilead, *The Panenmentalist Philosophy of Science*, Synthese Library 424,
https://doi.org/10.1007/978-3-030-41124-4_4

*discovery*, and the Wolf Foundation's announcement further states: "The discovery of Brout, Englert and Higgs was essential to the proof ... that the theory with massive gauge particles is well defined; and subsequent calculations in that theory, verified experimentally, culminating in the discovery of the massive W and Z particles."

The particles called "gauge bosons" are the carriers of the fundamental forces of nature. Massive particles are those that have mass. Not all sub-particles have mass: photons, the light-particles, for instance, are massless. In contrast, all material particles must have mass. They gain their mass owing to an interaction with the Higgs boson. Without the Higgs particle, physicists are incapable of explaining why particles have mass. Thus, in the discovery of that particle hangs the fate of the whole of the Standard Model.

Two, quite different, kinds of discovery are mentioned in the Wolf Foundation's announcement as cited above: (1) the discovery of the Higgs particle and (2) the discovery of *actual* particles, W and Z, the carriers of the weak nuclear force, as they were verified experimentally and their masses were measured (in 1983 at CERN). Discovery (2) depended on discovery (1), for it is the Higgs particle that should endow W and Z with mass. We can easily understand what a discovery of an actual particle or entity is. Nevertheless, such was not the case of the Higgs particle until July 2012[2]: until then this boson was not discovered or detected as an actual particle (though in 2011 there were already some tentative empirical signs of its actual existence).

Before 1983, the status of W and Z particles was similar to that of the Higgs boson before July 2012—namely, they were not known then as actual entities, namely, as actualities. As expected, the CERN site defined particles, prior to their discovery as actual entities, as "predicted" or "hypothetical and novel" (CERN 1983). As this Chapter is an ontological account and for reasons that will be explained below, I suggest replacing "predicted particles" and "hypothetical and novel particles," which are quasi-epistemological terms, with "possible entities or existents," which are modal terms that fit well my ontological account. As I will suggest below, these possible entities are real as much as actual entities are, though differently.

Another illuminating example of the two kinds of discovery is that of the positron, the first discovered antiparticle. In 1928, Paul Dirac discovered it (namely, inferred its existence) *on purely theoretical grounds*, whereas Carl Anderson discovered the actual positron in experiments performed in 1932. Dirac referred to this particle as a positively charged electron, whereas Anderson named it "positron." Dirac's Nobel Lecture clearly shows how his notable equation opened up the *possibility* for the existence of a positively charged electron, which "one can infer"

---

classified the plant and animal kingdom. It is the same mode of operation. My tool was group theory" (op. cit., p. 34; cf. p. 43).

[2] The CERN press release of 4 July 2012 announces: "... the ATLAS and CMS experiments presented their latest preliminary results in the search for the long sought Higgs particle. Both experiments observe a new particle in the mass region around 125–126 GeV." See the CERN site at http://press.web.cern.ch/press/PressReleases/Releases2012/PR17.12E.html

(Dirac 1933, p. 321) and which "appears not to correspond to anything known experimentally" (ibid., p. 323). Dirac characterized his discovery as an "inference" (ibid., p. 321) or "prediction" (ibid., p. 323).[3] Still, he was awarded the Noble prize "for the *discovery* of new productive forms of atomic *theory*." On grounds of such theoretical considerations Dirac also predicted the following: "It is probable that negative protons *can* exist, since as far as the theory is yet definite, there is a complete and perfect symmetry between positive and negative electric charges, and if this symmetry is really fundamental in nature, it *must be possible* to reverse the charge on any kind of particle" (ibid., pp. 324–325; my italics. A. G.). It is a prediction of other antiparticles, whose *possibilities* the theory *necessitates* or infers.

In his Nobel Lecture in 1936, Carl Anderson, the discoverer of the actual positron, stated: "The present electron theory of Dirac provides a means of describing many of the phenomena governing the production and annihilation of positrons" (Anderson 1936, p. 368). In this case, too, the second kind of discovery, that of the actual positron, depends on the first kind, which is Dirac's discovery of the possibility of a positively charged electron. Before Dirac's discovery of this *possibility*, such an electron had to be considered as impossibility. It was not until Anderson's discovery of actual positrons that scientists changed their attitude toward this possibility and did not ignore or exclude it any more.[4]

The story of the discovery of the neutrino is even more fascinating. In a famous letter of 4 December 1930, Wolfgang Pauli reported that he had "hit upon a desperate remedy for rescuing" the compatibility of the law of the conservation of energy with the statistically empirical data concerning beta decay. The remedy was "the *possibility* [*Möglichkeit*] that there *might exist* ... electrically neutral particles, which I wish to call neutrons" (Pauli 1994, p. 198; my italics, A. G.). In October 1933, Enrico Fermi reported on his hypothesis and theory concerning the beta decay and the existence of the neutrino (because 2 years after Pauli's discovery, James Chadwick discovered the neutron, Fermi suggested another name for the new possible particle—"neutrino," namely, "the little neutral one"). The experiments performed by Frederick Reines and Clyde Cowan in 1953 and 1955 detected an actual

---

[3] Mark Steiner argues that this is a modern kind of prediction: "Prediction today, particularly in fundamental physics, refers to the assumption that a phenomenon which is mathematically possible exists in reality—or can be realized physically" (Steiner 2002, pp. 161–162). Such a non-deductive, "Pythagoreanized" kind of prediction is of the kind according to which "possible implies actual" (ibid., p. 162), and "[in] the case of Dirac's prediction, then, to predict the positron took courage or faith in mathematics. And the equation which supported this Pythagorean prediction ... was 'derived' by purely formalist maneuvers" (ibid.). As the reader will see, the metaphysical view on which this chapter is based sees this discovery differently—to begin with, the possible does not imply the actual; instead, the actual depends on the possible, which thus conditions the actual.

[4] For examples of the ignoring and misinterpreting of positron tracks before Anderson's discovery, see Segrè 1980 (2007), pp. 191–193. On the grounds of the dependence of the discovery of an actuality on that of its possibility, it is reasonable to assume that had those experimentalists paid enough attention to Dirac's discovery, which was about the possible existence of the positron, they could have, even before Anderson's discovery, correctly interpreted their findings instead of misidentifying them.

neutrino directly and, thus, their work "verifies the neutrino hypothesis suggested by Pauli and incorporated in a quantitative theory of beta decay by Fermi" (Cowan et al. 1956, p. 103). In this case, we encounter, to begin with, a discovery of a possibility, then a quantitative theory that establishes it by calculations and, finally, the discovery of the actual particle (which is not the end of the story, for there were some later discoveries concerning neutrinos, and at present the standard model comprises three kinds of them). Again, the whole story begins with the discovery of a *possibility*[5] on which the discovery of the actual particle depends. From now on, I will call it "Pauli-Fermi's neutrino possibility," although it was Pauli who referred to the neutrino as a possibility (according to Fermi, too); Fermi, on the other hand, referred to it as a hypothetical particle (or "the hypothesis of the existence of the neutrino"), awaiting an experimental confirmation.[6] Yet when mentioning Pauli's idea, Fermi refers to the admitting of the existence of the neutrino as "a qualitative possibility" which squares facts concerning beta decay with the principle of the conservation of energy (Fermi 1933, p. 491). Unlike Dirac's discovery, Pauli's was of a *qualitative* possibility, namely it was independent of mathematical calculations and relied only upon theoretically physical considerations; whereas Fermi added the *quantitative* aspects to the discovered new *possibility*.

Bearing in mind these examples of discovery, it is not clear at all, at least philosophically, *what* is the nature of the discoveries by theorists such as Dirac, Pauli, Fermi, Brout, Englert, and Higgs, which were *not* discoveries of actual entities or facts. Experimentalists discovered the relevant actual entities years later, whereas in the case of the Higgs boson until quite recently there was no decisive evidence of its actual existence. It emerges that the aforementioned theorists discovered some new possibilities. But what is the nature of these possibilities and what is the connection between them and actual reality? Are they merely possible entities? Or, were these not entities at all? And, if not entities or facts, *what* did these theorists really discover? Or, perhaps, were they not discoveries at all but inventions? Perhaps, then, these theorists simply invented, created, envisaged, or stipulated hypotheses, conjectures, or predictions? Or, after all, perhaps they discovered some entities,

---

[5]Yet according to Reines, Pauli termed it also as a "postulate": "Pauli put his concern succinctly during a visit to Caltech when he remarked: 'I have done a terrible thing. I have postulated a particle that cannot be detected'" (Reines 1995, p. 204). I consider possibility (*Möglichkeit*) in Pauli's letter as a preferable version, for it is a direct and authentic statement.

[6]Pauli expressed doubts as to the actual existence of the discovered particle. Hence he wrote in his letter of December 4, 1930: "I admit that my remedy may perhaps appear unlikely from the start, since one probably would long ago have seen the neutrons if they existed" (Pauli 1994, p. 198). Having consulted Hans Geiger and Lise Meitner, he was more encouraged: "from the experimental point of view my new particles were quite possible" (ibid., p. 199). In contrast, the possibility of detecting such a particle was excluded by distinguished scientists such as Bethe and Peierls in 1934 (ibid.). Moreover, Niels Bohr pointed out in 1930 that no evidence "either empirical or theoretical ... existed that supported the conservation of energy in this case. He [Bohr] was, in fact, willing to entertain the possibility that energy conservation must be abandoned in the nuclear realm" (ibid., p. 203). It is a typical way of abandoning or even excluding possibilities that would be discovered later as indispensable for scientific progress.

existents, or facts, unknown as actual at the time of those discoveries? If so, what kind of entities, existents, or facts did they really discover?

My account in this Chapter is mainly ontological. Thus, the discovery of concepts or ideas is not my present concern. Any discovery has to be of something, of some existent or entity. Indeed, concepts can be considered as discovered mental entities, but the discoveries I would like to discuss are of entities that are independent of our mind, whereas psychical entities or concepts undoubtedly depend on our mind. Discoveries about ourselves, in the service of our self-knowledge and of knowledge in general—in philosophy, in psychoanalysis, and in other fields—are most valuable, but they are not my concern in this Chapter.

## 4.2   Creation or Invention

Although the Higgs boson was not discovered as an actual particle until July 2012, as a possible entity it was neither a creation nor an invention. Brout, Englert, and Higgs's theoretical considerations and mathematical calculations implied the discovery of a new particle that the Standard Model had lacked. Thus, the Higgs boson has completed the description of the behavior of all sub-atomic particles and fundamental forces in nature (except for gravitation). This comprehensive description should correspond to reality or nature existing independently of the theory. As taking a necessary part in such a description, the Higgs boson cannot be considered as an invention; it is a discovery. Neither Dirac's positively charged electron nor Pauli-Fermi's neutrino possibility were creations or inventions. They were, however, possibilities whose discovery led to the discoveries of actual particles. Similarly, scientists expected that the discovery of the Higgs boson as a possibility would lead to the discovery of the actual particle (namely, that this particle would be experimentally detected or empirically observed).

Creation or invention is quite different from discovery. Creation or invention produces its objects, which are entities that did not exist before, whereas discovery is of quite a different nature: it concerns what existed before, independently of the discoverer, as the discoverer does not create his or her discovery. In contrast, the products of creation or invention necessarily depend on the creator or inventor (in many of the cases, on the individual creator) and they could not exist without their discoverer(s), whereas the existence of the discovered entities or facts is independent of the discoverer in general and of any individual discoverer in particular. In the natural sciences and in mathematics, we can find some examples of several independent discoverers of the same discovery (the abovementioned example of Brout, Englert, and Higgs's discovery illustrates this perfectly).

## 4.3   Conjectures, Hypotheses, or Predictions

Conjectures, hypotheses, or predictions can be quite common and helpful in scientific discoveries. Major philosophers of science have devoted much thought to the contribution of conjectures and hypotheses to scientific knowledge (following Popper 1968). No less weighty appears to be the contribution of predictions to the discoveries of actual entities or facts.

Nevertheless, conjectures, hypotheses, and predictions are merely means to *discover* some entities or facts. The aim is the discovery, whereas the means to attain this end may be conjectures, hypotheses, or predictions. These are epistemological terms, whereas discovery is an ontological one. My question is: *What* kind of entities, existents, or facts did the abovementioned theorists discover? The answer should be in ontological terms, not in epistemological ones. Thus, their discoveries were not of a "hypothesized, conjectured, or predicted entities," which is a dubious hybrid of epistemological and ontological term (or a quasi-epistemological term). On these grounds, the Higgs boson, as a possible entity, was not a hypothesized, conjectured, or predicted entity; it was a discovered entity, participating indispensably in the Standard Model which describes mathematically the behavior of all subatomic particles and fundamental forces in nature (except for gravitation). Similarly, by means of his equation, Dirac discovered a new particle—a positively charged electron. Such discoveries make predictions possible.

Although it is accepted to characterize the theory that makes it possible to understand and explain the origin of mass—first introduced by Peter Higgs (in 1963) and, independently, by Englert and Brout (in 1964)—as hypothetical in respect of the Higgs field as well as the Higgs boson,[7] I would like to attempt to characterize it differently. Both the Higgs field and the Higgs boson have been discovered entities or existents, though until quite recently they were not known as actual. As the Higgs field possibly permeates the whole universe, it is a discovered possible fact about the whole universe. Instead of "hypothesized facts," I prefer to use "possible facts." Equally, Pauli-Fermi's neutrino possibility should be considered as a possible entity. It was a discovered entity, not simply a hypothesis or conjecture, even though Fermi described it as hypothetical.

As I will argue below, possibilities can be legitimately considered as possible existents or facts for which ontological terms are valid and which are independent of the mind or of the discoverer. Hence, possibilities are discoverable.

Because the aforementioned theorists had discovered possible entities, existents, or facts, they could predict the actual existence of such entities. Yet the prediction is not the discovery. It follows the discovery of the relevant possibility, which precedes and conditions the discovery of the actuality in question. In any case, prediction is an epistemological term, not an ontological one.

---

[7] For instance, Shears and Heinemann 2006, p. 3397.

## 4.4   Fictions and Thought-Experiments

Imagination plays a crucial role in constructing models which have contributed much to the making of discoveries. Model constructing may also employ fictions. Thought-experiments have been found quite useful for some major scientific discoveries,[8] and thought-experiments may consist of fictions. Such fictions, actually truthful fictions, serve us quite well in discovering real possibilities without which some of our most striking discoveries, if not all of them, could not be made.[9] The same holds true for the fictions involved in scientific models. These fictions are indispensable in serving scientists to make discoveries possible. We, thus, reach the same conclusion—imagination and fictions may serve scientists in achieving discoveries, both in theoretical and empirical or actual domains, but there is a major difference between these means and the discovered facts or entities, actual or possible.

## 4.5   Conventions

Is the discovery of possible entities simply a matter of convention? There were conventions about some alleged entities, for instance, phlogiston (in chemistry) and ether (in physics), and as soon as the conventions were discarded, no scientist believed any longer in the existence, possible or actual, of such entities.

Recently, Holger Lyre has raised doubt as to the reality of the Higgs mechanism as follows: "How is it then possible to instantiate a mechanism, let alone a dynamics of mass generation, in the breaking of … a kind of symmetry" which "is in fact a non-empirical or merely conventional one" and which does not possess any real instantiation, namely, realization in the world?[10] Entities that, to our knowledge, have no instantiations or realization, namely, actualization, in empirical reality are,

---

[8] For instance, Szilard's discovery of the nuclear chain reaction, Rowland and Molina's discovery of the loss of the atmospheric ozone layer, and Mullis's discovery of the polymerase chain reaction (Hargittai 2011a, b, c, d, pp. 244–245, 200, and 218–221).

[9] Regarding the vital role that truthful fictions play in discovering real possibilities, consult Gilead (2009).

[10] Lyre 2008, p. 121. Lyre follows John Earman's skepticism that gauge or gauge symmetry is simply a "descriptive fluff," whereas philosophers of science should ask, "What is the objective … structure of the world corresponding to the gauge theory presented in the Higgs mechanism?" (Earman 2004, p. 1239). Likewise, Lyre emphasizes that the symmetry in discussion is "a merely conventional symmetry requirement" (Lyre 2008, p. 121). Thus, he reached the conclusion: "no ontological picture of the Higgs mechanism seems tenable, the possibility of an as-yet-undiscovered process or a mechanism … notwithstanding. But … the Higgs mechanism 'does not exist'" (ibid., p. 128). According to Lyre, the "possible existence of the as-yet-undetected Higgs boson … is a purely empirical question" (ibid., p. 130). The *possible* existence of the Higgs boson has *not* been an empirical question at all; however, its *actual* existence was such a question. It appears that Lyre assumes that actual or empirical facts are the only existing facts subject to discovery.

in Lyre's view, mere conventions. Hence, he assumes, wrongly, that non-empirical entities, which are possible entities, are, as a matter of fact, merely conventions. According to such a view, if the symmetries involved in the theory of the Higgs mechanism are about such entities, there is no real discovery involved, and it is simply a convention that so far physicists have accepted with no philosophical or otherwise critical grounds. This assumption is wrong, for the discovery of possible entities, which at the time of the discovery were not then known as actual (or for their actual existence there was no empirical evidence), can be quite real from a philosophical or scientific point of view, as I will argue, and it may be free of any convention or independent of it. Such is the lesson that I learn from the examples of Dirac's possible positron and Pauli-Fermi's neutrino possibility. Furthermore, in many cases, as in these two examples, such discoveries challenge the accepted views or conventions. Time will tell about the fate of the Higgs boson. Yet I see no reason why its discovery as a possible particle, a discovery made independently by different scientists, would be considered as a convention at all.

## 4.6   Stipulation

Discovery and stipulation exclude one another.[11] The existence of the Higgs boson is not a stipulation; it is a discovery, whether of a possible entity or of an actual one. If the latter, it was discovered by means of the powerful Large Hadron Collider (LHC) at CERN. Similarly, the discovery of the possibility of a positively charged electron was not a stipulation that Dirac's equation required; it was, however, a discovery of a real possibility which was inferred by means of a theory in general and an equation in particular. The same holds for Pauli-Fermi's neutrino possibility. It was undoubtedly discovered, not stipulated. When one stipulates, one does not mean to discover something or to put it to empirical test.

## 4.7   Epistemic Aids and the Discovered Existents

Hypothesis, conjecture, prediction, fiction, thought-experiments, and the like all pertain to the epistemic aids for the discovery and should be discussed in the epistemology of discoveries. Yet the discoveries are not of these aids, the discoveries are of some *existents*, which are independent of these aids. It is impossible to discover something that does not exist, and it is meaningless to state that "one discovered something that does not exist," unless we would like to say that it was not a discovery

---

[11] Hence, Saul Kripke claims: "'Possible worlds' are *stipulated*, not *discovered* by powerful telescopes" (Kripke 1980, p. 44). Cf. "Generally, things aren't 'found out' about counterfactual situation, they are stipulated" (ibid., p. 49). Contrary to Kripke, I think that individual pure possibilities are discovered, whereas fictions about them are possibly stipulated or invented.

at all but simply an illusion or fiction. "Existence" has at least two meanings, only one of which is actual.

Existents pertain to the ontic realm, which is philosophically investigated in light of ontological considerations. It is clear that the Higgs boson has been a purely physical existent of a special kind; until quite recently, it was not known as an actual existent. An actual existent is spatiotemporally and causally conditioned and it is empirically, directly or indirectly, observable or detectable. Until quite recently, there was no empirical evidence of the actual existence of this boson, though the Standard Model necessitates its existence—unless the Higgs boson existed, there was no explanation for the mass that each body or material entity must have. If no empirical evidence for the actual existence of this boson were found, this would have been pulled the ground from under the empirical validity of the Standard Model as a whole. Hence, there is an inseparable, necessary connection between the existence of the Higgs boson and the validity of the Standard Model as a whole. Similarly, there is a necessary connection between Dirac's equation and the existence of a positively charged electron, which is a discovered possible existent, and the positron, which is an actual particle; just as there is a necessary connection between Pauli's neutrino possibility, Fermi's theory concerning it, and the discovery of the actual neutrino.[12] The Higgs boson (as a possible particle), Dirac's positively charged electron, and Pauli-Fermi's neutrino possibility are discovered possible existents. What is the nature of such discovered possible existents? Are they similar to pure mathematical entities and the facts about them?

There is a difference between pure mathematical entities and natural scientific possible entities such as the Higgs boson. Such possible entities or existents, unlike purely mathematical ones, are useless or insignificant in case that they have eventually no empirical or actual validity. Still, both kinds of existents, as we shall see, share something ontologically essential.

## 4.8 Calculation and Measurement

Possible entities are not subject to measurement but to calculation, whereas actual entities—actualities—are subject to measurement, for, unlike possible entities, actualities are subject to empirical observation, spatiotemporal location, and causality. For instance, a point in Euclidean geometry cannot be measured, whereas a dot, an actual point, is measurable. In the Standard Model there are 26 parameters, describing the strength of forces, particle masses and so on, which "must be measured experimentally and then added to the model" (Shears et al. 2006, p. 3396).

---

[12] In a similar vein, it is quite interesting to realize that Pauli wrote in a letter to Niels Bohr on February 15, 1955: "Einstein said to me last winter, … 'Observation cannot *create* an element of reality like a position, there must be something contained in the complete description of physical reality which corresponds to the *possibility* of observing a position, already before the observation has been actually made'" (Pauli 1994, p. 43; the italics are in the original).

While the mass of some particles is very accurately predicted by *calculating* the binding energy of their constituents, until quite recently there was no corresponding theory which could predict the mass of the fundamental particles themselves and that of the Higgs boson itself (Shears and Heinemann 2006, p. 3396). This has to wait for the experimental observations and *measurements*, which are and will be performed at CERN. The masses of fundamental particles are ascribed to the existence of the Higgs boson. This actual existence is waiting for more empirical evidence, which hopefully will also be achieved at CERN.

Calculations are *a priori* accessible and are primarily and directly valid for possible entities (such as the Higgs boson or a positively charged electron), to begin with, whereas measurements are only *a posteriori* accessible and are valid exclusively for actualities. Furthermore, calculations (such as Dirac's equation or Fermi's theory concerning beta decay and the neutrino) are associated with the necessity *about* the calculated facts (which are *possible* facts) and with the necessary relations among possible entities, whereas measurements have to do with actual entities and facts, which can be considered contingent (there is more about this below). Until quite recently, the predictions concerning the Higgs boson were based upon the correction of the calculations of the Standard Model, not upon measurements.

## 4.9 Abstract or Ideal Entities

Are possible entities or existents, such as geometrical entities and the facts about them, abstract or ideal entities? Although this is an accepted view about such entities, I consider it as wrong.

The received view is that geometricians in particular and mathematicians in general abstract from actual drawings of geometrical figures some ideal entities—"a circle," "a point," or "a line," for instance. By means of such abstractions they can make mathematical discoveries. In contrast, it is possible, following Kant or not, to show that mathematical discoveries are entirely independent of actual reality and empirical knowledge.[13]

---

[13] Giaquinto 2007, Ch. 4, "Geometrical Discoveries by Visualizing," shows how it is possible to make geometrical discoveries by visual means *in a non-empirical manner*. He thus relies on Kant in assuming synthetic *a priori* judgments in geometry (ibid., p. 50). Giaquinto's study is clearly epistemological, whereas I focus on the ontological aspects of discoveries. However, referring to Giaquinto's study, Daniel G. Campos discusses a similar view hold by Charles Peirce, for whom "reality is not circumscribed to what actually exists. 'Existing' and 'being possible' both are modes of 'being real'" (Campos 2009, p. 154). Campos relies at this point on Kerr-Lawson (1997). Kerr-Lawson, in turn, assumes that "no mathematical entities are existences in the fullest sense" (ibid., p. 79), as they are "hypothetical objects," and he somewhat connect this view with that of Putnam concerning "mathematics without foundations" (ibid., p. 84). In my view, in contrast, the entities that pure mathematics discovers, albeit purely possible, are real as much as actual existents are and they are not hypothetical objects. Both kinds of entities or existents are subject to discovery.

Unlike its representation or image as an actual dot, a point, for instance, has a position but no dimensions (according to the first definition in Book I of Euclid's *Elements*); it cannot be measured and yet it exists "in" the Euclidean space, namely it is subject to an *a priori* order. A line (according to the second definition in that book) is a "length without breadth," whereas any actual drawn line must have some breadth, however small. Actual drawn circles, lines, or dots are subject to our observation, whereas pure circles, lines, and points are not; only their manifestations, depictions, phenomena, images, or representations in actual space are. Pure geometrical entities are thus "ideal," but this does not necessarily make them abstractions from actual reality. Neither are they idealizations of empirical facts, for if they were, they should be idealized according to some ideal standards or paradigms, which, in turn, must be entirely independent of empirical facts and observation.

Actual drawn mathematical entities are actualities or depictions of purely possible mathematical entities. To identify actual mathematical entities we must rely upon such possible mathematical entities. Thus, to identify a dot as an actuality of a point we first must have access to the point as a possible entity. I argue this not on platonic grounds. I do not rely upon platonic paradigms or Ideas. I think about quite different entities, as I will explain below.

On such grounds, I do not consider mathematical entities as idealized abstractions from actual entities. Mathematical discoveries are, instead, of possible entities, which are not idealized abstractions from actual ones but they precede anything actual.[14]

Mathematical proofs are necessarily valid for all possible relevant cases, whether actual or purely possible. Such cannot be the case of abstractions, however idealized, for abstractions first rely or, rather, are contingent, on some actual cases, from which they are abstracted, whereas the mathematical proof must be valid for *every* relevant possible case. Relying upon some actual cases, not upon all relevant possible cases, must make the case empirical and hence, contingent, and instead of a deductive proof we would rely only upon an inductive one. To assume that the inference and logics involved are *a priori* cognizable is not sufficient to substantiate the proof as universally valid, entirely independently of actual contingent cases, for logics is purely formal, whereas mathematics is different from formal logics, as it deals with contents and not with logical forms only. Hence, what makes pure mathematics exempt from actual constrains is not only its logical aspect; what does so is equally its purely mathematical aspect.

---

[14] Discussing mathematical discoveries, Gian-Carlo Rota relates to mathematical possibilities and proposes that "a rigorous version of the notion of possibility be added to the formal baggage of mathematics" (Rota 1997, p. 191). In Rota's view, "[e]very theorem is a complex of hidden possibilities. … the proof of Fermat's last theorem foreshadows an enormous wealth of possibilities" (ibid., p. 195); or "proofs of theorems of Ramsey type are an example of a possibility that is made evident by an existence [non-constructive] proof, even though such a possibility cannot be turned into actuality" (ibid., p. 185). The happy expression "open up new possibilities for mathematics" (ibid., p. 190) in its various forms (ibid., pp. 191, 192, and 195) is a valuable leitmotiv in Rota's paper.

The discoveries of possible mathematical entities clearly show that they are primarily valid for possible facts that are associated with objective necessity, which cannot be ascribed to invention. That the sum of the angles of any Euclidean triangle is exactly 180°, for instance, concerns not only facts about any possible and actual Euclidean triangle; it also concerns the *necessity* about these facts. It is a necessary fact; there is nothing contingent about it. There is no Euclidean triangle that can be exempt from this fact. The mathematical discoverer must admit this necessity; he or she is not entitled to free his or her mathematical way of thinking from it to invent any Euclidean triangle that is not subject to this necessity.

Purely mathematical entities are not purely physical entities. So what about our Higgs boson as a possible particle? Are my considerations about the existence of purely mathematical entities valid for possible ("purely theoretical") particles such as the Higgs boson? After all, if alas, no empirical evidence of the actual existence of this boson had been found, this possible entity would have become useless or insignificant for physicists. Such cannot be the fate of purely mathematical discoveries. Nevertheless, like purely mathematical entities, the Higgs boson was *not* an idealized abstraction from any empirical data or actual facts. Dirac's equation, Fermi's theory of the beta decay, the Big Bang model, and the Standard Model are not such abstractions. Instead, they comprise discoveries of possible entities and their relationality (the general term concerns all the ways in which entities relate one to the other). The existence of these entities has been independent of actual physical reality and it conditions the discoveries of the actual facts for which the models are valid. The possible existence of the Higgs boson is a necessary condition for its discovery and identification as an actual entity, which must be left to actual reality and empirical evidence. The possible existence of Dirac's positively charged electron was a necessary condition for its discovery and identification as an actual entity—the positron. It was Anderson who found the empirical evidence for the existence of the positron as an actual entity. The same holds for Pauli-Fermi's neutrino possibility and the discovery of actual neutrinos by Reines and Cowan. In each of these cases, the theorist's discovery of the possible particle opens the way for the experimentalist's discovery or detection of the relevant actual particle.

My view concerning the discovery of possible entities rests neither on platonic nor on Kantian grounds. The way I consider possible entities, mathematical or purely physical, is quite different, for it rests upon the idea of individual pure ("mere," non-actual) possibilities and their necessary relations (in a general term—relationality).

## 4.10   Possible Existents as Individual Pure Possibilities

If possible existents are not idealized abstractions, then what are they? They are individual pure possibilities, which are real as much as actualities are, albeit differently. Regardless or independent of anything actual or of any actualization and exempt from any spatiotemporal and causal conditions, each individual possibility is pure. Individual pure possibilities are entirely independent of "possible worlds" as well as of any mind. The *concepts* of such possibilities are *de dicto*, but the possibilities themselves are possibilities *de re*.[15] As possibilities *de re*, individual pure possibilities are entirely independent of any mind and any concepts, and thus they are discoverable by us. We discover new individual pure possibilities, which are different from other pure possibilities, with some of which we are already familiar, and from known actualities as well.

To exist, any entity has first *to be* purely possible, to be a pure possibility. If an individual entity fails to exist as an individual pure possibility first, it cannot exist at all. Each existent, whether actual or not, has first to satisfy this *ontological* condition. Having this primary ontological condition satisfied and only then, the secondary ontological condition as to what are the spatiotemporal and causal circumstances under which this entity can or cannot actually exist, may or can be satisfied. Hence, the existence of any individual entity, whether actual or not, depends primarily on the existence of its pure possibility. Individual pure possibilities are thus the most fundamental existents.

No two pure possibilities can be identical—the law of the identity of the indiscernibles is necessarily valid for pure possibilities, which are exempt from spatiotemporal and causal conditions or restrictions. In other words, no two possibilities can be only numerically different, whereas two allegedly identical actualities can be only numerically different, for they exist at different places in the same time or at the same place in different times. Such cannot be the case of pure possibilities, which are exempt from any spatiotemporal restrictions; hence, the law of the identity of the indiscernibles is necessarily valid for them. Any "two" "identical" pure possibilities are, thus, one and the same possibility, and each pure possibility is an identity, too—the identity of the actuality that actualizes this pure possibility. Hence, in this Book, I use the expression "pure possibilities-identities." With no access to the relevant pure possibilities-identities, scientists and laypersons alike may be doomed to be blind to the identity of phenomena or entities they may encounter.[16]

Because no two pure possibilities can be identical, each pure possibility is necessarily *different from* all the others. On this basis, each pure possibility necessarily *relates to* all the others. As a result, the realm of pure possibilities shares a universal unifying or systematic relationality.

In 1928, Dirac's discovery of the positively charge electron was of a pure possibility-identity, which was a necessary condition for the discovery of the

---

[15] Thus, I do not follow Nicholas Rescher's conceptualism, replacing a "possibilism that is substantively oriented (*de re*)" by one that is "proportionately oriented (*de dicto*)." See Rescher (1999, 2003). For a critique of this powerful view see Gilead (2004a, b; Chap. 1 above). Nor I confine possibility to conceivability. There is much more to pure possibilities than conceivability, and the existence of individual pure possibilities does not depend on our mind.

[16] See, for instance, "We might marvel that Rutherford and Hahn did not grasp at the time the concept of isotopism, as they had discovered clear examples of isotopes; but when the mind is not prepared, the eye does not recognize" (Segrè 1980 [2007], p. 58). This is a good example of the indispensability of the discovery of a pure possibility-identity for the discovery and identification of the relevant actual entity or fact. At that time, Rutherford and Hahn did not consider the possibility that one and the same chemical element could have been two different physical entities. At the time, this was wrongly considered to be impossible.

actual positron. The same holds for Pauli-Fermi's neutrino pure possibility and the discovery of the actual neutrino.[17] Dirac's discovery of the pure possibility-identity of the positively charged electron and of those of other antiparticles predicted and paved the way to the discovery of the actual positron as well as other actual antiparticles. Until quite recently, the discovery of the Higgs boson was only the discovery of a pure possibility-identity. Without this fascinating discovery, physicists could not have predicted the actual existence of the Higgs boson, nor could they have explained how particles have mass, and how matter has been possible. As long as physicists had neither established evidence for the actual existence of the Higgs boson nor such evidence of its actual nonexistence, they still had well-established theoretical reasons to acknowledge its existence as a pure possibility. There appeared to be nothing to exclude it (despite some philosophical doubts). Thus, physicists thought that such a possibility *must exist* and should not be excluded; they knew *a priori* how to *identify* it; they understood and explained *why* it had to exist; they expected to discover its actual existence; and, thus, they *predicted* this discovery.[18] As a pure possibility-identity, the Higgs boson has been a necessary existent, owing to the relationality within the scope of the Standard Model.[19] This model necessitates the pure possibility-identity of the Higgs boson, and this necessity is independent of actual physical reality and empirical physical observations or experiments. In contrast, the physical utility and significance of this possibility depends, nevertheless, on actual physical reality and empirical observations or experiments, namely, on an empirical validity. The same holds true for the positron: the relationality of pure possibilities-identities involved in Dirac's theory and equation requires or necessitates the *possibility* of a positively charged electron regardless of actual reality and empirical observations or experiments.[20]

---

[17] Frederick Reines entitled his Nobel Lecture (for the detection of the neutrino) in December 1995—"The Neutrino: From Poltergeist to Particle." In light of my view in this Chapter, I would like to rephrase this title thus: "The Neutrino: From Pure Possibility to Actual Particle." It is striking how both Pauli and Fermi were closely attached to Reines's detection of the neutrino. It was Fermi who was more an experimentalist and from whom Reines took advice since 1951 about the "possibility of the neutrino detection" (Reines 1995, p. 202; 206–208), and Pauli was the first to be informed by Reines's telegram about the detection: "We are happy to inform you that we have definitely detected neutrinos from fission fragments by observing inverse beta decay of protons" (ibid., p. 214). Pauli responded: "Everything comes to him who knows how to wait" (ibid.).

[18] As CERN Director General, Rolf Heuer, put it in July 2011, "We know everything about the Higgs boson except whether it exists."

[19] Although this model may have possible alternatives, the necessity under discussion holds true also, though differently, for the Higgs bosons in alternative models. Riccardo Barbieri, Lawrence J. Hall, and Vyacheslav S. Rychkov found it justified to consider possible alternative roads for physics beyond the Standard Model. See Riccardo Barbieri et al. (2006). For another possible alternative, see T. Gregoire et al. (2004). Also consider Shears and Heinemann 2006, pp. 3402–3403, for the Supersymmetry Model's prediction of five kinds of Higgs boson as well as for other possible alternatives.

[20] See Dirac 1928, p. 612. It was crucial that Dirac negated the attempt to exclude the very (pure) possibility of a positively charged electron. In this way, he opened up new possibilities for particles

Similarly, Fermi's theory and calculations about the beta decay and the neutrino *necessitate* Pauli-Fermi's neutrino possibility. Given that conservation of energy is valid for the beta decay, this particle *must* exist! Though "if you didn't see this particle in the predicted range then you have a very real problem" (Reines 1995, pp. 203–204). This reminds me very much of some quite recent thoughts about the Higgs boson.

Contrary to Steiner's interpretation (2002, p. 162), in Dirac's case possibility does not imply actuality; instead, the actual is an actualization of pure possibilities, existing independently of our mind (and likewise in the case of the neutrino). By means of our theories, mathematical or physical, we gain access to these possibilities. The necessity in discussion is not necessarily a deductive relationality; there are many kinds of pure possibilities, each is necessarily different from the others, and each necessarily relates to the others. Dirac's discovery is about *physical* pure possibilities, which necessarily relate to one another and which are described *mathematically*. Mathematical description has been indispensable for any physical discovery since Galileo's days until our own. Whether Dirac was a "Pythagorean" (in Steiner's terms) or not, this does not necessarily reflect on his discovery of the positively charged electron, as long as we consider it as a pure possibility in the view that I present in this Chapter.

The relationality of mathematical pure possibilities in any mathematical proof is necessary, independent or regardless of any contingency and actuality or actualization. Unlike physically possible entities, the significance and strength of mathematical entities is independent of actual reality and empirical observations or experiments. Yet, physics, theoretical or applied, cannot exist without a strong reliance on mathematical language. Physics thus depends on mathematical pure possibilities and their necessary relationality. Pure mathematics and pure physical theory enable the discoveries of the pure possibilities-identities without which experimentalists cannot make the discoveries of the relevant actualities.

## 4.11   How Is *a priori* Accessibility to Pure Possibilities Possible?

Our intellect and imagination are good enough to allow us access, however limited (as we are limited beings), to the realm of pure possibilities. Pure possibilities are certainly different from actualities. Observing actualities, we always can free our thought from actual constraints and think about—discover—pure possibilities, which are different from them. Our imagination and intellect help us to do so in many occasions. For some reason or other, we pay more attention to our capability of abstraction, and we are inclined to forget that we can abstract because we can

---

physics. On the way that excluding possibilities may result in blocking scientific progress and, in contrast, how saving possibilities contributes to this progress, see Gilead 1999.

think about pure possibilities that are different from the actualities with which we are already familiar. Moreover, we can think about pure possibilities that are quite different from the pure possibilities with which we are already familiar. On these grounds, we are capable of discovering new pure possibilities. In other words, we have accessibility to new pure possibilities even though and because they are different from all the actualities as well as of all the pure possibilities with which we are already familiar. We do not need empirical observations and experiments to have access to new pure possibilities. We can thus rely upon our thinking, intellect, and imagination to gain such access, which is certainly good enough to put forward our mathematical and pure scientific theories. Our accessibility to the realm of pure possibilities relies upon the universal relationality of all pure possibilities as well as actualities, insofar as they are actualization of the relevant pure possibilities-identities. Every possibility, pure or actual, including the possibility of one's mind, is different *from* the others and, hence, it relates *to* the others. This provides us with sufficient accessibility to the realm of pure possibilities, and this accessibility is *a priori*.

## 4.12   Panenmentalism and Its Uniqueness and Originality

Until quite recently, I was not aware of the applications of panenmentalism to the discoveries discussed in this Chapter. Philosophy of science is one of the domains included in this systematically comprehensive metaphysics. This domain is the context in which this Chapter is embedded. Panenmentalism is a theory about individual pure possibilities and their universal relationality.

Panenmentalism is entirely different from any kind of possibilism known to me and it opposes actualism. The philosophical view that does not admit individual pure ("mere", non-actual) possibilities altogether or at least as existing independently of actual reality—is called "actualism," whereas the view that does acknowledge such possibilities I term "possibilism." Actualism is generally allowed to use the idea of possible worlds and possible world semantics.

To the best of my knowledge, no actualist theory, including the most recent ones, admits the aforementioned absolutely independent existence of individual pure possibilities *or*, more traditionally, even any existence of them (consult, for instance, Bennett 2005, 2006; Nelson and Zalta 2009; Contessa 2010; Menzel 2011; Woodward 2011; Vetter 2011; Stalnaker 2004, 2012).

Challenging actualism, panenmentalism is a possibilism *de re*, according to which pure possibilities are individual existents, existing independently of actual reality, any possible-worlds conception, and any mind (hence, they are not ideal beings). To the best of my knowledge, panenmentalism differs from any other kind of possibilism. Claiming that, it is not in my intention to argue that it is preferable to the other kinds; I say only that it is a novel alternative to them.

The following are the main features in which panenmentalism differs from other kinds of possibilism: First, panenmentalism is strongly realist about individual pure possibilities, which are thus independent existents rather than mere "beings" or "subsistents." Over this point, panenmentalism disagrees with Meinonigians, Neo-Meinonigians (to begin with Richard Routely [Sylvan]; see Gilead 2009, pp. 23–27, 33–38, 46–47, 83–91, 109–113, and 121) and their followers (such as Graham Priest, Nicholas Griffin, Terence Parsons, and Edward Zalta). Second, it dispenses with the idea of possible worlds, which almost all the possibilists known to me have adopted. As I have explained above, among other reasons, this idea is quite problematic for various reasons: for instance, it is not clear enough, and there are many controversies about it with no universal or long-standing consent; the problem of the epistemic accessibility from one world, especially from the actual world in which we live, to any other possible world does not appear to have a satisfactory solution; and if we can dispense with this idea and find a satisfactory, clearer and simpler, alternative to it, we should take this possibility into consideration. Third, panenmentalist pure possibilities are not abstract objects or entities, neither are they potentialities, for abstractions (as abstracted out of actualities or actual reality) and potentialities depend on actualities which are ontologically prior to them, whereas pure possibilities are ontologically prior to and entirely independent of actualities. Four, though using truthful fictions, panenmentalism, acknowledging the full, mind-independent reality of pure possibilities, differs from any kind of fictionalism, especially modal fictionalism (Gilead 2009, pp. 80–83; this difference holds also for Kendall Walton's make-believe theory). Five, as mind-independent, pure possibilities are not concepts, hence panenmentalism is possibilism *de re* and not conceptualism or possibilism *de dicto*.

If some readers may think that the panenmentalist pure possibilities allegedly remind them of Edward Zalta's "possible objects" or "blueprints" (Zalta 1983; and McMichael and Zalta 1980) or of Nino Cocchiarella's "possible objects" (Cocchiarella 2007, pp. 26–30; Freund and Cocchiarella 2008), such is not the case at all. First, these possible objects rely heavily on possible-worlds conceptions. Second, according to Cocchiarella's conceptual realism, framed within the context of a naturalistic epistemology, abstract intensional objects "have a mode of being dependent upon the evolution of culture and consciousness" (Cocchiarella 2007, p. 14), whereas panenmentalist pure possibilities are entirely independent of such evolution and of its naturalistic context as well and are *a priori* accessible. Third, following Meinonigians and Neo-Meinonigians, both Zalta and Cocchiarela consider "actual" and "exists" as synonyms, while, in their view, possible objects are merely "beings." In contrast, panenmentalism treats both pure possibilities and actualities as existents, though in different senses of the term "existence" (distinguishing between the existence of pure possibilities and that of actualities—the former is spatiotemporally and causally conditioned, while the latter is entirely exempt from these conditions).

Although David M. Armstrong adopts a special kind of possibilism (such as "possibilism in mathematics") and is committed to mere possibilities, namely, those without instantiation (Armstrong 2010, pp. 89–90), this is not a possibilist view in my terms: in Armstrong's view, these possibilities do not exist (ibid., p. 90), as he states that the only existence is spatiotemporal. Hence, his hypothesis is that there are no objects outside space-time (ibid., p. 5). Furthermore, though as a "one-worldler," Armstrong rejects other possible worlds (ibid., p. 16), the mere possibilities that he adopts explicitly supervene on the actual (ibid., p. 68). Thus, they are not pure possibilities in my terms (that is to say, entirely independent of anything actual). Finally, if mere possibilities are not existents, in what sense are they discoverable?

Since Yagisawa's modal realism heavily relies on the conception of merely possible worlds in which there are mere possibilia (Yagisawa 2010), I do not follow his view, either. The same holds for his distinction between "being" and "existence" or between "reality" and "existence." With panenmentalism, all individual pure possibilities are full-fledged existents—not only "real" ones. As for the problem of transworld identity, it does not exist for panenmentalism, avoiding the idea of possible worlds altogether.

With panenmentalism, pure possibilities are the possibilities-identities of actualities. Each actuality has a pure possibility-identity, which cannot be shared with other actualities. This makes panenmentalism a unique kind of nominalism. Universal terms and laws rest upon the relationality of individual pure possibilities. Our accessibility to the realm of pure possibilities is *a priori*, whereas actualities are only *a posteriori* cognizable. Thus, our knowledge of actualities can be empirical only. Panenmentalism as a whole is thus neither empiricist, nor rationalist; yet it is rationalist about our knowledge of pure possibilities, and empiricist about our knowledge of actualities.

Necessity pertains to the existence of individual pure possibilities and to their relationality. Necessity also pertains to the inseparable connection between any pure possibility-identity and its actuality. There is no necessity at all about actualization. Thus, not all pure possibilities, albeit actualizable, are actualized, and the so-called "principle of plentitude" is *not* valid for actualities.[21] The contingency about each actuality is strongly compatible with the *a posteriori* and empiricist nature of our knowledge of actualities. The necessity about pure possibilities and their relationality can be *discovered* by means of logical, mathematical, and other theoretical considerations (including truthful fictions), but the discovered possibilities and their relationality are entirely independent of these considerations or means. As for our knowledge concerning the actualization of such discoveries, it is entirely subject to empirical observations and experiments.

But, if all actualities are contingent, what is the point in predicting actual existents on the grounds of pure possibilities-identities and their relationality? The crucial point is that only on the basis of such predictions can scientists empirically recognize,

---

[21] In contrast, Arthur Lovejoy's famous principle—"Possible implies actual" (Steiner 2002, p. 162) or "Any genuine possibility actualizes at some moment in an infinite time" (Bangu 2008, p. 249)—has been considered as inspiring the praxis of modern physics. For instance, Helge Kragh associates this principle, in its version as Gell-Mann's "totalitarian principle"—"Anything which is not prohibited [namely, possible] is compulsory"—with Dirac's reasoning (Kragh 1990, p. 272). On the aforementioned panenmentalist grounds, I see Dirac's discovery as well as the other discoveries discussed in this chapter in quite a different light.

identify, understand, and explain the predicted actual entities. The discovery of the actual positron, of the actual omega minus particle, of actual particles such as W and Z, of some predicted actual elements in light of the eka-elements in the Periodic Table, and many other discoveries of actual entities are fine examples of demonstrating this crucial point. The *a priori* acquaintance with pure possibilities-identities made the discovery of the relevant actualities really possible. In contrast, excluding some possibilities on whatsoever grounds has hindered scientific progress (for instance, in the case of isotopes, the advent of quasicrystals [Gilead 2013] and others).

Pure possibilities are not ideal entities, which depend on our mind. We discover pure possibilities just as we discover actualities, though truthful fictions may help us greatly in discovering pure possibilities, which are independent of our mind. Thus, panenmentalism is not Kantian either. Pure possibilities and actualities are "things in themselves," not phenomena. This does not render our knowledge absolute or exempt from failure; on the contrary, although our accessibility to the realm of pure possibilities is *a priori*, our knowledge of it is quite limited. We know quite a little about pure possibilities and even more so about their *universal* relationality. It is inevitable that we are also subject to mistakes and errors about existents, possible or actual. After all, on grounds of "lazy" or convenient conventions, received views, preconceptions and so on, we quite habitually exclude vital pure possibilities, which are indispensable for our discoveries and scientific knowledge, and thus hinder scientific progress and fail in our aiming at truths. Nevertheless, because my discussion in this Chapter focuses on ontological considerations about discoveries, I do not discuss these major epistemological problems in this occasion.

Again, individual pure possibilities are not members of any possible world. Panenmentalism is exempt from possible-worlds semantic or metaphysics. As is well known, the idea of possible worlds has served actualists who have denied the existence of individual pure possibilities, which are entirely independent of actual reality in general and of actual individuals in particular.

## 4.13 A Metaphysical Platform

As all individual pure possibilities universally relate to each other, there is a metaphysical platform for embedding all there is in a universal system. On this platform, physical pure possibilities and their relationality also rest. The Standard Model reveals not only the symmetries that govern physical reality as a whole but also discovers how the breakings of these symmetries, which made it possible for particles to gain mass and to be material particles, are restored. Symmetry plays a crucial role in modern physics, not for aesthetic reasons and not necessarily for mathematical reasons,[22] I think, but because symmetry is a universal and unifying relationality

---

[22] Steiner argues that predictions by the use of symmetries "are of the (nonreductive) 'possible implies actual' variety because symmetry conditions define more what cannot occur rather than what must occur" (Steiner 2002, p. 162). According to panenmentalism, in contrast, because any

of the multiplicity in nature under simple common law.[23] Note that the classical function of symmetry has been of "harmonizing" the *different* entities into a unified whole; whereas the modern concept relies *also* on equal entities, but still the relationality among the entities, whether different or equal, is maintained.[24] Panenmentalism bases the relationality of pure possibilities on their differences, as no two pure possibilities can be identical.

According the Standard Model, the massless photons can reach any point in the universe; they can spread themselves infinitely. Whenever the symmetry in the universe breaks, the omnipresent photons "mend" this and restore the symmetry. According to the panenmentalist metaphysical platform, photons thus *actualize* the basic universal relationality in the physical universe, for the photons communicate each distinct part of this universe to all the rest. The actualized relationality is the symmetry, apparent or hidden, that governs physical reality. We may say that the Standard Model thus discovers this symmetry in the two senses of discovery which this Chapter explicates: (1) the discovery of the relationality among all the physical individual pure possibilities (of particles and forces) and (2) the discovery of this symmetry as an actual fact about physical reality.

## 4.14    Two Kinds of Discovery

In sum, there are two kinds of discovery: (1) discoveries of possible entities, which are individual pure possibilities-identities, and of their relationality; (2) discoveries of actualities. The second kind of discovery depends on the first kind. To discover new actualities we have to discover their pure possibilities-identities first or, at least, not to exclude these possibilities but to admit them, knowingly or unknowingly. One

---

actuality is contingent, no scientific prediction can be about "necessary" actual existence. The necessary relationality of pure possibilities does not imply actual necessity, whereas the inseparable connection between any pure possibility-identity and its actuality is necessary, though there is no necessity about the existence of any actuality. Given these restrictions, well-established predictions on grounds of symmetry may be very helpful scientifically, for instance, in the case of the discovery of the omega-minus particle. For an opposite view, questioning even the scientific status of such a prediction, which appears to be merely an educated guesswork, consult Sorin Bangu concerning the discovery of omega-minus (2008, pp. 256–257).

[23] For this reason, symmetry has occupied the attention of physicists until the present: for instance, Pierre Curie's interest in crystals' symmetry; classical crystallography and the quasicrystals; symmetry's role in the special and general theory of relativity; in quantum mechanics; and in the Standard Model. One of the most illuminating insights of Pierre Curie was about the importance of symmetry in determining which phenomena are possible. An anonymous reviewer to Springer Nature reminds me at this point of "the Woodward-Hoffmann rules in organic chemistry about the symmetries of molecular orbitals of the reagents in deciding whether a reaction would be allowed or forbidden."

[24] Cf. Brading and Castellani (2008). As for permutation symmetry, it is a moot point whether the law of the identity of the indiscernibles is valid for quantum physics (ibid.), whereas panenmentalism applies it to every individual pure possibility.

of the major hindrances in the path leading to scientific and other discoveries is our inclination to exclude possibilities from the outset. For instance, the discovery of quasicrystals was greatly hindered by the supposition that such crystalline structures were theoretically and empirically impossible.

Were the positron and other antiparticles not purely possible in the light of purely physical theory, all that we know today in physics could be entirely different and some major discoveries of some actual antiparticles would not be possible from the outset. The same holds true for the Pauli-Fermi neutrino possibility and the application of the principle of the conservation of energy on the subatomic reality. Were the Higgs field and the Higgs boson not purely possible from the outset, namely, in the light of the purely physical theory of the Standard Model, physicists could not attempt to discover their actualities, and our understanding, explaining, and knowledge of the universe would have been much less than they are today.

# Chapter 5
# Mathematical Possibilities and Their Discovery

**Abstract** This Chapter attempts to answer the following questions: What is the ontological status of mathematical entities? What enables our epistemic access to such entities? Do we discover mathematical entities or simply invent them? Mathematical entities do not exist in space, time, or as links in a causal chain. Nor are they abstracted from actual-physical reality. Instead, mathematical entities are individual pure possibilities existing independently of our mind, of possible worlds, and of anything actual. Our intellect (by means of proofs, calculations, and the like) and imagination are reliable enough to give us sufficient epistemic access to such entities. Mathematical entities can be discovered; they cannot be invented. Nevertheless, truthful fictions may help us discover such entities as well as their general relations.

## 5.1 The Questions

This Chapter attempts to answer the following questions: What is the ontological status of mathematical entities? What enables our epistemic access to such entities? Do we discover mathematical entities or simply invent them?

## 5.2 Pure Mathematical Entities Are Not Actualities

Undoubtedly, we may think about or imagine pure mathematical entities (or objects) that, to the best of our knowledge, never have existed and are not expected to exist in the future as actualities, as actual-physical entities. As we shall see, "existence" has two meanings—purely possible and actual; to be, to exist, any being has to be purely possible first. As I will attempt to show below those mathematical entities are not fictions but real, true existents, entirely meaningful or significant, at least mathematically speaking. In other words, actual existence is, at least, not a necessary condition for the existence of mathematical entities. We can do mathematics that is entirely exempt of any actual conditions. Henceforth, I use and mention "actual" and "physical" interchangeably because both of these terms are equally valid for existence which is subject to spatiotemporal and causal conditions. There is no

© Springer Nature Switzerland AG 2020

A. Gilead, *The Panenmentalist Philosophy of Science*, Synthese Library 424, https://doi.org/10.1007/978-3-030-41124-4_5

actuality that can be exempted from these conditions. The same holds for any physical existent. Hence, there is no actualization except for the physical one. We thus assume that pure mathematical entities do not exist under spatiotemporal or causal conditions and, as such, they are not subject to any empirical observation or experiment.[1] They are not empirical or actual entities. Nevertheless, they are accessible to our intellect and imagination.

To argue so, we first have to deal with an expected objection: Although mathematical entities are not actualities, they are still abstractions, idealizations, or descriptions based on our experience, by means of observations or of experiments, of actual reality. I meet this objection in the following thought-experiment. Take, for example, Euclidean geometry. Imagine that the actual reality in which we exist would not have been subject to even one definition, axiom, postulate, or proposition included in this geometry, and yet this geometry would have been possible with no change at all in its validity, consistency, and coherence, namely, of its *truth*.[2] It could have been applied to entirely non-actual or counter-actual reality, and yet would have lost nothing of its validity or strength. As a matter of fact, even in such a case this geometry would not change at all. The earth that it would have "measured" would be purely possible or counter-actual. We can imagine such a possibility quite clearly and refer to it as a real or truthful one. The gist of the matter is that even in such a case, Euclidean geometry would have been invariable. The same holds true for any pure geometry, Euclidean or non-Euclidean, and for any mathematical pure theory. Pure mathematical figures, structures, and bodies, any pure mathematical entity or number, are all independent of actual reality. Furthermore, no mathematical proof has to rely upon actual or empirical data; it can rest only upon logical and purely mathematical considerations.

Indeed, however closed are mathematics and physics and however valuable are the mathematical ideas that physics has contributed, pure mathematics is independent of physics, whereas physics depends on pure mathematics and cannot do without it. Furthermore, pure, theoretical physics itself consists of non-actual possibilities. I conclude then that any pure geometry or mathematical theory is entirely independent of actual reality.

What requires revision is not necessarily this or that kind of mathematics but the wrong idea that some of these theories, let alone one of them, embrace all mathematical possibilities (such as the belief, universally accepted until the middle of the nineteenth century, considering Euclidean geometry as the only possible and

---

[1] Cf. Cory Juhl's analysis of the distinctiveness of mathematical entities (Juhl 2012).

[2] The truth under consideration need not correspond to actual-physical reality or actualities. Mathematical truth consists of formal logical validity, mathematical consistency, and mathematical coherence, while all the theorems are based on proofs and calculations. As for the correspondence valid for mathematical truths, I will discuss it below. In any event, we do not need the Quine-Putnam Indispensability Argument (about which consult Putnam 1972 and Colyvan 2011) to realize in what conditions pure mathematics is true and mathematical entities are real existents. Cf. Cory Juhl's view that indispensability arguments fail to show that empirical data bear on the mathematical (Juhl 2012).

complete mathematics). It has been quite wrong to assume that any one of these theories can exhaust the whole realm of mathematical possibilities or even some of its domains (think, for instance, of the geometry of 230 possible crystalline space groups, which crystallographers assumed as complete until the 1980s). The realm of mathematical possibilities is never closed; instead, it should always be kept open to further, even quite different discoveries in the field.

On the grounds of my arguments in this Chapter, I strongly suggest the reader to reconsider the view that "[e]verything there is is *physical*. … universals, too, are physical … the universals which exist are all real physical properties and relations among physical things. … the world around us, the world of space and time, does contain mathematical objects like numbers" (Bigelow 1988, p. 1). Following David Armstrong, John Bigelow calls this *a posteriori* realism, according to which mathematical objects are not mere abstractions, "existing separately from the physical things around us" (ibid.). The panenmentalist *a priori* realism about mathematical individual pure possibilities and their general relationality (namely, the ways in which they relate to one another), which I will discuss below, is an attempt to challenge such views. Whenever, in this Book, I mention "pure possibilities," I have "individual pure possibilities" in mind. The same holds certainly true for mathematical pure possibilities.

## 5.3   Mathematical Entities Are Individual Pure Possibilities

One of the major mathematical questions, if not *the* major mathematical question is: What is mathematically possible and what is impossible? This question is equal to the following: What does or does not mathematically *exist*? Both questions and the possible answers to them are entirely independent of actual or empirical reality. To state that a round square does not exist means that a round square is mathematically impossible, and vice versa.

If mathematical entities are neither actualities nor abstractions or idealizations based upon them, mathematical entities have to be pure possibilities only, and pure mathematics is, in fact, the exploration of mathematical *pure possibilities* and the ways they relate to one another, namely, their relationality. As Kant already realized, unlike logical objects, mathematical entities are particular and not general. In panenmentalist terms, mathematical entities are individual pure possibilities.[3] The

---

[3] Although David M. Armstrong adopts a "possibilism in mathematics" and is committed to mere possibilities, namely, those with no actual instantiation (Armstrong 2010, pp. 89–90) but with a possible one (Armstrong 2004, p. 117), this is not a possibilist view in my terms. I cannot agree with Armstrong as he maintains that these mathematical possibilities are not existents, as he states that they do not exist (Armstrong 2010, p. 90) and that the only existence is spatiotemporal. Hence, his hypothesis is that there are no objects outside space-time (op. cit., p. 5). Furthermore, though as a "one-worldler," Armstrong rejects other possible worlds (op. cit., p. 16), the mere possibilities that he adopts explicitly supervene on the actual (op. cit., p. 68). Thus, they are not pure possibilities in my terms (namely, entirely independent of anything actual). Finally, what is the criterion for the possibility of instantiation of mere possibilities in space-time?

ontological dependence and conditioning of all existents upon individual pure possibilities should be distinguished from a logical, modal, or epistemological one. The existence of individual pure possibilities does not depend on anything actual, and such existence is an ontological necessary condition for the existence of any individual entities, whether actual or possible. Undoubtedly, actual entities—actualities—exist, though contingently, whereas the primary existence of their pure possibilities is indispensable for the existence of these actualities. On these grounds, too, individual pure possibilities independently exist. These possibilities determine the identity and discernibility of their actualities, as the law of the identity of indiscernibles is certainly valid for individual pure possibilities (as I have argued above).[4] As the existence of pure possibilities is not conditioned or limited by anything actual, there are pure possibilities—purely possible individual existents—that are not actualized, that are non-acual. All these are good enough reasons to be a realist about individual pure possibilities and their relationality.

Let me demonstrate the need for such realism with a mathematical example. Suppose that, for the first time, Arthur hears about irrational numbers, such as $\sqrt{2}$, $\pi$, and $\varphi$ (the golden ratio). He already knows what an integer, a ratio, and a square root are. Nevertheless, he absolutely refuses to accept that there can be a real number, which cannot be calculated as a ratio of two integers. "Such numbers do not exist at all; their existence is impossible from the outset," he challenges his good friend, Mary. She, in turn, tries to help him change his mind by showing him some geometrical examples, drawn on a piece of paper, all of whose relevant data she measures and, in calculating the outcomes, shows that the results are such irrational numbers. Nevertheless, Arthur adamantly maintains his resistance: "Each of these results does not really exist; indeed, they must fail to exist. Something must be wrong with your *measurements* concerning these diagrams. There is no number such as 3.14159265…, which is indeterminate. Such an absurd entity fails to exist or to be named as a number. Everything that exists in the real world of pure numbers—and there is such an entirely rational realm—is exact and *determinate* and there should not be any approximation; such approximations or indeterminacy are the results of ignorance, error, or simply wrong *measurements*, not of the reality of pure numbers and their true calculations. Hence, it is not true that such irrational 'numbers' really exist." "But," Mary protests, "I have just shown you some *actual* palpable examples, drawn just in front your eyes, in which the measurements and calculations clearly show that there must be such numbers! Now, examine again *this* isosceles right triangle, the length of each of its sides is one centimeter. Anyone who is familiar with the Pythagoras Theorem, can surely answer the following question without further measurements: What is the length of the hypotenuse of *this* triangle?

---

[4] Remember that spatiotemporal conditions are not valid for pure possibilities but only for actualities, two "identical" individual pure possibilities are, in fact, one and the same possibility. In contrast, we can discern between two "identical" actualities only because they cannot exist at the same place in the same time. Only the individual pure possibility of an actuality safeguards the identity of this actuality and its discernibility from any other actuality under any circumstances whatsoever. Hence, panenmentalism considers individual pure possibilities as pure possibilities-identities.

The answer is obviously $\sqrt{2}$, which cannot be expressed as a ratio of two integers." Notwithstanding her objections, Arthur continues to argue: "No. It is impossible. We fail to measure it exactly. I don't expect any approximation or indeterminacy in such an exact science as *pure* mathematics. Something must be *im*pure about your actual examples. In pure mathematics, there are no such things. In pure mathematics, they are simply impossible. Don't give me actual, physical examples like these. Show me how such so-called 'irrational numbers' are *purely possible*, how they are possible *prior to* any actual example, quite *independently* of any actual-physical illustration. Show me that they truly, independently exist." "Fine," answers Mary, "Let us turn to another example, taken from the realm of 'pure numbers' as you call it. It is a simple equation: $x = \frac{1 + \sqrt{5}}{2}$. Let me calculate, and you will meticulously check my calculation. The result is, mind you, 1.6180339887..., whose approximation is 1.618034 or 1.618. As there is nothing actual-physical about this equation and about this result as well, and neither of them depends on anything actual-physical, they are surely pure enough. Suppose that there were no actual-physical entities corresponding to that number, it is still surely valid and truly exists as purely mathematical. Even if you are entitled to say that my measurements of the physical-actual, that is, the drawn examples that I have just shown you, are not accurate enough and because of this I have reached wrong results, in this last example there is still no need for any measurement; it is a matter of pure algebraic *calculation* only. What is wrong with it? The same holds true for all the cases that I illustrate as actual-physical. Ignore completely their actual-physical embodiment. Instead, think of them as mathematical pure possibilities, as ratios of pure numbers. In each case, you can follow my *general* proofs based on calculations, which are entirely independent of any particular palpable case, any actual-physical example." After a period of deep and long thought, Arthur has no choice but to answer her: "Indeed, there is nothing wrong with your *calculations*. You have made your point. I can see now that each of your examples is eventually about mathematical pure possibilities and the ways in which they relate to one another—their relationality (ratios, in your words). Because they are purely possible, they cannot be actually impossible, and now I cannot argue that they are inaccurately measured. We don't measure mathematical pure possibilities; we can only calculate them in considering their relationality. I was wrong in thinking that irrational numbers should be impossible. Now I realize, on purely mathematical grounds, that they *exist* as mathematical pure possibilities, and, *as a result* of this, they *can or may* also exist as actual-physical possibilities."

Establishing the existence of individual pure possibilities, what makes the propositions about such possibilities *true*? In other words, what are the *truthmakers* of pure possibilities? A received way to treat mathematical truthmakers is that suggested by David Armstrong (Armstrong 2004). According to his view, such truthmakers depend on actualities or the actual reality as a whole. Discussing counterfactuals and relying upon the modal truism concerning possibility and contingency, Armstrong argues that if $p$ contingently exists, not-$p$ is necessarily possible. This truth about the possibility of not-$p$ depends on the contingent truth that $p$ exists. According to my view, in contrast, the truth about pure possibilities, including mathematical one, does not depend on anything actual, let alone on instantiation, possible or actual, in actual reality. Hence, what are the truthmakers of mathematical pure possibilities? The truths about them are relational or contextual and are subject to *proofs*. For instance, in the context of, or relating to, the Euclidean geometry some axioms and theorems are true and some should be considered false. No correspondence to actualities is required to make them either true or false. Formal logic of some kind is also indispensable, though not sufficient, for any mathematics, and logical truths have to do with consistency and coherence and not with correspondence to actual-physical reality. The proposition that there are triangles whose sum of the angles is greater than or less than 180° (or, that such triangles do exist) is logically consistent, for such a possibility is not involved with any contradiction, and it is coherently, contextually true following the relationality of non-Euclidean geometrical pure possibilities. This relationality is subject to proof. Generally speaking, when it comes to the theory of truth concerning pure possibilities, coherence or relationality plays a crucial role, whereas correspondence is indispensable for the truths about actualities. As for the correspondence of true propositions to individual pure possibilities and their relationality, this correspondence first depends on the inclusion of the relevant possibilities while their exclusion is invalid (or vice versa). Thus, to acknowledge the existence of these possibilities, we must first guarantee that they are not excluded, and this can be made only on the basis of coherence within a particular realm of pure possibilities. Hence, true propositions about non-Euclidean entities correspond to the individual pure possibilities and their relationality, which a non-Euclidean geometry validly includes and not excludes. The validity under discussion is subject to proof(s).[5]

Let me refer to an example which shows how proof, discovery, truth, and the exclusion (or inclusion) of the existence of some pure possibilities are inseparable in mathematics: Fermat's Last Theorem states that for any integer $m > 2$, there are *no* positive integers—$a,b,c$—such that $a^m + b^m = c^m$. Until 1994, there was no proof for this theorem. In 1994, Andrew Wiles proved it, which means that he successfully *discovered* that this theorem must be *true*,[6] that is, that *there are no* positive integers that can satisfy that equation. It is a valid proof, because it presents an argument that is logically complete and exhaustive,[7] namely, it covers all the relevant pure possibilities. There is no need for this proof to rely upon any actuality, otherwise its unconditional universality would have not been safeguarded. Assuming that each of

---

[5] Mathematical consistency and coherence are inescapably subject to proofs. This dependence on proofs clearly distinguishes between fictional and real existence. Because mathematical truths are subject to proof, they cannot be considered as fictional but rather as discoveries. On the connection between mathematical *reductio* proofs and discoveries see Jacquette 2008. As Jacquette shows, "*reductio* reasoning in mathematics constitutes a method of discovery rather than mere confirmation of prescient truths" (op. cit., p. 251). The discovery in discussion is of new truths (p. 253).

[6] "The point is that Wiles's successful proof is the way we discovered that it really is true" (Day 2012, p. 49).

[7] "The final criterion is whether the proof presents an argument that is logically complete and exhaustive" (op. cit., p. 50).

the variables of this equation is an individual pure possibility, the proof thus excludes the existence of such pure possibilities within the equation—within the context or the relationality that the equation constitutes. Finally, the discovery of this mathematical truth, corresponding to ("about") mathematical reality, rests upon logical necessary connections between prior known truths and the proven, new truth of the theorem.[8] The mathematical reality in discussion consists of mathematical individual pure possibilities and their relationality.

Mathematical individual pure possibilities do not exist in any actual space and time and are devoid of any causal efficacy or influence. Pure mathematical space and time are the ways in which some mathematical possibilities relate to one another, in terms of structure and order, for instance. There is nothing actual about them, as long as pure mathematics is concerned or as far as the relevant applied mathematics is not applied to actual reality but to another kind of pure possibilities, say, pure physical possibilities which a physical theory discovers.

Panenmentalism (or panenpossibilism) is devoted inter alia to the philosophy of mathematics, in this philosophy, the relationality of mathematical individual pure possibilities plays a crucial role.

For instance, the Riemann hypothesis is about the relationality, in terms of order, of the prime numbers, which are numerical fundamental pure possibilities. Marcus du Sautoy describes prime numbers as "the most fundamental objects in mathematics" or as "the very atoms of arithmetic" (Du Sautoy 2003, p. 5). I would describe them, instead, as "the most fundamental pure possibilities in mathematics (or arithmetic)." Du Sautoy's "timeless numbers that exist in some world independent of our physical reality" (ibid.) is closer to that description (though panenmentalism does not rely on the concept of "possible world"). His "objects with no obvious physical reality" (op. cit., p. 31) should be replaced by better alternatives—mathematical pure possibilities.

Note that there is nothing Platonic in this view of mine. Unlike transcendent and universal Platonic mathematical Ideas, mathematical pure possibilities are immanent individuals (particulars) whose relationality is general. Although mathematical pure possibilities, like Platonic mathematical Ideas, are entirely atemporal and atopical, there is no further similarity of these quite different concepts. Contrary to the Platonic view separating noumena from phenomena, panenmentalism does not separate the two kinds of reality—that of pure possibilities and that of actualities. Such separation would lead to unsolvable problems (*aporias*). As for modern "platonism" in the philosophy of mathematics, it, too, appears to consider mathematical objects or entities as non-physical, non-causal, and non-spatiotemporal. Nevertheless, my view is different from this "platonism" in that such objects or entities are *non-actual, individual pure possibilities*.[9]

---

[8] Op. cit., p. 7.

[9] It is also different from Maddy's mathematical realism (Maddy 1980). On her view, sets exist independently of human thought and "are taken to be individuals or particulars, not universals" (op. cit., p. 163, note 1). In her version of mathematical realism, numbers "are taken to be properties of set" (op. cit. note 3). Possibilities, let alone pure ones, are not mentioned in her paper. In

Furthermore, with panenmentalism, the pure possibility of any actuality is inseparable from it, and the actuality is a spatiotemporally conditioned and limited *part* of this pure possibility. How can a part of a pure possibility be actual and spatiotemporally conditioned? It is because such a part, subjected to these conditions, is what has been actualized of the pure possibility, which comprises *all* which is possible for that entity, whether it is purely possible or actually possible. Every actuality is possible, too, although it is not purely possible. The inseparability of any actuality from its pure possibility does not blur the categorial *distinction* between pure possibilities and their actualities, according to which the latter are spatiotemporally and causally conditioned and they depend on the former, whereas the former are entirely exempt from these conditions and are absolutely independent of the latter.

Panenmentalism is different in principle also from idealism of any sort and from possible-worlds theory as well. It is a unique kind of possibilism *de re*, acknowledging the independent existence of individual pure possibilities and their relationality without relying upon any possible world or actual reality. This possibilism *de re* is especially valid for mathematical entities or objects. Contrary to some doubts,[10] the distinction between possibility and actuality is valid for mathematical objects. They are not beyond this distinction.

Note that the Poincaré conjecture, for instance, raises a question concerning possibility(ies): "Is it *possible* that the fundamental group of a manifold *could* be the identity, but that the manifold *might* not be homeomorphic to the three-dimensional sphere?"[11] It is a crucial question about pure possibilities quite analogous to other crucial questions, such as: Are quasicrystals mathematically possible? Are crystalline (pure) structures of fivefold or tenfold symmetry possible at all, contrary to the expectations of almost all the crystallographers in 1982?[12] In these examples, taken from pure mathematics, "possibility" and "existence" are interchangeable.

My view is quite different from that of Meinong, Meinongianism, or Neo-Meinongianism about non-actual objects.[13] Neither is my view Kantian or Fregean, even though I, too, consider purely mathematical objects as non-actual.[14] Unlike Hilary Putnam, I do not explicate mathematical existence with the set-theoretic

---

recent books (Maddy 2003, 2007; Leng 2011), too, she does not mention "possibilia," "mere possibilities," "pure possibilities," "possibilism," or "actualism" which may be relevant to her view. Instead, she mentions "pure sets" (for instance, in Maddy 2003, p. 59).

[10] Parsons 1982, p. 496; cf. Parsons 1990, p. 333; Parsons 2008, pp. 1–40; and Parsons 2008, pp. 7–8, followin Kant. Though mathematical objects are not located in space and time, and though they do not stand in causal relations and cannot be perceived by the senses (Parsons 1982, p. 491)— for the abovementioned reasons, I also do not follow Parsons's view that they are abstract, though he is quite right in denying them any spatiotemporal and causal limitations.

[11] O'Shea 2007, p. 135, citing *Oeuvres de Henri Poincaré* (Paris: Guthiers-Villars, 1952, vol. 6, p. 498); the italics are mine. A. G.

[12] See Gilead 2013.

[13] See Gilead 2009, pp. 35–40; 45–47; and 83–91.

[14] Cf. Parsons 1982, pp. 494–496.

notions of possibility and necessity.[15] According to Putnam's "modal picture," one need not suppose "eternal objects" (following "platonism") but, instead, this "mathematics has *no* special objects of its own, but simply tells us what follows from what."[16] Thus, according to his view, mathematical statements involve modalities but not special objects.[17] Nevertheless, in fairness to Putnam, it should be noted that he himself states: "Even if in some contexts the modal-logic picture is more helpful than the mathematical-object picture, in other contexts the reverse is the case,"[18] and that he attempts "smoothing the transition from the modal-logic picture to the mathematical-object picture."[19] In any case, this Chapter is incompatible with Putnam's idea that "mathematics is essentially *modal* rather than existential."[20] Of course, neither is this Chapter compatible with Putnam's following idea: "Not only are the 'objects' of pure mathematics conditional upon material objects; they are, in a sense, merely abstract possibilities."[21] Finally, even if there is no sharp separation between mathematics and logic,[22] there is, nevertheless, a sharp *distinction* between them. The irreducible plurality of pure possibilities and the differences between them are essential for panenmentalism. Thus, mathematical pure possibilities are irreducibly different from logical pure possibilities, notwithstanding the strong connections and "intimate" relations between these two kinds of pure possibilities. The differences under discussion are epistemological as well as ontological. Pure possibilities, such as mathematical pure possibilities, are irreducible existents.

## 5.4   Individual Pure Possibilities and What Mathematics Really Is

One of the more ardent antagonists of the platonist approaches to mathematics is, undoubtedly, Reuben Hersh. His *What Is Mathematics, Really?* (Hersh 1997) calls the reader's attention to the fact that mathematics is social, cultural, and historical and that the right way to know what mathematics really is, is to look at the actual ways in which mathematicians do mathematics. In Hersh's view, to relate to mathematical entities as if they were existing outside of space and time, as "abstract entities," independent of history and of social praxis, education, conventions,

---

[15] Putnam 1979, especially p. 49; cf. p. 47: "Mathematics as Modal Logic" and "Mathematics as Set Theory."

[16] Op. cit., p. 48.

[17] Op. cit., pp. 48–49; cf. pp. 58–59: "Introducing modal connectives ... is not introducing new kinds of objects."

[18] Op. cit., p. 57.

[19] Op. cit., p. 59.

[20] Op. cit., p. 70.

[21] Op. cit., p. 60.

[22] Op. cit., p. 2.

culture, and generally of human beings—is simply wrong. His philosophy of mathematics is thus called humanistic.

As I have just explained, I am not a platonist, especially concerning the question, What is mathematics, really? I have no doubt that actual reality, the actual achievements of mathematicians (who, obviously, are human beings), and the history of mathematics are strongly connected with mathematical entities, objects, or existents. In this sense, too, I am not a platonist at all. Nevertheless, these entities, objects, or existents are *distinct* from anything actual or physical and are independent of causal and spatiotemporal circumstances and, as distinct and independent, they exist independently of anything actual, historical, and empirical. Hence, mathematical existents are discoverable and are not the creations of our ingenuity (to be discussed in Sect. 5.7 below). Nevertheless, the independence and distinction under discussion do not imply detachment or separation. We should differentiate between distinction and separation, which are not—by any means—one and the same. As actual, physical reality and the history of mathematics are inseparable from the mathematical existents, which are individual pure possibilities. No actuality can be separated from its individual pure possibility. As I have argued above, this inseparability distinguishes my approach from any platonic (or Platonic) one, in which the separation and separate existence of mathematical existents have played a crucial role. And as I have mentioned above, another essential distinction between my approach and the platonic ones is that mathematical objects, entities, or existents are individual pure possibilities.

According to Hersh, mathematical objects are not physical objects (with which I agree); but, rather, they are "a distinct variety of social-historic objects. They're a special part of culture" (Hersh 1997, p. 22). As such, they are actual (in contrast and as I have argued above, my view is that physical and actual entities are the same and any actualization is physical). I see this quite differently from him. Our social-historic objects and our culture as a whole are inseparable from mathematical individual pure possibilities and their relationality. Nevertheless, these possibilities exist independently of our culture, history, and social life. I believe that Hersh is mistaken about the inseparability in question. Mathematics, as it *really* is, does not depend on anything actual, whereas reality—physical, historical, social, or cultural—actualizes, inter alia, mathematical possibilities without which physical objects and natural sciences could not exist at all. The connection between actualities and individual pure possibilities is necessary, for each actuality is an actualization of an individual pure possibility.

Hersh writes: "Mathematics is the example par excellence of a practice inseparable from its theory, its body of knowledge" (Hersh 1997, p. 207). In a panenmentalist translation, as a practice, mathematics *actualizes* mathematical individual pure possibilities and their relationality, which are accessible to the theory or which are discovered or "captured" by means of the theory. Both criticizing and explaining Wittgenstein's ideas of mathematics (though accepting his anti-platonic approach), Hersh gives a fine example in which $12 + 1 = 14$ makes sense (Hersh 1997, pp. 207–208): Suppose that one is living at a building whose tenants are superstitious about number 13 and hence in their building there is no floor numbered 13; the floor above the twelfth floor is the fourteenth floor. Now, in a panenmentalist translation, in numbering the floors at that building, the individual pure possibility of number 13 was not actualized, and the reasons for this were "cultural." This clearly shows that mathematics is not a matter of convention, arbitrariness, circumstantial, and the like, whereas the actualization by human beings is culturally, socially, and historically conditioned. In this way, a panenmentalist can show what is wrong with Wittgenstein's approach to the status of mathematics more clearly and more simply than the way in which Hersh does this. It also shows what mathematical praxis is all about—discovering or actualizing mathematical individual pure possibilities and their relationality.

If, according to Hersh, a philosophy of mathematics should look at what mathematicians actually do (namely, within a historical-social-cultural or humanistic context), I would like to consider *what are the mathematical individual pure possibilities and their relationality* that mathematicians *really actualize* by their mathematical activity within a human, social, cultural, and historical context. Thus my concern focuses on the question of the *identity* of mathematical existents. As for Hersh's humanistic-historical-social context, it is an actual context, actualized in physical facts and events. Pure possibilities, in contrast, are not conditioned or confined by anything actual or by causal and spatiotemporal conditions whatsoever.

To demonstrate my divergence from Hersh's view, I will take an example that Hersh analyzes, namely an exercise in Pólya's heuristics (Hersh 1997, pp. 3–7). This example concerns an answer to the intriguing question, Is there a 4-dimensional cube? (Hersh 1997, p. 3; hereafter 4-cube). To answer it, we have first to answer the question, How many parts or ingredients has a 4-cube? My conclusions with regards to this example are quite different from Hersh's.

Hersh considers that example as taking part in an "inquiry into mathematical existence" (Hersh 1997, p. 4), with which I fully agree, but he also adds that "by guided induction and intelligent guessing, you'll count the parts of a 4-cube" (ibid.). I will try to demonstrate that the very same task of discovering the answers to those two questions can be performed not by induction or by guessing but only by referring to mathematical individual pure possibilities and the ways in which they necessarily relate to one another. For the fulfillment of this task all we need is to know what are the individual pure possibilities of the parts or ingredients— interior, faces, edges, and vertices—of a cube (which is the mathematical individual pure possibility that consists of such ingredients) and to be able to distinguish between these individual pure possibilities. We inquire the ways in which all these possibilities relate to one another. This inquiry ends with an adequate answer to the question: Is there a 4-cube? Let us answer this question step by step, not necessarily in the way Hersh suggests.

We all know what a 3-dimensional cube (hereafter 3-cube) is and we have had experience with many such cubes, all of which are physical-actual. Of course, we can know how many parts any 3-cube must have and we can count them without thinking about or perceiving any actual-physical cube. Just thinking about the purely possible *structure* of a 3-cube, we can quite easily count or calculate its ingredients. Each 3-cube necessarily has 27 parts or, rather, ingredients: 1 interior (which supervenes on its being 3-dimensional), 6 faces (each of which is 2-dimensional), 12 edges (each of which is 1-dimensional), and 8 vertices (each of which is 0-dimensional). The sum of all these ingredients is 27. This answer *exhausts all* the relevant individual pure possibilities and their relationality concerning the parts of *every* 3-cube.

In the same vein, suppose someone asks: Does a 3-cube exist? We need not observe actual reality to provide one with an answer. We would rather translate this question into the following one: *Is* a 3-cube possible? Or, *How* is a 3-cube possible? If a 3-cube is a mathematical individual pure possibility, to demonstrate or to prove that such a cube exists, even necessarily exists, is to show that it is possible and how it is possible. Now, my answer to that question is as follows: a 3-cube is possible if and only if it consists of exactly 27 purely possible ingredients, which are: 1 interior, 6 faces, 12 edges, and 8 vertices. Because of these ingredients and the way they relate to one another, a 3-cube is possible, which means that it exists. I will also explain later why it *necessarily* exists.

Take the individual pure possibilities of 8 vertices, 12 edges, 6 faces, and 1 interior, which the other ingredients enclose, and these are all you need to *construct* a purely possible 3-cube. There is only one way in which all these ingredients are combined to construct such a mathematical object. As each of these ingredients is a mathematical possibility, this is the only possible way in which all of them *relate to* one another to construct a geometrical unit that is a 3-cube. This *relationality* enables this mathematical individual pure possibility as a whole. In this way a pure individual mathematical possibility of 3-cube exists. Thus, it is an existent. To construct it, to make it possible, no actual or empirical fact is required. All that is required are the abovementioned mathematical individual pure possibilities and the way in which they relate to one another.

Now, following Hersh to some extent, there are some interesting implications of the structure or relationality of a 3-cube, namely, of the way that its ingredients relate to one another.

Let us consider each of the faces. Each face is, as Hersh shows, a 2-dimensional cube (hereafter a 2-cube). Each face consists of 9 ingredients: 1 face, 4 edges, and 4 vertices, the sum of which is 9. Moreover, each of the edges is a 1-dimensional cube (line segment, hereafter 1-cube) consisting of 1 edge and 2 vertices, which make 3 ingredients. As I have just said, all these cubes are included in the 3-cube or implied by it. The relationality between these three kinds of individual pure possibilities pertaining to the 3-cube is quite interesting. Numerically, we have a series here: of 1, 3, 9, 27 ingredients. This is a clear relationality of these three sums of pure possibilities. As this relationality clearly demonstrates, the 3-cube (hexahedron) is to the 2-cube (square) as the 2-cube is to the 1-cube (line segment), and as the 1-cube is to the 0-dimensional cube (point, hereafter 0-cube).

What about the fifth possibility, that of a 4-cube, which nobody could observe in the empirical reality that is three dimensional? On the basis, and only on the basis, of the relationality of the abovementioned four individual mathematical pure possibilities of cubes, combined or internally related in the individual pure possibility of a 3-cube, we can easily *calculate* the number of the purely possible mathematical ingredients of which we can construct the individual pure possibility of a 4-cube. Following the abovementioned series, it must be 27 times 3, which is 81. We thus have the following series: 1, 3, 9, 27, 81. Indeed the relationality of the sums of the ingredients in each kind of cube clearly demonstrates that the 4-cube (tesseract) is to the 3-cube (hexahedron) as the 3-cube (hexahedron) is to the 2-cube (square), as the 2-cube is to the 1-cube (line segment), and as the 1-cube is to the 0-cube (point).

To construct the individual pure possibility of a 4-cube we have to take the individual pure possibilities of 16 vertices, 32 edges, 24 faces, 8 cells, and 1 volume. This answer *exhausts* the individual pure possibilities and their relationality concerning every 4-cube.

To solve the problem that Hersh mentions concerning the 4-cube, we need no observation, no experiment, or experience, no actual-physical facts. All we need are mathematical individual pure possibilities and the way(s) in which they relate to one another. Furthermore, the indisputable truths that a 3-cube consists of exactly 27 parts or ingredients and that a 4-cube consists of exactly 81 parts or ingredients are entirely independent of any historical-social-cultural conditions, restrictions, or circumstances. The answers to the questions concerning these two kinds of cubes are historically, socially, culturally, or geographically absolutely independent. Every mathematician anywhere, at any time, and in any culture would give the same answers to these questions. Regardless and independent of any historical, social, and cultural circumstances, any human being has full access to these mathematical pure possibilities and their relationality as long as he or she can distinguish between these different possibilities and discover their relationality.

So far, I have mentioned possibilities. What about the *necessity* which is involved in their relationality? The sums of the ingredients are not only possible, they are also necessary. Each necessary relation must be possible, too. Necessity is a kind of possibility. There are no other possibilities to sum up the ingredients or parts of these cubes. You have simply to calculate, to construct, or to analyze to know that there are no other sums, other possible sums, of the ingredients of these cubes, as much as there are no other ways in which they relate to one another. Hence, the individual pure possibilities in discussion as well as their relationality are not only possible; they are necessary, too! Indeed, it is impossible to construct any 3-cube with fewer or more than 27 ingredients or parts. It is possible to construct it only from these ingredients and not from others. Thus, it is necessary to construct it by means of *these* 27 ingredients and not others.

First of all mathematical possibility entails mathematical existence. As Hersh rightly puts it: "In mathematics, nonexistence usually is a matter of impossibility" (Hersh 1997, p. 84). Thus, in mathematics, existence usually is a matter of possibility. Which possibility? A mathematical individual pure possibility whose existence is necessary. As I will explain below, mathematical individual pure possibilities are necessary existents.

## 5.4.1   Hersh and a Failure in Proving That There Are Square Circles

### 5.4.1.1   Mathematical Proofs, Showing, and Telling

Mathematical rigorous proofs worth their name *tell* us that such and such *must* be the case; they do not simply show you that such and such a case *actually is*. An actual example, unless in *reductio ad absurdum*, should not be considered as a mathematical rigorous proof. Thus, it is not sufficient to show one or some examples of a square circle for rigorously proving that square circles are mathematically possible. Unlike other kinds of possibilities, mathematical possibilities, mathematical existents, and mathematical necessities are the same. There are, thus, one and the same mathematical modality. If you prove that a square circle is possible, it means that a square circle mathematically exists and, moreover, if the proof under discussion is really a rigorous one, this existence is also necessary. In contrast, using one or more actual examples to show that square circles exist, unless this employs a *reductio* proof, does not rigorously *prove* that such circles necessarily exist. This is what is required of a rigorous mathematical proof.

Note that showing some actual examples or illustrations does not increase our understanding, which only mathematical rigorous proofs can do (some call them "good proofs").[23] Showing teaches us that such and such is the case; not *why* such and such *must be* the case or that the case is necessary. For this, we need a mathematical telling.

I am bearing in mind that since the nineteenth century, mathematicians have been "seriously (and irreconcilably) divided over what constitutes a proof in mathematics" (Kleiner 1991, p. 308) and that since then, the term "mathematical rigor" has become more and more equivocal.[24] Nevertheless, in this Chapter, I refer to a

---

[23] "'Proof, in its best instances, increases understanding by revealing the heart of the matter' note Davis and Hersh ... . 'A good proof is one which makes us wiser,' echoes Yu. I. Manin ... ." (Kleiner 1991, p. 309).

[24] Or, in other words, "standards of rigor have changed in mathematics, and not always from less rigor to more. The notion of proof is not absolute. Mathematicians' views of what constitutes an acceptable proof have evolved" (Kleiner 1991, p. 291). Mathematicians have suggested methods of proving alternatives to rigorous proofs, for instance, dispensing with rigorous proofs for testing primality (i.e., discovering new prime numbers), Michael Rabin suggested a probabilistic method, using randomization within the computation and producing "the answer with certain controllable miniscule probability of error" (Rabin 1980, p. 129). Rabin implemented this method to finding primes by computer. Using computers for proving mathematically is a well-known method to be attempted as a surrogate for rigorous mathematical proofs, but it is involved with serious difficulties and doubts (see, for instance, MacKenzie 1999, which should be consulted for the history of computerized four-color theorem and the negotiability of mathematical proof). As for the evolution of the type of proofs that mathematicians use—proof by deduction, by mathematical induction, by transfinite induction, by exhaustion, by contradiction, and by construction—consult Bramlett and Drake 2013, pp. 20–33. As they conclude, "rigor in proof has been and still is an issue in the presentation of mathematical concepts and ideas. It is a fine balance that one must obtain to construct a proof that is sound and correct while also being understandable by others" (op. cit., p. 32).

condition which I consider necessary for any rigorous mathematical proof. In any event, I relate to mathematical rigor generally, not to "mathematical absolute rigor." The history of mathematics has taught us an obvious lesson: we should not question the solid fact that mathematical proofs are not absolutely immune from challenges, questionings, doubts, and revisions.

### 5.4.1.2   Hersh's Examples

Using intriguing examples, Reuben Hersh *shows* that there are some actual cases of square circles (Hersh 1997, pp. 265–267). The first actual example that he gives is taken from a practice prevalent amongst some taxi companies.

Hersh is not concerned at this point with the classic Greek problem "proved in modern times to be unsolvable, of constructing with ruler and compass a square with area equal to that of a given circle" (op. cit., p. 265). Let us begin with definitions: a circle is a plane figure in which every point has a fixed distance, which is the radius, from a fixed point, i.e. the center; and a square is a quadrilateral with equal sides and equal angles (op. cit., pp. 265 and 266). The following is the example, based on actual facts, that Hersh suggests to solve the problem of a square circle:

> Suppose I live in a flattened, building-less war zone. Transportation is by taxi. Taxis charge a dollar a mile. There are no buildings, so they can run anywhere, but for safety, they're required to stick to the four principal directions: east, west, north, and south.
>
> People measure distance by taxi fare. If two points are on the same east-west or north-south line, the fare in dollars equals the straight-line distance in miles, traveling only east-west and north-south.
>
> The taxi company has a map showing the points where you can go for $1. These points form a square, with corners a mile north, south, east, and west of the taxi office. In the taxi metric, *this square is a circle*—it's the set of points $1 from the centre.
>
> Yes, a square circle! Inconceivable, yet here it is! (op. cit., p. 266)

In some cases, the so-called "inconceivability" should not serve as an obstacle for mathematicians. Most of the time, inconceivability is only a matter of common sense or a dogmatic way of thinking. Many people consider the idea of, say, transfinite numbers or that of "Hilbert's Hotel" as inconceivable, and yet they are conceivable on the grounds of Cantor's mathematical approach. Thus, the inconceivability under discussion is not a shortcoming that one is entitled to attribute to Hersh's proof.

Another square circle to which Hersh refers, is a regular Euclidean circle (as just defined above) on which equally spaced points are inscribed, while the circumference is divided into four equal sides, and they all meet at the same angle, 180 degrees (op. cit., p. 266). Alternatively, considering the equator of the earth as a huge circle, it has four equal straight sides and four equal angles. In this case, too, we have a

---

Consult MacKenzie 2005 for "at least two notions of proof—non-mechanical rigorous argument, and mechanized formal truth" from the viewpoint of a sociologist of science (MacKenzie 2005, pp. 2343–2344). As MacKenzie states, "While most mathematicians are in practice committed to rigorous-argument proof, formal proof has become the canonical notion of 'proof' in modern philosophy" (op. cit., p. 2344).

square circle (ibid.). Finally, think of Cartesian coordinates and of a graph of a regular Euclidean circle. The graph is exactly the unit square of the "taxicab metric" that Hersh mentioned above. Now, for $p$ finite, the graph is a larger square, with horizontal and vertical sides, i.e. the four lines, and so on (op. cit., p. 267). As a result, "we have infinitely many unit circles of various shapes. The first and the last are square!" (ibid.).

*Does Hersh succeed in squaring the circle*? Does he succeed in accomplishing an allegedly *impossible* mission? If such a successful accomplishment demands a rigorous proof, the answer is *no*. All he suggests are actual illustrations or examples in which the space—the area—that a square occupies is equal to that which a circle occupies; their areas are equal. Rigorous proof in this case demands that in *any case* of a circle we are capable of finding out a square that is equal in area to that of a circle. Nevertheless, neither Hersh nor anybody else has so far succeeded in doing *that*. So far, there is *no algorithm to calculate this*. Of course, such an algorithm must be fully justified and understood; it cannot rest merely upon mechanical or unsurveyable grounds. Hence, a computerized algorithm that cannot be fully surveyed and understood by able mathematicians is not the algorithm that we are looking for.

In these examples, one does not prove that square circles exist in the mathematical sense of "existence," i.e., necessarily exist, necessarily possible. One case or a few such cases proves nothing rigorously. In any of these cases, we have no *algorithm to calculate whether the relevant areas are equal or not in every case of a circle or a square*.

In other words, Hersh's actual examples do not satisfy the demands of mathematical rigorous proof. They refer to some *actual* possibilities of square circle but not to its necessary existence. While squares and circles are geometrical necessary existents, and we have all we need to prove this rigorously, we have no proof of the necessary existence of square circles. Rigorous proof *tells* about some necessity; it does *not just show* some tangible or actual cases.

Nevertheless, Hersh's approach to the problem of square circle has some merits of its own. One of the familiar ways to solve the problem of how a square circle is impossible and yet meaningful and comprehensible, is to employ the idea of sortal properties or predicates (for instance, Feldman 1973).[25] According to this approach, being a square circle is a uninstantiated property. Hersh clearly shows that such is not the case, for there are instances of square circles; there are some actual examples of them. Yet, as he also shows, the fact that there are some actual cases of square circles does not render the concept "square circle" conceivable or understood. To show that there are instances of a square circle does not provide us with a rigorous mathematical proof of its existence. I deem as a merit of Hersh's examples that they obliquely imply that the approach relying on instantiated/uninstantiated properties is not sufficient to rigorously solve the problem in question. To refer to an instant or

---

[25] "'Round square thing,' 'colorless red thing,' 'part of a perfectly honest man,' etc. are all sortals. Some of these could be removed by requiring that a sortal predicate be consistent" (Feldman 1973, p. 272).

more does not solve rigorously the problem under question. Hence, in both of these respects, Hersh enlightens us with his intriguing examples.

In fairness to Hersh, he also argues thus: "Our inherited notion of 'rigorous proof' is not carved in marble. People will modify that notion, will allow machine computation, numerical evidence, probabilistic algorithms, if they find it advantageous to do so. Then we are misleading our pupils if in the classroom we treat 'rigorous proof' as a shibboleth" (Hersh 1993, pp. 395–396). Nevertheless, if we wish to maintain both demands of adequate proof—rationally convincing and explaining—what I see as panenmentalist requirements of mathematical rigorous proof should be met.

Note that in the context of our discussion, "rationally" should be added to "convincing," as other kinds of "convincing" can be merely authoritative: for instance, "As Professor Certainty considers it, this proof must be convincing;" or, "The mainstream view is that such a proof is convincing." No genuine mathematician would subscribe herself to such kinds of "convincing" seriously. Now, if "a proof is just a convincing argument, as judged by competent [or qualified] judges" (Hersh 1993, p. 389), who judges the judges? What are the criteria to decide who is a "competent judge"? What are the rational grounds on which the competent judge decides that such and such is a proof worth its name? Is a distinguished Professor of Mathematics at an exalted University a "competent judge" concerning mathematical proofs? After all, as John Locke rightly claimed, there is no error to be named which has not its own Professors. Indeed, cultural or accepted standards are always questionable and challengeable. Hersh, too, emphasizes the fact that there is "deep disagreement about standards of rigorous proof" (Hersh 1993, p. 394). If Hersh's goal is "to impart an understanding" (following op. cit., p. 397) of the possibilities of squaring the circle, his aforementioned actual examples fail to do so. All the more, as he adds "first by some important [?] examples, then by general concepts" (ibid.). Note that according to Hersh, "a proof is a complete explanation" (ibid.). In any event, just referring to actual examples does not appear to me to serve as an adequate explanation, let alone as a complete one.

### 5.4.1.3 The Panenmentalist Possibility of Square Circles

Individual pure possibilities are possibilities that are mind-independent, primordial existents,[26] which may not actually exist, that is, they are entirely independent of actual circumstances or of any actuality, exempt from any spatiotemporal and causal conditions or restrictions, and, as individual, they are wholly concrete and not

---

[26] Primordial in the sense that their existence is the basic, primordial necessary condition for the existence of anything. In other words, if individual pure possibilities not exist, their actualities could not exist either. If it were a purely impossible actuality, it could not exist at all in whatsoever circumstances. "Existence" has two meanings: (a) "actual existence," that is, in space and time, and (b) possible existence, such as the existence of numbers, which do not exist in time and place.

abstract entities. Individual pure possibilities are entitled "panenmentalist possibilities."[27]

Without individual pure possibilities, the identity, distinctness, and the very existence of anything, whether actual or possible, could not have been possible. If something is purely impossible, such an entity cannot exist at all. Individual pure possibilities are not haecceities: as each such possibility comprises all that is possibly open to an entity under one and the same identity, it is not a sort of an essence, let alone an actual essence; thus, an individual pure possibility is not an haecceity. An individual pure possibility is a nonactual or unactualized ("mere") possibility, only to the extent that something must be left of any individual pure possibility that remains pure and not actualized, despite that actualization of the relevant entity. The reason for this is that anything actual could have been different and yet remains one and the same entity. For instance, this computer screen in front of your eyes could have been one and the same entity, would it have been not in your office, right now, and not manufactured by a specific manufacture but by another one belonging to the same company, and yet it would be the same individual entity. Any actuality is simply the spatiotemporal and restricted part of its individual pure possibility which is, thus, always more comprehensive or "wider" than its actuality.

Panenmentalist possibilities are highly relevant to the ontology of mathematics, as numbers and geometrical figures, as pure mathematics deals with them, are in fact mathematical pure possibilities. Instead of "mere, or non-actual possibilities," I use the term "pure possibilities." In the same way that pure mathematics is distinguished from applied mathematics, individual pure possibilities, mathematical or otherwise, are distinguished from their actual possibilities, i.e., actualities.

Are square circles possible as individual pure possibilities? Is not a square circle an entirely and absolutely unactualizable possibility? So far, we have seen that their actualizability is not absolutely or entirely excluded (as Hersh's actual examples *show*). The question is: Are they *necessarily* actualiz*able*?

Metaphysically speaking, a square circle is a pure possibility of which we know nothing yet, a possibility that, to date, we could not conceive or comprehend (despite the abovementioned actual examples). With panenmentalism, only one *basic* absolute impossibility exists, namely, it is absolutely impossible for two pure possibilities to be identical. The reason for this is that the principle of the identity of the indiscernibles is absolutely valid for individual pure possibilities, as they are exempt from any spatiotemporal conditions or restrictions. In contrast, actualities—actual possibilities—are necessarily subject to these conditions. Hence, when one discerns no difference between two actualities, say, two drops of water, one can distinguish between them numerically (i.e., that there are two of them instead of one) only

---

[27] As panenmentalism needs no possible-worlds conceptions, the attempt to solve the problem of a square circle on the grounds of impossible worlds is not accepted on panenmentalist grounds or on those that do not follow any possible/impossible-worlds conception. In contrast, see Bjerring 2014, p. 327: "Since there are impossible worlds in which the mathematically impossible happens, there are impossible worlds in which Hobbes manages to square the circle." In such an impossible world, such a possibility holds true.

because at the same time they exist in different places (or at different times in the same place). Such can never be the case of individual pure possibilities, as they exist neither in time nor in place. Thus, one has to distinguish between them only on the basis of the principle of the identity of the indiscernibles, which means that in any case in which "two" individual pure possibilities are indiscernible, "they" are, in fact, one and the same individual pure possibility. Thus, an individual pure possibility can be identical only to itself.

Illogical or non-logical possibilities exist, since unlike Leibniz's, Kant's, and many other philosophies, with panenmentalism, the realm of pure possibilities as a whole cannot be restricted to logical possibilities alone. To argue otherwise means to consider logic as metaphysically prior to metaphysics, which is absurd, and, even worse—to believe that our present or current logical knowledge exhausts all logical possibilities. Such was the wrong assumption of logicians of the past who assumed that all logical possibilities were known to them, that logic exhausted itself, as it were. Nowadays, we are familiar with an impressive variety of logics. Do we not laugh today at logicians of the past, who were familiar only with a binary (two-valued) logic, who excluded the possibilities of any other, non-binary logic? The same holds for mathematical possibilities. From a Euclidean geometrical viewpoint, parallel lines sharing the same point(s), except for at an infinite distance, are just impossible, whereas a non-Euclidean geometry fortunately opens up such possibilities for us. To assume impossibilities may lead other thinkers (or the same thinkers under different intellectual circumstances, or after some intellectual progress) to reconsider the possibilities of such "impossibilities" and to discover that they are eventually possible.

Yet, you may argue, a square circle is a logico-mathematical impossibility, as most of the logico-mathematicians so far have assumed and, perhaps, as all logico-mathematicians in the future should assume. Such is the case, the argument goes, since without the assumptions and the procedures demonstrating that such a circle is absolutely impossible, neither logic nor mathematics is at all possible. Nevertheless, such an argument relies on the tacit assumption that the realm of pure possibilities as a whole is logical and that the illogical or the non-logical is impossible. This especially holds, of course, for the current state of logical knowledge and it cannot hold so for all logical possibilities, some of which we do not even dream about at present. Such an argument arbitrarily confines the realm of purely logical possibilities, which should not be confined to any current logical knowledge or awareness. In such a confined spirit, Kant and many others have assumed that the law of contradiction is the necessary condition for each possibility.

We examine the realm of pure possibilities through many different lenses. Each of these lenses has limitations, owing to the impossibilities that it discerns or considers. Each lens is "blind" to some possibilities, considering them as not existing or just as impossibilities, whereas under another lens the same "impossibilities" would be considered possible, even indispensably possible. We do not know yet of a pure possibility of a square circle, but this does not imply (or enforce upon us) the conclusion that such a possibility is absolutely or metaphysically impossible. Considering the realm of pure possibilities as a whole, a pure possibility of a square circle may be possible; it may exist there, and *if* so, it exists there necessarily as do all the other pure possibilities. All we need is a rigorous proof to demonstrate that.

Since each pure possibility necessarily relates to any other pure possibility, which must necessarily be different from it (to differ *from* implies to relate *to*), this relationality[28] may involve contradictions. Contradictions are not impossible under the panenmentalist sky. So is the case of any paraconsistent logic. Priest, Tanaka, and Weber, emphasizing the "inevitable presence or the truth of some contradictions," write about the possibilities of contradictions in such logics as follows: "If there are true contradictions (dialetheias), i.e., there are sentences, $A$, such that both $A$ and $\neg A$ are true, then some inferences of the form $\{A, \neg A\} \vDash B$ must fail. For only true, and not arbitrary, conclusions follow validly from true premises. Hence logic has to be paraconsistent" (Priest et al. 2013). There are various examples of productive and useful inconsistencies from which one can drive non-trivial, non-arbitrary, and true conclusions following validly from true premises and, hence, are subject to paraconsistent logic. For instance, in mathematics, the best available theory concerning infinitesimals is inconsistent: the calculation of derivative infinitesimals in the infinitesimal calculus of Leibniz and Newton had to be both zero and non-zero (op. cit.); and in physics, Bohr's theory of the atom implies major contradictions.[29] These two cases are subject to paraconsistent logics.

Each pure possibility may relate to "its" impossibility, namely, to the possibility that contradicts it. *To relate to* and *to refer to* are not identical. At least for the time being, we cannot refer to a pure possibility of a square circle as we are not familiar or acquainted with such a possibility and, as such, we cannot even conceive it. Since each pure possibility must relate to all that is different from it, including to what contradicts it, we can relate to the possibility of a square circle. One would argue that the pure possibility, the very possibility, of a square circle is impossible, because the very idea of a square circle is contradictory (being a square and being a circle contradict each other). Nevertheless, panenmentalists would answer to that argument: contradictions can be not sufficient to exclude some possibilities. Take for example parallel lines according to Euclidean geometry—parallel lines sharing a common point, which does not lie in the infinite distance, are involved with contradiction. Yet, *this apparent contradiction disappears as soon as the total realm of*

---

[28] Relationality comprises all the ways in which individual pure possibilities necessarily relate to each other. As any individual pure possibility is necessarily different *from* any other individual pure possibility, it necessarily relates *to* this possibility. While pure possibilities are individual and concrete, their relationality is general or universal. Panenmentalism is nominalist about pure possibilities and leaves generality and universality to their relationality only. There are no general or universal pure possibilities. For these reasons, panenmentalism does not follow the (wrong) idea that being a square circle is a (general or universal) uninstantiated property or attribute.

[29] According to this theory, "an electron orbits the nucleus of the atom without radiating energy. However, according to Maxwell's equations, which formed an integral part of the theory, an electron which is accelerating in orbit must radiate energy. Hence Bohr's account of the behavior of the atom was inconsistent. Yet, patently, not everything concerning the behavior of electrons was inferred from it, nor should it have been. Hence, whatever inference mechanism it was that underlay it, this must have been paraconsistent" (op. cit.).

*pure possibilities, which is not confined to the boundaries of Euclidean geometry, is considered.* In this realm, there is a secure region for non-Euclidean geometries. In this region, the pure possibilities of parallel lines sharing a common point, the pure possibilities of triangles the sum of whose angles is more than or less than 180°, and so forth certainly exist.

As much as non-Euclidean geometries or modern kinds of logic do not cancel out any of the Euclidean or classically logical possibilities but, instead, add some vital alternatives to them, and thus enlarge the known scope of possible geometry or logic, we may use the known logical or mathematical laws or rules to *relate* to quite different logico-mathematical possibilities that are *not* under these laws or rules. By relating thus, we do not cancel these laws or rules, but reveal their limitations and confine or limit their validity, which is a necessary result of acknowledging that the realm of metaphysical possibilities is wider and more comprehensive than that of some logico-mathematical possibilities, let alone of the currently known ones.[30]

To keep the realm of pure possibilities open for us, we must reduce the number of pure impossibilities as much as we can. Panenmentalism, like any other metaphysical theory or "lens" has its limitations. On panenmentalist grounds, we are entitled to exclude, for instance, the possibility that machines can think (Gilead 1999, pp. 137–158). This exclusion is not absolutely valid for the entire realm of pure possibilities. Instead, it is valid on panenmentalist grounds in a way similar to that in which Euclidean geometry excludes the possibility of two parallel lines sharing the same point. Thus, through panenmentalist lenses, conscious or thinking machines are just impossible, as much as, through Euclidean lenses, non-Euclidean possibilities are impossible. We should not confine the entire realm of pure possibilities to any restriction of any of its parts or to what is confined to the range perceived by particular "lenses." I have argued for one principal restriction only as metaphysically valid for that realm as a whole—no two pure possibilities can be identical. I believe that this absolute basic restriction holds for any possible metaphysics.

In sum, the pure possibility of a square circle, though being considered impossible through the lens of any currently known logic or mathematics so far, is an actualizable pure possibility. Although we cannot refer to this possibility, by all means we can relate to it as contradicting the currently known purely logico-mathematical possibilities. At least to that extent, this "impossible" possibility is meaningful and significant. Such relationality keeps the realm of logico-mathematical possibilities open.

As for the rigorous proof for the existence of square circles, from a panenmentalist viewpoint, such a proof has to point out the necessary existence of such a pure possibility. There is no contingency about pure possibilities; their existence and relationality are necessary, whereas actualities exist contingently only. There could have been nothing actual instead of anything actual. Nonetheless, individual pure

---

[30] The following analogy is fascinating: a "way to develop a many-valued paraconsistent logic is to think of an assignment of a truth value not as a function but as a *relation*" (Priest et al. 2013).

possibilities necessarily exist and there is nothing contingent about them and about their relationality: as exempt from any spatiotemporality, no pure possibility is changeable (change and temporality are inseparable, and where there is no temporality, no change exists). Actualities are dependent on uncountable actual circumstance, a dependence that implies an infinite regress. When it comes to actualities, things could have been always different and there is no necessity about them. All actualities are contingent, whereas all individual pure possibilities and their relationality are necessary. In this sense, panenmentalism takes a Humean approach to causality and empirical reality as contingent only.

All the above-mentioned examples that Hersh suggests for the possible existence of square circles are contingent. As a self-declared anti-platonist, Hersh rests mathematics only on actual historical and cultural grounds. In fact, he confines mathematical possibilities to actualities—to what mathematicians have actually done. Panenmentalism, in contrast, though non-platonist, is clearly possibilist. That is, with panenmentalism, mathematical individual pure possibilities exist entirely independently of actual circumstances, historical or cultural, though they are discovered in the course of the history of mathematics and in particular cultural contexts and circumstances. When a panenmentalist seeks a rigorous proof for the necessary existence of square circles, if they can possibly exist, such a proof cannot depend on actual examples or on any other actualities.

Generally speaking, in fact, Hersh's approach is actualist. From a panenmentalist viewpoint, it appears that pure mathematics, in contrast to applied mathematics, cannot be understood and explained on actualist grounds. Panenmentalism can suggest adequate metaphysical grounds for the ways in which pure mathematics discovers its discoveries, proves its proofs, and makes its fascinating progress in enlarging for our knowledge and understanding the realm of mathematical possibilities. To consider a square circle as purely impossible implies excluding some mathematical pure possibilities that may be discovered as mathematically vital.

## 5.5  An Alternative Indispensability Argument

Even though my view concerning the ontological status of mathematics does not need the Quine-Putnam Indispensability Argument or other indispensability arguments to show in what conditions mathematical objects are real existents and pure mathematics is true, I would not like to ignore this important argument entirely. After all, since Putnam's view (Putnam 1972), following Quine's "ontological commitment" (Quine 1960),[31] philosophers of mathematics have invested many efforts

---

[31] As is well known, Quine acknowledges "numbers as objects" (Quine 1960, p. 245). This view explicitly excludes the existence of "unactualized possibles" (ibid.). My attempt in this section to suggest an Alternative Indispensability argument is clearly incompatible with that of Putnam-Quine. Quine's examples of "the possible hotel on that corner" or "the possible new church on that corner" (ibid.)—which he used to show the doubtfulness of such objects because of the "perplexity

in discussing indispensability arguments, interpreting, criticizing, revising, and improving them. The following is Mark Colyvan's representation of the paradigmatic "Quine-Putnam Indispensability Argument:"

(P1) We ought to have ontological commitment to all and only the entities that are indispensable to our best scientific theories.

(P2) Mathematical entities are indispensable to our best scientific theories.

(C) We ought to have ontological commitment to mathematical entities. (Colyvan 2011)

As my view is realist about individual pure possibilities and their relationality as well, I obviously do not endorse the view that ontological commitment is valid *only* for what is indispensable to scientific theories. Nonetheless, individual pure possibilities, especially mathematical ones, and their relationality are indispensable to all best scientific theories. Taking this indispensability into consideration, an Alternative Indispensability Argument may be as follows:

(P1) We ought to have ontological commitment to all the entities that are indispensable to our best scientific theories.

(P2) Pure Mathematical entities are indispensable to our best scientific theories.

(P3) Pure mathematics explores individual mathematical pure possibilities and the ways they relate to one another (namely, in sum, their relationality).

(P4) It explores these entities or objects and their relationality quite independently of any empirical evidence or of physical reality, actualities, and spatiotemporal and causal conditions as well as of possible worlds and of our mind.

(C) We ought to have an ontological commitment to mathematical entities—individual pure possibilities and their relationality—existing quite independently of physical reality, actualities, spatiotemporal and causal conditions, possible worlds, and our mind.

In each case in which mathematical possibilities are actualized, they are spatiotemporally and causally restricted. This renders them, as actualized (and not as pure) physical-actual objects. Actual-physical reality actualizes some individual mathematical pure possibilities and their relationality. This reality thus transforms the nature of their existence—from that of pure possibilities to that of actualities. The indispensability of mathematical pure possibilities and their relationality is quite obvious: Without them, physics and our knowledge of the actual-physical reality would be simply impossible, and without them, there were not actual quantities, structures, and measurements. This indispensability, too, means that individual mathematical pure possibilities are real existents to which we ought to have ontological commitment. Furthermore, if our physics explores physical reality as it is in itself (and not as a phenomenon), these mathematical pure possibilities are indispensable also to the existence of physical reality.

Nevertheless, what about mathematical theories that have no known physical application? First, this is not valid for my view: Applicability, whether actual or

---

over identity" with which they are inescapably involved—do not hold true in my view, according to which pure possibilities are not spatiotemporally conditioned or located. I am not alarmed by what Quine calls "vagaries of unactualized possibles" (op. cit., p. 246). As I mentioned above, panenmentalism successfully applies the principle of the identity of the indiscernible to individual pure possibilities. In fact, this view attempts to challenge Quine's stricture that "if we waive location, there supervenes a perplexity of identity" (Quine 1960, p. 252).

possible, is not a necessary condition for the existence and truth of mathematical entities as individual pure possibilities and their relationality. Second, in fact, we cannot know in advance which mathematical theory would be found as physically applicable. What we know in advance is that there is no physical theory that is exempt from purely mathematical significance, that there is nothing physical that is exempt from mathematical application. Whatever is purely mathematically possible is not physically impossible; it can be a physical possibility too, whether actualized or not. To know, understand, and predict natural phenomena we must have in store as many mathematical possibilities as possible (including their relationality). In any case, when new discoveries are made, it is also, and first and foremost, because scientists are aware of mathematical pure possibilities that so far were not known as physically-actually applied. Because we can never tell in advance which mathematical theory can and may be found as physically applicable, the Alternative Indispensability Argument can be valid for *any* pure mathematical theory.

For this reason, too, this Alternative Argument is compatible with the thought-experiment that I suggested above, according to which Euclidean geometry would have been true and valid if it had been counterfactual, for even in a case in which a mathematical theory appears to be counterfactual, it does not mean that it could not have been actualized. The actual state of affairs is inescapably contingent,[32] whereas the dependence, epistemic and ontic, of actual-physical reality on mathematical individual pure possibilities and their relationality is necessary or indispensable. Thus, these possibilities are indispensable to the physical-actual reality and its scientific knowledge, not the other way around. Nevertheless, the existence of and the truths about mathematical pure possibilities are independent of any of the above indispensability arguments, though such arguments may contribute something to the acknowledgment of those existence and truth.

As for Field's arguments for the *dispensability* of mathematics, according to which mathematical entities or objects are replaceable by structures of physical space (Field 1980), it does not cut under my panenmentalist view of mathematics, for this view does not inevitably depend on indispensability arguments to prove the existence of pure possibilities. Neither do Field's arguments cut under the indispensability arguments, following Armstrong,[33] which adopt Aristotelian realism about mathematical objects or entities and are exempt of any "platonism" or Quinean commitment (Newstead and Franklin 2012, following Armstrong 2004).

---

[32] According to Armstrong's Possibility principle, "For each contingent truth, a shadow truth companies it: the possibility of its contradictory. It is a 'mere' possibility only. Given $p$, and given that it is contingent, the truth <it is possible that not-$p$ > is entailed" (Armstrong 2004, p. 83). My view is quite different from his: First, in my view, *everything* actual-physical is contingent. Second, my view of individual pure possibilities is by no means confined to counterfactuals, as it states that *any* actuality must have a pure possibility. Finally, to be real, true existent, no individual pure possibility needs actualization.

[33] According to this view, "the existence of a mathematical entity, according to my Possibilism, is no more than the possibility of instantiation of that sort of structure somewhere in space-time" (Armstrong 2004, p. 124; cf. 111). I, notwithstanding, do not consider Armstrong's view as real possibilism.

## 5.6 The *a priori* Epistemic Accessibility to Mathematical Pure Possibilities

Why are pure possibilities accessible to our imagination and intellect, without relying upon empirical means or actual data? Considering all the possibilities with which we are acquainted, whether actual or pure, we are capable in principle of thinking about different, even quite different, pure possibilities. Such is the case because each pure possibility necessarily relates to all the other possibilities, pure or actual, *from which it is different*. As I have argued above, difference implies relationality—"to be different *from*" necessarily implies "to relate *to*." As no two individual pure possibilities can be identical but each is different from the other, each necessarily relates to the other. Individual pure possibilities are not discrete entities; as a rule, they relate to one another. Hence, Euclidean pure possibilities necessarily relate to non-Euclidean pure possibilities. Nevertheless, this relationality can be known to us only if some particular mathematical possibilities are not excluded. Should mathematicians exclude, for one reason or another, the possibility of any negation of the Euclidean parallel postulate, it would be impossible for them to accept non-Euclidean possibilities. Once such an exclusion is removed, the way is open to follow the relationality in question, and mathematicians gain access to such possibilities. Such relationality relies only upon the difference between the pure possibilities in discussion and not on the state of the mathematical knowledge or on actual reality. Hence, it is subject to discovery.

Regardless and independently of any physical (actual) dot, line, triangle, or any other actual figure, we can easily and directly think about or refer to point, line, triangle, or any other figure, Euclidean or non-Euclidean, which is a mathematical pure possibility. Such possibilities are entirely exempt of any actuality or actual restriction. Thus, for instance, we think about and refer to points, which have positions but no dimension, or to lines, which are lengths without breadth. Contrary to actual, physical dots or lines, these pure possibilities are not restricted by any actual spatiotemporal or causal conditions and they are entirely accessible to our imagination and intellect. In principle, we can refer to any pure possibility simply on the ground that it is different from any other pure possibility as well as from any actual one, which we can take into consideration. In this way, we can think about and refer to *counter*factual or *counter*-actual possibilities, to any possible negation of a Euclidean pure possibility (namely, to a non-Euclidean possibility), and so on, simply because it is different from other possibilities, pure or actual. Even on a basis of quite a limited number of pure or actual possibilities, we can think about or refer to an open-ended number of pure possibilities. Pure possibilities are *a priori* accessible to our imagination and intellect, considering the limitations of these capabilities. The metaphysical fact that no two individual pure possibilities are identical and that each of them necessarily relates to the others guarantees this accessibility.

## 5.7  Pure Possibilities Are Discoverable, Not Inventible

Mathematical pure possibilities are not invented; instead, they are subject to discovery. These pure possibilities can be discovered and their existence is entirely independent of our imagination or reasoning, though they are accessible enough to our intellect and imagination. Mathematical pure possibilities are subject to necessity that is independent of our will, received ways of thinking, expectations, or subjectivity. Hence, mathematical pure possibilities are not ideal entities; they exist as objective entities, entirely independent of any mind, human or otherwise.[34] Mathematical necessity of any kind is subject to our discovery, not to our wishes or imagination, thought it may be accessible by means of our intellect and imagination alone.[35] Mathematicians may *discover* the *existence* of a whole class of functions (for instance, Poincaré's discovery of a new class of Fuchsian functions[36]) *contrary* to their expectations, or contrary to what they *had believed possible*. Indeed, some inventions may surprise the inventors, especially to the extent that unexpected consequences are concerned but, unlike inventions, discoveries and their consequences are enforced on the discoverer, who cannot help but to accept them, however unexpected they may be or however impossible they may have been thought to be. This also clearly shows that mathematical objects or entities are discoverable rather than inventible. Similarly, the prime numbers are discoverable, not inventible, and their existence is independent of the mathematicians' expectations and conjectures about them. Once a mathematician accepts some mathematical conditions and postulates or some pure mathematical possibilities, the consequences are *enforced* on him or her. Unlike any inventor, he or she is not free to change these consequences, for there is a logical and mathematical *necessity* about them. The surprising outcomes of inventions are contingent; there is no necessity about them. They are always subject to the inventor's change of mind.

Another example is that of the discovery of imaginary numbers in the nineteenth century. Despite their name, their existence has been always independent of the discoverer's imagination. These numbers were not fabricated or artificially created.

---

[34] And, contrary to Mary Leng's stance, these are not merely logical consequences of our mathematical assumptions (Leng 2011, p. 65). There is much more to mathematical discoveries than logical consequences, as any reduction of mathematics to logic (or of mathematical possibilities to logical ones) must fail. Thus, contrary to Leng's view, mathematical objectivity has to do with an independent realm of mathematical objects or entities.

[35] Hence, Grisha (Gregory) Perelman, who proved the Poincaré conjecture, relates mathematical possibilities, necessity, and proof of Hamilton's mathematical *discovery* (Perelman 2002).

[36] O'Shea 2007, pp. 115–118. O'Shea also characterizes Gregory Perelman's achievement associated with Poincaré conjecture as "a stumbling discovery relating to the shape of our universe" (op. cit., p. 2). O'Shea mentions many other mathematical discoveries (such as the discovery of non-Euclidean geometries, ibid., p. 4). And he is one of the mathematicians or philosophers of mathematics relating to the exploration of mathematical possibilities, for instance: "Precision allows one to reason sensibly about objects outside ordinary experience. It is a tool for exploring possibility; about what might be as well as what is" (op. cit., p. 21). For similar ideas see Rota 1997, pp. 191, 192, and 195.

Instead, they were real mathematical entities, waiting for their discovery or for the acknowledgement of their existence. While the eighteenth century mathematicians did not admit the (real) existence of such numbers, the nineteenth century mathematicians were brave enough to admit it (Du Sautoy 2003, pp. 68–69). Though Gauss displayed their existence by means of a pictorial proof (op. cit., p. 70), this constructed means served him, in fact, to discover a new kind of mathematical possibilities, namely, of imaginary numbers. The pictorial proof served him as an invented constructed tool to discover these entities—mathematical pure possibilities—analogously to microscopes and telescopes by means of which we can observe actual entities and phenomena, whose existence could not be discovered otherwise. Gauss's pictorial proof, which is an image, by no means reflects on the ontological status of these numbers, which are entirely real (in the ontological sense), not imaginary at all; not stipulated but discovered. As we will realize, imagination is one of the helpful vehicles in which to tour mathematical reality and to make new discoveries—to discover new mathematical pure possibilities, which are different from the known ones and from actual possibilities as well. Euler discovered that imaginary numbers would open unexpected paths, new possible ways, in mathematical reality. Imaginary numbers, like those of other kinds, are mathematical pure possibilities, really existing independently of our mind. Before the discovery of imaginary numbers, mathematicians would regard any number that would satisfy the equation $x^2 = -1$ as quite impossible, namely, as if not existent. The discovery showed, nevertheless, that such numbers *must be (purely) possible*, or, in other words, that, purely mathematically, they *must necessarily exist*. The discoverer(s) acknowledged and saved such possibilities, whereas other mathematicians attempted to exclude them, because they believed that they could not exist. After all, it is quite difficult to think about the existence of the pure possibility of such a number, as long as one believes that no $x^2$ can be equal to $-1$. Such an equation was certainly unexpected. In concluding, discoveries may quite possibly be contrary to the expectations or beliefs of the discoverer, whereas inventions are not contrary to the plans of the inventor.[37]

Still, the reader may have some doubts: Perhaps all these discoveries are simply reflections of some necessary patterns of our thinking, of the Human Reason (to use a Kantian term), and thus, in the final account, they are mind-dependent (or Human Reason-dependent)? Such discoveries are not of phenomena, let alone of fantasies or illusions, which are mind-dependent. The reality of these discovered entities is fundamental and irreducible to other sort of reality. Even the attempts of reducing

---

[37] Jody Azzouni draws our attention to the possibility that the "the implications of a stipulated set of rules can surprise us because they are, to a large extent, epistemically opaque. …We should beware of taking the psychological impact of algorithmic independence as a symptom of ontological independence: a set of axioms and inference rules … yields surprising implications" (Azzouni 2000, p. 231). In my view, however, the unexpectedness and surprise concerning the discoveries of ideal numbers, non-Euclidean possibilities, quasi-crystalline geometric possibilities, and other examples in this Chapter are due to the dogmatic exclusion of possibilities that mathematicians believed that they could not *exist*, namely, that they are impossible.

pure mathematics to logic have been failed (while logical pure possibilities in turn are independent of our mind and thus, like mathematical possibilities, are applicable to actual reality, even if we do not follow any Kantian assumption). There is nothing else beyond mathematical discoveries. Since pure possibilities are fundamental beings or existents, they are not phenomena of anything and they are mind-independent. In a sense, they are "things-in-themselves," which can be discovered by our intellect, calculations, and imagination. There is nothing ontologically more fundamental than individual pure possibilities, including the mathematical ones. Whenever physical reality actualizes mathematical pure possibilities and their rela-tionality (subjecting them to spatiotemporal and causal restrictions), it can do so because they are first of all mathematically, purely possible, quite independent of actual reality. On the other hand, if the discoverer finds actual objects, data, or states subject to measurement and yet are considered mathematically (or theoretically) impossible (such as the case of Dan Shechtman's discovery of a "quasicrystal" on 8 April 1982)—to make such actual discoveries intelligible and to clear them of the fault of experimental error, one has to prove and explain them mathematically, to prove their purely mathematical existence-possibility, quite independently of the actual data.

Another doubt or worry concerns mathematical stipulations. Unlike mathemati-cal pure possibilities (and their relationality), stipulated entities are not discoverable and they depend on our mind. My view about mathematical entities is immune, I think, to the worry that Jody Azzouni expresses so well in the following: "You can-not simultaneously treat mathematical objects as ones which are genuinely indepen-dent of us, in the sense that their existence and their properties are not matters to be stipulated by us, and at the same time, fail to provide an explanation of how what we claim about them is dependable with respect to *them*" (Azzouni 2000, p. 242). As pure possibilities, mathematical entities or objects are accessible enough to our intellect and imagination and are discoverable by means of these capabilities. What we claim, logically or otherwise, about these entities or objects is strictly valid for *them as they are in themselves* (as I have argued), notwithstanding their complete ontological independence of us. Contrary to individual pure possibilities, Kripkean possible worlds are stipulated, not discovered (remember that panenmentalism is entirely free of the concept of "possible worlds"). Hence, as individual pure possi-bilities, mathematical entities or objects are not stipulated. They are discoverable instead.

Reality as a whole consists of pure possibilities and of actualities. Pure possibili-ties are as real as actualities, though in a different way. "Existence" has two mean-ings—purely possible and actual. To be, to exist, any being has to be purely possible first. Ontologically speaking, pure possibilities are the primary or fundamental beings, whereas actualities supervene on them, not the other way round.

## 5.8   Some Examples of Discoveries of Mathematical Pure Possibilities

According to Euclidean geometry, owing to the Fifth Postulate (concerning parallel lines), the following possibilities do not and could not exist[38]:

1. Given a line $l$ and a point $P$ not on the line, there is more or less than one line through $P$ in the plane determined by $l$ and $P$ that does not intersect $l$.
2. The sum of the angles in a triangle is more (or less) than 180° (or two right angles).
3. The ratio of the circumference to the diameter of a circle is not the same for all circles, no matter how large or how small.
4. Given any triangle, there are no arbitrary large and small triangles with the same angles and whose sides are in the same proportion to one another.
5. The Pythagorean Theorem is invalid.

In contrast, according to a non-Euclidean geometry, each of these possibilities *does exist and it is as real, valuable, and significant as the Euclidean possibilities.* These five possibilities became quite new discoveries, however paradoxical, since the middle of the nineteenth century. On the grounds of Euclidean geometry all these five are simply and undoubtedly impossible. In other words, Euclidean geometry has excluded such possibilities, whereas non-Euclidean geometries have discovered them as real and saved them as mathematical pure possibilities.

As Davis and Hersh put it, "A whole series of geometers tried to prove the parallel postulate by showing that its negation led to absurdities. But they were led not to absurdities but to the discovery of 'fantastic' geometries that had as much logical consistency as the Euclidean geometry of 'the real world'."[39] In panenmentalist terms, because any two pure possibilities are necessarily different from one another, they necessarily relate to one another. Hence, Euclidean possibilities necessarily relate to non-Euclidean possibilities, though this relationality was quite unknown until the middle of the nineteenth century. Each of the abovementioned five negations thus adequately reflects the result of the relationality of a Euclidean possibility, following the parallel postulate, to a non-Euclidean possibility, which is the negation of the Euclidean possibility. Like the non-Euclidean pure possibilities, the relationality under discussion is also a fruit of discoveries, quite surprising or unexpected before the middle of the nineteenth century.

Likewise, Davis and Hersh mention that following the discovery (or the proof of the legitimacy) of a non-Cantorian set for which the axiom of choice or the continuum hypothesis is false, one has to use the axioms of restricted set theory "to construct a model in which the negation of the axiom of choice or the negation of the continuum hypothesis can be proved as theorems" (op. cit., p. 233). In panenmentalist terms, pure possibilities of which a Cantorian set consists relate to their negations

---

[38] This is the non-Euclidean list that is contrary to the Euclidean list in O'Shea 2007, p. 73.
[39] Davis and Hersh 1998, p. 232.

in a non-Cantorian set. The discovery under discussion requires the relationality of mathematical pure possibilities without any reliance on experience, empirical data, or actual reality.

## 5.9　Mathematical Language, Conceptual Possibilities, and the Applicability of Mathematics

There is a strong temptation to argue that mathematics affords various languages to describe the world. Galileo's Book of Nature written in mathematical language is one of the most fruitful ideas in clarifying the indispensable role of mathematics in the history of physics and natural science. Why should I, then, posit mathematical pure possibilities existing independently of our mind as well as of the mathematical language(s) in which we describe nature? Why not assume that mathematical possibilities are simply conceptual possibilities, as Howard Stein did (Stein 1988, p. 252)? This is nicely compatible with the view that mathematics is a language in which we may describe reality or nature (or that it is a general or formal language about the languages by means of which we describe reality or nature).

Nevertheless, such a view has to overcome a substantial obstacle—why should nature or reality be subject to such a language, which, in many cases, has been constructed independently of actual reality and nature? On Kantian grounds, we can find a well-established answer to this question but if we do not follow a Kantian way of thinking about this problem, we have to suggest quite another answer. If indeed mathematical possibilities are mind-independent, we are left with a clear answer—in fact, nature or actual reality actualizes, among other pure possibilities, mathematical pure possibilities. It is an empirical fact, which a very long history of observations and experiments has shown to be undeniable. If this is a contingent fact, like any other actualities, there is nothing wrong with it. Unlike pure possibilities and their relationality (which are necessary), actualities—empirical facts—are contingent, although necessarily subject to spatiotemporal and causal circumstances. It is a matter of contingency or contingent facts that anything actual exists; and, at least to some extent, it is a contingent fact that nature or actual reality actualizes mathematical possibilities. We depend entirely on empirical findings whenever we would like to know whether a pure mathematical theory holds true for nature or actual reality. And these findings are contingent, even when they appear to follow a necessary law or rule. Nevertheless, nature could not actualize anything but individual pure possibilities and their relationality, among which there are mathematical pure possibilities and their relationality. Hence, there are actualities that are of mathematical pure possibilities and their relationality and, because of this, in fact, the Book of Nature is written in mathematical language(s). Actualization thus depends or supervenes on individual pure possibilities and their relationality, not the other way round. Unlike the existence of actualities themselves, this dependence or supervenience is inescapable or necessary. As for mathematical conceptual possibilities, of which mathematical languages consist, they simply reflect the accessibility of our mind to the mathematical pure possibilities and their relationality that nature actualizes. Only in this way can mathematical language be valid for actual reality.

Concerning the applicability of mathematics, I have to say more. Any actuality actualizes its own individual pure possibility and each physical entity actualizes the mathematical part of its individual pure possibility. As spatiotemporal objects (whose existence is contingent, although their spatiotemporal and causal conditioning is necessary), all actual entities necessarily have mathematical properties (for instance, there is no spatial entity that has not a geometrical form, just as there are no temporal and spatial entities that are not subject to numbers). As actualities, physical objects are spatiotemporal entities and, as such, they inevitably share some mathematical properties, which are actualizations of mathematical pure possibilities and their general relations. Indeed, all that we know about physical reality clearly demonstrates that there is no physical entity devoid of some mathematical properties. Thus each physical entity is, to some extent, an actuality of a mathematical pure possibility, which is epistemically and ontically independent of any actuality or empirical reality. Physical objects ontically and epistemically depend on their mathematical pure possibilities but not vice versa. Each physical entity is an application of an individual physico-mathematical pure possibility, and this makes the applicability of mathematics to actual, empirical reality, *necessary*.

## 5.10 Is It Valid to Quantify over Mathematical Pure Possibilities?

Geoffrey Hellman mentions "our unwillingness to quantify over possibilia."[40] Claiming that "possibilia are not recognized as objects,"[41] Hellman takes, wittingly or unwittingly, an actualist stance, whereas possibilism is a sound alternative to actualism or naturalism, especially as far as purely mathematical objects or entities are concerned. There is no necessity to adopt an actualist stance. Ascribing full-fledged existence to pure possibilities, as much as to actualities, panenmentalism must acknowledge two kinds of existential and universal quantifiers—over actualities and over pure possibilities. Once we acknowledge that, we have no special difficulty in quantifying over pure possibilities,[42] especially over mathematical pure possibilities, such as: "The sum of the angles of *every* (possible) Euclidean triangle is exactly 180 degrees, no less and no more;" "There (possibly) *exist* triangles of which the sum of their angles is more (or less) than 180 degrees." These two

---

[40] Hellman 1989, p. 17. On the problems of quantifying over possibilia and commitment to possibilia, see Hellman 1993, 2003, pp. 147 and 149; Restall 2003, p. 82; and Burgess 2005, p. 83.

[41] Op. cit., p. 59. For possibilist alternatives, see the examples of "ontologically radical strategies" that, according to Jody Azzouni, take "modal idioms as primitives and mathematical reference as, strictly speaking, concerned with possibility rather than actuality" (Azzouni 1994, p. 7).

[42] See Gilead 2009, pp. 39 and 45, including the context in the relevant sections.

propositions are true, quite regardless or independently of the fact that Euclidean and non-Euclidean geometries are applicable to actual reality or nature. If Euclidean and non-Euclidean geometries were not applicable to actual reality or nature, these geometries, as mathematically *purely possible*, would still have remained valid. Their existence, validity, or truth does not depend on their applicability in general and to actual reality in particular. Such is the case because, among other things, it is entirely valid to quantify over mathematical pure possibilities.

## 5.11  Proofs and the Reality of Mathematical Pure Possibilities

To reveal the reality of their objects, mathematicians cannot rely upon experiment or observation; instead, they rely upon proof.[43] Proofs are secured mathematical ways to discover new mathematical objects, which are mathematical pure possibilities, and the ways they relate to one another. Moreover, fictions are not subject to proofs, otherwise they would not have been considered as fictions but as truths or real existents. Thus, pure mathematical proofs inescapably entail mathematical discoveries of non-actual real existents. There are other mathematical means of discovery about which we will elaborate below. Like logical proofs, mathematical proofs reveal the necessity about their entities. Like all other pure possibilities, mathematical possibilities and their relationality are strictly necessary. Mathematical discoveries are about mathematical necessities, which are also existents as pure possibilities. Discoveries concerning actual, empirical reality are not of necessary phenomena but of contingent ones. Mathematical discoveries are quite different—mathematical proofs are the most reliable way to discover necessary mathematical pure possibilities and their necessary relationality. Mathematical proofs combine possibility, necessity, reality (or existence), and discovery in such a way that the discovered facts in mathematics—the facts about prime numbers, for instance—are such that they will never "change in the light of future discoveries" (Du Sautoy 2003, p. 32). After all, tangible reality is not more real than the reality of mathematical objects and their relationality, all the more so, as the stability or durability of the latter is greatly firmer than that of the former.[44] The proof reveals the necessity about

---

[43] Cf.: "In other scientific disciplines, physical observation and experiment provide some reassurance of the reality of a subject. While other scientists can use their eyes to see this physical reality, mathematicians can rely on mathematical proof, like a sixth sense, to negotiate the invisible subject" (Du Sautoy 2003, p. 32). Cf.: "The eyes of the Mind, by which it sees and observes things, are the demonstrations themselves" (*Ethics* 5, the demonstration of proposition 29; Spinoza 1985, p. 608). This sentence is about the nature of reason (*ratio*) to conceive things as real and necessary "under a species of eternity" (namely, atemporally). Notwithstanding, Spinoza was an actualist and he could not accept the whole idea of pure possibilities.

[44] Compare with Alain Connes's view as follows: "I think that I'm fairly close to the realist point of view. Take prime numbers, for example, which, as far as I'm concerned, constitute a more stable reality than the material reality that surrounds us" (Changeux and Connes 1989, p. 12; cf. du

mathematical pure possibilities, about mathematical facts, and their relationality as well. It also reveals that mathematical truths, unlike truths about actualities, are not subject to temporality, locality, and causality. Thus, mathematical truths, unlike those about actualities, are immutable and atemporal. The laws of nature may "lie" (as Nancy Cartwright suggests in Cartwright 1983), whereas mathematical laws never lie; what the latter describe, they also explain. As long as one is realist about pure possibilities and their relationality, the laws of mathematics, unlike the laws of nature, cannot lie, cannot deceive us. In mathematics, unlike in actual nature, there is no gap between "phenomena" and reality. Mathematical objects are substantially real; they are not phenomena. As I will argue below, even mathematical fictions are truthful, for they help mathematicians discover mathematical facts, truths, and necessities.

All this does not render mathematics dogmatic. Owing to its special kind of necessity, which is a necessity about individual pure possibilities and their relationality, the non-Euclidean necessary pure possibilities do not exclude the Euclidean necessary pure possibilities, and vice versa. These two kinds of possibilities are different and in some cases negate each other, but they do not exclude the other kind of possibilities, rather the contrary—because these two kinds are different, they necessarily relate to each other. They all exist in the same realm (but in different domains, the Euclidean and the non-Euclidean)—the whole realm of mathematical pure possibilities and their relationality. Hence, mathematical necessity and mathematical pluralism are compatible, whereas mathematical dogmatism and mathematical necessity exclude each other, for mathematical possibility (which is ample with variety) and mathematical necessity are entirely compatible. This leaves the whole realm of mathematical possibilities ever open. No mathematical theory or domain can exhaust the whole mathematical realm.

## 5.12  Mathematical Fictions, Creativity, and Discoverable Truths

Mathematicians have been endowed with ample imagination and they have certainly been creative. Like artists, they have been capable of creating amazing fictions. Nevertheless, these fictions are truthful—they are of immense help to mathematicians in discovering mathematical truths subject to proofs. Mathematical theories may be such fictions, indicating the creativity and inventiveness of their authors.[45] Nevertheless, these theories, once they are subjected to proofs, reveal or

---

Sautoy 2011, p. 23; and Polkinghorne 2011). Connes explicitly mentions mathematical reality, which mathematicians explore and discover.

[45] Cf. Rosen 2011a, pp. 14–15. Unlike mathematical objects, mathematical theories, such as the calculus, are subject to temporality and come into existence; hence, according to Rosen, mathematical objects are discoverable, whereas mathematical theories are inventible (he names them "abstract artifacts"). I do not accept Rosen's qualified realism about mathematical objects (such as

discover facts and truths about mathematical objects, which are mind-independent.[46] These facts, truths, and necessities are independent of the theories and of the imagination, creativity, and inventiveness that enabled these discoveries, which are of mathematical individual pure possibilities and their general or universal relationality. Mathematical fictions are epistemic tools, whereas mathematical pure possibilities and their relationality enjoy a substantial ontological status.

Algebraic tools helped mathematicians discover various mathematical entities or objects, such as zero, negative numbers, positive numbers, and real numbers. These are real discovered entities, which exist independently of our mind, any truthful fiction, actual reality, and empirical data. The discovery of such entities requires no empirical or quasi-empirical means, no matter what were the historical, actual circumstances in which they were actually discovered. For instance, suppose that the postulate of the correspondence between the points on the line and the real numbers is a truthful fiction, an illuminating mathematical artifact. Remember the example of the abovementioned Gauss's pictorial proof. Though it is not rigorous as recursive proofs are, it is no less helpful, sometimes indispensable, for discovering new mathematical pure possibilities and their relationality. As with recursive proofs, such fictions are not empirical or quasi-empirical. Instead of relying only on our intellect, they rely also upon our imagination, both of which take us beyond actual or empirical reality. By means of such fictions, mathematicians could make fascinating mathematical discoveries, regardless of their applications to physics. Because of the discovery of mathematical pure possibilities and their relationality, such actual applications are possible; not the other way round. Contrary to Putnam (Putnam 1979, pp. 64–65), no quasi-empirical, let alone empirical, means have to be involved in the possibility or capability of such discoveries. The (actual) history of such discoveries is quite irrelevant to our discussion, for the historical circumstances are contingent, as is any actuality. Human beings are capable of relating to the purely possible (for instance, to the counterfactual or counter-actual) as much as to the actual. We are not only natural or instinctual creatures; we are equally rational and imaginative creators, inventors, and discoverers, capable of transcending or liberating ourselves from the actual, of creating new realities, and, especially, of discovering new pure possibilities.

Mathematics has made most fruitful uses of mathematical metaphors—its amazing capability of translating or transferring from one domain to quite a different one (for instance, from numbers theory or arithmetic to geometry and vice versa, or

---

numbers) (Rosen 2011b), which ignores the irreducible, fundamental existence of mathematical objects as mathematical individual pure possibilities.

[46] Cf.: "What about the question of whether mathematical objects really exist? ... There are some things out there that are independent of our existence or act of imagining them. Prime numbers, simple groups, elliptic curves. It is not mathematicians who made these things" (Du Sautoy 2011, p. 23). Du Sautoy gracefully oscillates between mathematical creation and discovery, whereas, on my view, mathematicians invent or create truthful fictions, but what they discover are mind-independent mathematical pure possibilities and their necessary relationality.

from geometry to algebra and vice versa) has contributed greatly to further mathematical discoveries.

Thus truthful fictions, mathematical metaphors, and, first and foremost, proofs—all of which are not quasi-empirical, let alone empirical—have contributed to the mathematical discoveries of further pure mathematical possibilities and their necessary relationality.[47]


## 5.13   Mathematical Discoveries Are of Pure Possibilities, Not of Actualities

Consider, for instance, the following propositions: (1) There are Euclidean spaces of every dimension; (2) There are non-Euclidean spaces of every dimension; and (3) There are spheres, Euclidean or non-Euclidean, of every dimension. The existence of such spaces and spheres, or of any other geometrical figure or solid of every dimension, is not drawn, inferred, or abstracted from empirical data or from actual reality. Their existence is of pure possibilities—possibilia—and their relationality. Some of these possibilities may be actualized or applied to actual phenomena. Nevertheless, their existence is entirely independent of actual or empirical reality. Mathematical discoveries are of mathematical individual pure possibilities and their general relationality, whether they are physically actualizable or not.

Hilary Putnam and other philosophers would not accept the possibilist view that this Chapter suggests, probably on the grounds of prevalent philosophical prejudices against realism about individual pure possibilities and their relationality. Unlike Putnam (for instance, Putnam 1979, p. x), I do not think that Euclidean geometry has been overthrown owing to the discovery of non-Euclidean ones. On the contrary, the new discoveries have enriched and expanded our knowledge of mathematical pure possibilities. I do not accept Putnam's characterization of geometry as "the theory of physical space" (op. cit., p. 77). As I argued above, even if physical, actual space would not have existed as we are acquainted with it, Euclidean geometry, let alone non-Euclidean ones, would have been invariably saved. The practical uses of geometries, in science and for "mundane" measurements of the land, simply indicate the actual applicability of pure mathematics, which, in itself, is entirely independent of actual reality. We accept axioms as evident or true, not because they rely upon actual reality, which provides us with probabilities and inductive generalities only and not with what may satisfy such axioms; we accept them, instead, because we prefer their possibilities to others (in the event of us

---

[47] Thus, unlike mathematical fictionalism, which denies the existence of mathematical objects or entities (Leng 2010; cf. Pincock et al. 2012; Balaguer 2011), my view concerning truthful fictions (Gilead 2009), which can help in discovering mathematical pure possibilities and their relationality, requires the existence of these possibilities as mind-independent entities or objects.

knowing enough about the other possibilities).[48] Before 1979, Putnam conjured that it was possible that Riemann hypothesis would not be proven but verified "by a quasi-empirical method" (op. cit., p. 62). Time will tell if he was right but so far the progress that has been made toward a possible proof has not relied upon such a method; instead, it has relied, among other things, on translating geometry into "abstract algebra."[49]

Notwithstanding Putnam's reservations and criticism,[50] mathematical pure possibilities are *a priori*, non-empirically accessible, provided that we do not deny their independent existence and their relationality as well and that pure possibilities and necessities are entirely compatible.

Alain Connes enlightens this from an angle similar to mine (Changeux and Connes 1989, pp. 17–18). Whenever we have to rely solely on calculations, we have to leave behind "all contact with physical reality" (actual reality, in my terms). For Connes, "this is what mathematical reality is all about: there exists, quite inexplicably, a coherence independent of our system of sensory perception ... a coherence that entirely surpasses the coherence yielded by sensible intuition, the direct intuition of phenomena." But what provides us with metaphysical grounds and an explanation for this special kind of reality is an assignment that philosophers, not mathematicians, take upon themselves. Panenmentalism is one of the ways to fulfill such an assignment.

As for calculations, they take part in proofs. Indeed, Riemann and Hilbert preferred insightful thinking to *Rechnung* (computation, calculation), but this did not mean that they excluded calculations altogether from taking part in proofs (Laugwitz 2008, pp. 302–304). Moreover, if Detlef Laugwitz is right in ascribing to Hilbert the view that mathematical objects are concepts, and if concepts are mind-dependent, this would ignore the ontological status of mathematical entities or objects as mind-independent pure possibilities, whereas calculations, no less than insightful thinking and imagination, serve mathematicians in discovering mathematical objects that are mind-independent mathematical pure possibilities. In any case, like proofs, calculations are *a priori* accessible as well as primarily and directly valid for pure possibilities, whereas measurements are only *a posteriori* accessible and are valid exclusively for actualities. Furthermore, like proofs, calculations are associated with the necessity *about* the calculated facts (which are pure possibilities) and with the necessary relationality among them, whereas measurements have to do with actual entities or facts, which are contingent. Calculations are intellectual performances, and they are entirely independent of sensual, tangible, or actual reality.

---

[48] Subject to our aim—discovering further new mathematical possibilities. We thus prefer axioms which are more fruitful than the others in proving more interesting true theorems and, hence, in discovering new individual mathematical pure possibilities and their relationality.

[49] To judge from Du Sautoy 2003, p. 306.

[50] Putnam 1979, pp. viii–x and 60–70. In contrast, Giaquinto 2007, Ch. 4, "Geometrical Discoveries by Visualizing," convincingly shows how it is possible to make geometrical discoveries by visual means *in a non-empirical manner*. He thus relies on Kant in assuming synthetic *a priori* judgments in geometry (op. cit., p. 50).

Nevertheless, the reality of mathematical pure possibilities and their relationality, the reality that is subject to proofs and calculations, is as real as actual reality. Distinguishing mathematical reality from physical reality, Riemann considered mathematical objects only (O'Shea's 2007, p. 102). These objects or entities are subject to calculations and, as O'Shea puts it, everything you can calculate ("compute") could not be more real (ibid.).

## 5.14   How Does Panenmentalism Challenge Parsons's Recent View on Mathematical Objects?

Charles Parsons applies the distinction between Meinongian being (referring to an existential quantifier) and existence (namely, actual existence spatiotemporally and causally determined) also to mathematical objects (Parsons 2008). In the light of this distinction, the proposition, "there is a natural number between 1 and 3," (which is 2) is valid for the *being* of a mathematical object or entity, *not for any actual existence* (actuality) of it. What about nonexistent mathematical objects, such as the round square? Parsons's answer to this question is as follows: "surely fiction can speak of nonexistent mathematical objects, though of course the necessity of mathematics implies that the existence of such objects is *impossible*. Although fiction is doubtless constrained by considerations of coherence, it is certainly not limited to recounting the possible" (op. cit., p. 25). At this point, I believe, Parson is wrong. Literary fiction is based on the assumption that the persons, objects, situations, and events depicted in it are *possible*, otherwise no one would have any interest in reading such a fiction and in welcoming it with expectations and excitement. This is a moot point between panenmentalism and Parsons. Indeed, panenmentalism can clearly demonstrate how the round square is mathematically possible after all (Gilead 2009, pp. 73–75).

The case appears to be that some of the ideas that I have expressed above have also been introduced by Parsons's book. Take, for instance, his definition of mathematical objects: "Taken at face value, mathematical language speaks of objects distinctively mathematical in character: numbers, functions, sets, geometric figures, and the like. ... are distinctive in being abstract. ... an object is abstract if it is not located in space and time and does not stand in causal relations" (op. cit., p. 1). Indeed, mathematical pure possibilities are individual entities that do not spatiotemporally and causally exist. They *really* exist in a different manner, namely, in the realm of individual pure possibilities and their relationality which are independent on any actuality. Thus, in the proposition "There is a mathematical object or entity— a natural number—between number 1 and number 3," the phrase "there is" does not signify merely an existential quantifier; contrary to Parsons, I think that this phrase does not hold merely for a being but also for a *real existent*, although a nonactual one! We should distinguish between the existence, fundamentally real, of individual pure possibilities (including their relationality) and the existence of their actualities.

Any individual pure possibility is not a concrete spatiotemporal object, yet it is absolutely real. In fact, individual pure possibilities are more real than their actualities, as these possibilities are ontologically and epistemologically prior to their actualities and render them possible. While any actuality is dependent on its individual pure possibilities, these possibilities are independent of their actualities. Individual pure possibilities are fundamental, most basic and thus prior, existents. Any mathematical pure possibility is real, because it is a real mathematical individual entity existing independently of its physical embodiment, namely, an actuality, and prior to this actuality. Individual pure possibilities can exist with no actuality and quite independently of any actuality, whereas no actuality can exist without its pure possibility (which panenmentalism considers as the identity of an actuality) on which it necessarily depends. Panenmentalism thus endows individual pure possibilities and their relationality with ontological priority. Moreover, we know purely mathematical objects (to be distinguished from applied mathematical objects, which are physical actualities) in an *a priori* manner and independently of any empirical evidence. To sum up, individual pure possibilities and their relationality maintain ontological as well as epistemological priority. With regard to my understanding of it, such is certainly not the view of Parsons.

Parsons does not ignore the connection between mathematical objects and possibilities. Discussing Kant's philosophy of mathematics, Parsons mentions the fact that "There are indications that he [Kant] would treat mathematical existence under the category of possibility; mathematical examples occur in the elucidation of the category in the Postulates [*Critique of Pure Reason* A220–221/B268; A223–224/ B271]. ... There is some affinity between Kant's view and those now called modalist, ... . This affinity is limited by the fact that Kant conceives possibility as 'real' possibility, not certifiable by mathematical arguments" (op. cit., p. 7). Parsons further writes: "Kant's refusal to apply the category of actuality to mathematical objects suggests another distinctive feature of such objects, that the distinction between potentiality and actuality, particularly between possible and actual existence, does not apply to them" (op. cit., pp. 7–8).

In contrast, panenmentalism clearly distinguishes between individual pure possibility and potentiality—potentiality rests upon actual state of affairs, whereas pure possibility is independent on anything actual. Parsons cites Kant's disciple—Johann Schultz—as follows: "'In mathematics possibility and actuality are one, and the geometer says *there are* (*es gibt*) conic sections, as soon as he has shown their possibility *a priori*, without requiring as to the actual drawing or making them from material'" (op. cit., p. 8, footnote 9). Of course, Schultz was right, and I would happily adopt that view of his, except for the clear panenmentalist distinction between a mathematical individual pure possibility and its actuality under physical circumstances. Thus, if Kant did not really distinguish between number 2 as a mathematical pure possibility and its actuality in physical circumstances, panenmentalism must challenge his view (as well as that of Parsons).

Following Meinong, Parsons appears to think that "the primary meaning of 'existence' is what we are calling actual existence, so that in the most basic sense, mathematical objects do not exist" (op. cit., p. 25, footnote 33). This is precisely what panenmentalism opposes. Mathematical objects do exist, to begin with, as mathematical individual pure possibilities and their relationality, hence, they *may be actualized as physical* objects, events, circumstances, and the like.

Parsons uses the term "pure abstract objects" (op. cit., p. 36) and states: "it appears that natural numbers are pure abstract objects" (ibid.). Yet, the case appears to be that "mathematical pure possibilities" is a much better and clearer phrase (see my objections above to wrongly considering mathematical objects or entities as "abstract"). Using the term "pure abstract objects," Parsons has Platonic Forms in mind: "Perhaps the earliest clear reference to pure abstract objects is the discussion of the Greatest Forms in Plato's *Sophist*" (ibid., footnote 59). In contrast, first, pure possibilities are individuals whose general nature lies only in their relationality. Second, although these possibilities are exempt of anything spatiotemporal or causal, they are not, contrary to Platonic Forms, transcendent. It would be a misconception to ascribe any platonist approach to panenmentalism in general and to individual pure possibilities in particular.

## 5.15 Conclusion: The Answers to the Abovementioned Three Questions

Mathematical entities are individual pure possibilities existing independently of our mind, possible worlds, and anything actual. They do not exist in space, time, or as links in a causal chain. Nor are they abstracted from actual-physical reality, either. Our intellect (by means of proofs, calculations, and the like) and imagination are reliable enough to give us epistemic access to such entities.

Mathematical entities can be discovered; they cannot be invented. Nevertheless, truthful fictions may help us discover such entities as well as their general relationality.

# Chapter 6
# A Panenmenalist Approach to Molyneux's Problem and Some Empirical Findings

**Abstract** My panenmentalist approach to Molyneux's problem rests upon the assumption that the newly-sighted person has an innate epistemic access to the individual pure possibilities of the actual sphere and cube, irrespective of any specific geometry and empirical data. If the new sight of such a person were normal, he or she could, immediately or after some training, distinguish between the sphere and the cube and relate them to his or her tactile previous acquaintance with these objects, while still being blind. This innate capability of cross-modal sensory identification is well understood in terms of my panenmentalist approach and some recent empirical tests as well. The example of synesthesia and that of "visual-to-auditory sensory substitute devices" (SSDs) demonstrate that our brain's capability of cross-modal sensory identification is innate.

## 6.1 Molyneux's Problem

On July 7, 1688, William Molyneux wrote a letter to John Locke, posing the following problem:

> A Man, being born blind, and having a Globe and a Cube, nigh of the same bigness, Committed into his Hands, and being taught or Told, which is Called the Globe, and which the Cube, so as easily to distinguish them by his Touch or Feeling; Then both being taken from Him, and Laid on a Table, Let us Suppose his Sight Restored to Him; Whether he Could, by his Sight, and before he touch them, know which is the Globe and which the Cube? Or Whether he Could know by his Sight, before he stretch'd out his Hand, whether he Could not Reach them, tho they were Removed 20 or 1000 feet from Him? (Molyneux 1688)

Locke, referring to Molyneux's letter, phrased the problem in *An Essay Concerning Human Understanding* (II, 9, vii).

Molyneux's problem, which has troubled philosophers and scientists from 1688 until the present, has been phrased differently as follows: before gaining sight, the blind person distinguished between the sphere and the cube simply by touching them. Could he now, having his sight given to him, not only distinguish between the two actual objects but also relate the sphere and the cube that he can now see to the sphere and the cube that he could only touch before? Can he now identify the visual sphere and the cube with the touched ones respectively? This problem is about the cross-modal (or the meta-modal) sensory identification of objects. There are some ways—modes—in which one can perceive the same objects: by sight, touch, and the like. The trouble is that these modes are *essentially different* one from the other. What is perceptually common to a tangible object and a visual one? On what grounds can the newly-sighted person (as well as we) *know* that it is one and the same object, under two different sensory modes, and not two different objects? If there are such grounds at all, are they innate or empirically acquired?

Both versions of Molyneux's problem have been vigorously considered by philosophers and scientists for about 350 years. Even if scientists can settle or solve the problem empirically, it still certainly has had a vital philosophical significance. For instance, it has been related to the major philosophical question about the origin of our knowledge: is knowledge innate or acquired by means of experience only?

The current Chapter relates to three senses of the concept "modality": (1) The philosophical sense in which the concepts and categories of the possible, the actual, the necessary, the contingent, and existence are considered; (2) mathematical modalities in which the same mathematical entities are treated in different geometries, for instance, Euclidean and Non-Euclidean ones; (3) sensory modalities in which the same objects are differently perceived by different senses, such as tactile, visible, or audible ones.

## 6.2    Purely Geometrical Objects and a Philosophically Modal Approach to Molyneux's Problem

Relying upon a classical empiricist basis, on Lockean empiricist grounds, Molyneux and Locke suggested a negative answer to Molyneux's time-honored problem. If what everybody can know about objects supervenes on data received by one's senses, there is no possibility for the newly- sighted man in Molyneux's thought-experiment, as long as he cannot touch them, to distinguish between a sphere and a cube. On such classical empiricist grounds, the newly-sighted person also cannot identify the object that he previously touched with the same object that now becomes visual to him. He cannot perceive it cross-modally as one and the same object.

Can a rationalist approach solve this problem? I doubt it, for the point in the treatment of the problem by Molyneux and Locke is that empirical data are indispensible for treating it, whereas any rationalist approach assumes the contrary, namely, empirical data are reducible to rational principles. Nevertheless, it is quite clear that as much as actual sphere and actual cube are concerned, and as actual objects are subject to experience and to empirical data, these data are indispensible or irreducible. Experience or empirical knowledge is vital to get any information about actual objects. To know something about the actual sphere and cube laying before his eyes, the man in Molyneux's thought-experiment must use his sight, in any case in which he cannot, for one reason or another, touch these objects. Knowledge about facts cannot do without experience and empirical data, which are received, directly or indirectly, by means of our senses.

The problem is that, on Lockean empiricist grounds, until the moment that he gains sight, the newly-sighted in the experiment could not see the sphere or the cube and he had no visual data whatsoever of them. How can he recognize the sphere and the cube and distinguish between them, if he can now see them for the first time but, for the time being, cannot touch them?

Nevertheless, could this person while blind conceive *purely* geometrical objects without relying upon any sense but only upon his intellect and imagination and without applying these objects to anything empirical?

It is possible that purely geometrical objects have not been abstracted from empirical data but were discovered without relying upon the senses. We conceive such objects in an *a priori* manner. Purely geometrical objects were possible and intelligible, even if none of them—figures, lines, and points—have been physically embodied or applied at all and all the physical objects in our world were entirely different from these imagined purely geometrical objects. After all, there are purely geometrical figures and structures that have no correspondence with the empirical bodies and shapes of which we have experience.

Now, before gaining his sight, the man in Molyneux's thought-experiment could refer to purely geometrical spheres and cubes, independently of any experience. His purely geometrical knowledge about them was entirely independent of his tactile experience no less than of his missing sight. Before gaining sight, he adequately distinguished between purely geometrical cubes and spheres, irrespective of any specific geometry. In this way, he gained his *concepts* or *ideas* of sphere and cube. This is, obviously, not compatible with the ways in which we gain all our ideas or impressions according to Locke or any other classical empiricist.

The question of the correspondence of pure mathematics with actual reality, a reality that we are acquainted with by experience only and by means of our senses, this question is a big and most intricate one. The application of pure mathematics to empirical reality, an application without which physics were not possible, is a question that many minds have attempted to solve. A philosophically modal approach to this question is strongly relevant to the approach in the current Chapter to Molyneux's problem.

My philosophically modal approach to this problem rests upon a very simple distinction—the distinction between individual pure possibility and its actuality. My approach does not rely at all on the notion of possible worlds but only upon those of individual pure possibilities, their general relations, and their actualities as well.

The *a priori* epistemic access to any individual pure possibility is, in principle, open to our intellect and imagination, for we can usually imagine or think about any individual pure possibility that is different from any other individual pure possibility about which we think or which we imagine. To have such an access, we need not rely upon any experience or perception or upon an abstraction from it, but only upon our free imagination and our intellectual capability to relate to what is beyond what is given to us in experience and especially beyond what is actual.

Actual entities are concrete objects, whereas individual pure possibilities are specific entities. As their name indicates, individual pure possibilities are not general or abstract. In my terms, abstract "objects" or "entities" are abstracted or generalized from actual, concrete entities. For instance, "the table" is a general, abstract object, which is abstracted or generalized from actual, concrete tables with which we are empirically acquainted. Hence, my use of the term "abstract object" is different from the received use of these terms to which many other philosophers currently subscribe.[1] I will return to this point below.

There are individual pure possibilities that are psychical or mind-dependent. Nevertheless, in the current Chapter, I refer to individual pure possibilities that are clearly mind-independent and thus our mind has to discover rather than create them. Such pure possibilities are, for instance, numbers and geometrical objects.

Any actual entity, any actuality, which is subject to empirical study, is possible, otherwise it would not exist at all. To be actual, any entity must be possible *at the outset,* and this is a prior condition, prior to any spatiotemporal and causal conditions, which make such entity possible *as actual, too.* Strip any actual entity of all its spatiotemporal and causal conditions or restrictions and of all the circumstances under which it exists in fact, and you consider then its individual pure possibility. In such a case, you do not consider it as an actuality. This pure possibility distinguishes the entity under discussion from any other entity, possible or actual, because no two individual pure possibilities can be identical[2] and each individual actuality has exclusively an individual pure possibility that cannot be shared with any other actuality. This pure possibility is conceived by means of our intellect and imagination instead of by our experience and senses, in which actualities are perceived.

Any actuality and its individual pure possibility are not identical, though they are necessarily connected or united. They are not identical as the actuality is necessarily spatiotemporally and causally conditioned and restricted, whereas the pure possibility is absolutely exempt from such conditions and restrictions. It is impossible to separate an actuality from its individual pure possibility. Had it been severed from this possibility, the actuality under discussion could not exist at all. Thus, an actuality and its individual pure possibility are inseparable, yet they are distinct from each other; they are not identical.

---

[1] Nevertheless, following Donald Cary Williams, Keith Campbell (1983) considered tropes as abstract particulars, which have been abstracted from empirical reality. According to Williams and Campbell, we single out tropes ("characters"), which are particulars, spatio-temporally existing. Thus, Campbell claimed: "Tropes are brought before the mind by an act of abstraction. That is the sense, and the only sense, in which they are abstract entities" (Campbell 1983, p. 130). Marina Folescu considers universals as "general notions, formed by abstracting away from the tropes we encounter" (Folescu 2016., p. 641). Cf. Knox 2016, pp. 50 and 52 (also employing "abstracting away from"). Cf. Lowe 1995, p. 514, following Locke, considering abstraction as separation in thought.

[2] "Two" identical pure possibilities are one and the same possibility. In contrast, two allegedly identical actualities (say, two identical drops of water) exist in the same time at two different places (or at the same place in two different times) and hence an observer can always distinguish between them. In contrast, because pure possibilities do not exist at place and in time, no two "identical" pure possibilities can exist, and the law of the identity of the indiscernible is strictly and unqualifiedly valid for pure possibilities. Hence, two "identical" pure possibilities, which nothing, spatiotemporal or otherwise, may discern between them, are indeed one and the same pure possibility.

The objects or entities of pure mathematics are mathematical individual pure possibilities some of which are actualizable (applicable) in physical, actual reality.[3] The newly-sighted man in Molyneux's thought-experiment, provided that his sight is normal,[4] *should* be able to identify the cube and the sphere and *should* well distinguish between them by means of his sight alone resting upon his *a priori* acquaintance with the geometrical pure possibility of the sphere and that of the cube. He *should* understand that what he sees in fact are actualities of these two different pure possibilities, whose actualization is subject to spatiotemporal and causal conditions. Because both Molyneux and Locke were classical empiricists, they could not leave room for any such an *a priori* acquaintance. As we will realize, their empiricist view on this matter is not necessarily commonsensical, for, on a commonsensical basis, Thomas Reid argued to the contrary and explained why the newly-sighted man could immediately distinguish between the visual cube and sphere.

Note that my philosophically modal approach to Molyneux's thought-experiment is neither rationalist nor empiricist. The use of sight or that of touch is a necessary condition to recognize an *actual* sphere and an *actual* cube and to distinguish between them. This makes this approach not rationalist. In contrast, to *identify* an actual cube or sphere, regardless of spatiotemporal and causal conditions and of actual circumstances as well, and to tell the essential difference between them, one has to have, also as a necessary condition, an *a priori* knowledge of their individual pure possibilities. This makes my approach not empiricist.

As classical empiricists, Molyneux and Locke answered negatively to the question whether the man in the thought-experiment could tell the cube from the sphere and identify each of them. My philosophically modal answer is affirmative, provided that the man who gains sight has some prior knowledge of purely geometrical objects or any knowledge concerning the pure possibilities of geometrical figures

---

[3] Nevertheless, there are also purely geometrical entities, such as in Banach-Tarski's paradox, which are geometrical individual pure possibilities and yet *physically impossible*. This means that they are physically inapplicable. Such purely possible but physically impossible objects are the result of decomposition of, say, a ball, into a finite number of pieces (five of them suffice!) and, only having each of the pieces translated and rotated, recombining them into two identical balls in the same size of the original one. In geometry, "translation" means changing the positions of these pieces in space. There is another version of this paradox: the mathematician may recombine the same pieces into a much bigger object. Referring to this version, mathematician Karl Stromberg writes: "This seems to be patently false if we submit to the foolish practice of confusing the 'ideal' objects of geometry with the 'real' objects of the world around us. It certainly does seem to be folly to claim that a billiard ball can be chopped into pieces which can then be put back together to form a life-size statue of Banach. We, of course, make no such claim. Even in the world of mathematics, the theorem is astonishing, but true" (Stromberg 1979, p. 151). You may translate "ideal objects" and "real objects of the world around us" into the terms of my approach, namely, "individual pure possibilities" (or "purely geometrical objects") and "actualities" respectively. Indeed, we should never confuse pure possibilities with actualities, whether these possibilities are physically applicable (actualizable) or not (as in the case of Banach-Tarski's paradox).

[4] That is, provided that he has not lost his inborn neural-cerebral capability to recognize shapes and figures.

and provided that he has not lost his inborn neural capability to recognize shapes and figures under whatsoever sensual modes.

Molyneux's question also concentrates upon the cross-modal connection between tactile and visual acquaintance with a sphere and a cube. The geometrical pure possibilities of these objects allow different mathematical cross-modal representations or descriptions of the same possibility. The Euclidean geometry demonstrates that geometrical figures can be represented by numbers and their relations and vice versa, namely, numbers and their relations can be geometrically represented. Analytic geometry translates visual figures into algebraic equations and thus reduces any visualization of the figures under discussion to quite a different description. Anyone acquainted with such geometrical pure possibilities does not experience any difficulty in understanding that such possibilities are the same under different descriptions, representations, or terms—under different purely mathematical *modes*. This holds true also for the actualities of these possibilities. As a result, cross-modality (or "meta-modality") in the realm of sensory experience and actual reality is parallel to the cross-modality in the realm of mathematical pure possibilities.

Geometrical cross-modality between different geometries, Euclidean and non-Euclidean alike, is relevant to Molyneux's problem. Our capability to recognize and identify a triangle, for instance, whether it is a Euclidean or a non-Euclidean one, is analogous to our capability to recognize and identify cross-modally a cube, for instance, as being touched or seen. The geometrical modalities are different from each other and yet we recognize and identity a pure possibility of a cube, sphere, or square and the like under the modality of any geometry. The same holds for the analytic-algebraic modality of a geometrical figure and its Euclidean modality. It is the same figure or structure under two different descriptions or geometrical modalities.

As for sensory cross-modality, the tactile description of, or acquaintance with, a cube or a sphere is of the same cube or of the same sphere that the visual description or acquaintance captures. Each of these descriptions and ways of acquaintance captures partial actualization of one and the same individual pure possibility.

We have an empirical epistemic access to actualities but how can we epistemically access to mathematical pure possibilities? Our imagination and intellect are sufficient for this aim. We know that $5 + 2 = 7$, not because we know that, for instance, five chairs and seven chairs are twelve chairs. On the contrary, we know that five and seven chairs are twelve chairs *because* we *primarily* know that the mathematical pure possibilities of 5 and 2 are 7. We know that, not by relying upon experience and not dependently on anything actual, but only by thinking about or imagining these pure possibilities and their general relations. The same holds true for purely geometrical entities or for geometrical pure possibilities. We need no actual, empirical reality to have the pure possibility of a sphere or a cube. These are constructions or structures that our mind grasps without relying upon any experience or acquaintance with actual reality. Such pure possibilities are accessible to the imagination and intellect of any human being.

Such a view is compatible to some extent with that of Thomas Reid, analyzing the case of the blind mathematician, Nicholas Saunderson (Reid 1997 [1764], section 6. 11),[5] considering the innateness of mathematical knowledge (in Reid's term—the originality of mathematics, which is not acquired). Nevertheless, Reid did not refer this to a modal metaphysical question concerning actualities and their pure possibilities, and this is a big difference among others between my approach to the problem and Reid's (without mentioning Molyneux by name though there is no doubt that Reid discusses Molyneux's problem at this point).

As for the processing of numeral symbols (as well as letters), a recent most interesting empirical brain study demonstrates that "specificity in the ventral 'visual' stream can emerge independently of sensory modality and visual experience, under the influence of distinct connectivity patterns" (Abboud et al. 2015, p. 1). This empirical new study is compatible with the views of Reid and myself, *mutatis mutandis*, about the innate origin of our mathematical knowledge.

Our epistemic accessibility to mathematical pure possibilities and their general relations may sound to the reader as if it were a kind of Kantian "innate geometry", which Brigitte Sassen analyzes, discussing Kant and Molyneux's problem (Sassen 2004, p. 473).[6] No, I have no innate geometry in mind but, rather, innate geometrical pure possibilities, which are modally conceived in a different way ("mode") by each specific geometry, Euclidean and non-Euclidean. These innate possibilities are *cross-modal* geometrical pure possibilities, independent of any specific geometry, Euclidean or non-Euclidean.

What I have in mind is not Kant's notion of *a priori* mathematics, but quite another kind of mathematical apriority. Kant did not refer to mathematical individual pure *possibilities* that empirical reality actualizes but, rather, to the transcendental conditions that make experience possible. Referring to space and time, he discussed the *a priori*, transcendental conditions of perception. These conditions pertain only to the human mind (hence, they are ideal in Kantian terms). In contrast, my view is about mathematical possibilities that are *entirely mind-independent* and which are actualized in empirical reality *as it is in itself* and *not* as a phenomenon only, *not* as an object only for the human knowledge. This empirical reality, unlike Kant's consideration of it, is not a phenomenon, whose forms of intuition and those of the intellect are mind-dependent (better, Human Reason-dependent). We have access to the aforementioned mathematical possibilities by means of our intellect and imagination, which are the tools in serving mathematicians in deciding which mathematical entities *exist*, for *mathematical possibilities* are *mathematical existents*.

---

[5] Cf. Van Cleve 2007, Hopkins 2005, and Copenhaver 2010, pp. 298–299.

[6] As Sassen explains, Kant's contemporaries, J. G. Feder and H. A. Pistorius, believed that this kind of geometry was incompatible with Cheselden's empirical findings about a newly-sighted boy (which I will discuss below). They thought that Molyneux and Locke were justified in answering Molyneux's question negatively, because Cheselden's empirical findings had proven this. In contrast, I think, like some others, that these findings did not answer Molyneux's question at all, for they, like some other current findings, have not demonstrated that newly-sighted persons have gained with a normal, full-blown sight capability.

The way I consider mathematical objects and their independence of empirical and actual reality, is certainly not the way that many, probably the great majority, of current philosophers of mathematics consider these objects. My panenmentalist approach is of a special kind of possibilism, which does not rely upon the notion of possible worlds, which so many philosophers still accept.[7] Many of the current philosophers, even the great majority of them, admittedly or not, are actualists of one kind or another. Actualists deny the existence of pure possibilities that are entirely independent of actualities or of actual reality in general. Actualists certainly would not accept my view in the current Book.

An empiricist, who is also an actualist, would argue against me that the only way that we can acquire mathematical or any kind of knowledge is by experience and by *abstraction* from actual and empirical data. According to such an empiricist, we have no epistemic access to "mere[8] or pure possibilities" (which actualists assume to depend on actual reality) but only by way of abstraction from actual and empirical objects. Mathematical objects or entities for such an empiricist are simply abstract entities.

My answer to this challenge should be long and detailed. I will suggest now only a very concise and partial answer to this challenge and only to the extent that it is relevant to my approach to Molyneux's problem.

First, I would like to clarify that individual pure possibilities are not the so-called "abstract individuals" either.[9] Again, my terms of abstraction and "abstract objects" are different from the prevalent ones.

---

[7] My modal approach is realist about individual pure possibilities, whereas platonic theories of possibilities (to which Forrest 1986; Bigelow and Pargetter 1990; Tugby 2015, and Ingram 2015 are subscribed) are realist about possible universal properties that are uninstantiated but not about possible particulars or individuals. It is questionable whether all these Platonic theories are fully-fledged possibilist, if at all.

[8] Note that individual pure possibilities are not simply mere, non-actual possibilities, or mere possibilia, for individual pure possibilities *can* be actualized (unless their applicability to physical-actual reality is impossible, as in the examples mentioned above) and the parts of many of them are, in fact, actualized. It is important to re-emphasize that though any actuality is inseparable from its individual pure possibility, they are not identical and the distinction between them is, ontologically and epistemologically, maintained.

[9] Such as are mentioned in Linsky and Zalta 1995, pp. 534–536 and 554 (cf. Linsky and Zalta 2006, pp. 77, 88, 90–92). Even though these mind-independent abstract individuals are free from any spatiotemporal and causal restrictions or conditions and even though possibilia and possible worlds pertain to these "individuals", they are quite different from the individual pure possibilities mentioned in the present Book. These possibilities, in contrast, are the primal necessary ontological conditions for all actualities (i.e., individual pure possibilities are metaphysically fundamental), and no actuality is separable from its individual pure possibility. Furthermore, my approach considers properties as pertaining to the general relations amongst individual pure possibilities, which is entirely different from the manner in which Linsky and Zalta treat properties and abstract individuals. Linsky and Zalta's "Platonized naturalism" relies heavily upon Kant, Frege, Meinong (borrowing the comprehension principle), Russell, and, to some extent, on Quine, whereas my approach is entirely independent from any of these grand theories. Finally, my approach is strictly possibilist and realist about individual pure possibilities, whereas Linsky and Zalta occasionally rely upon actualist considerations and they subscribe to a conception of possible worlds. It is up to

Furthermore, abstraction should be carried out according to some concepts, criteria, methods, or models, none of which should be abstracted from any empirical or actual entities, otherwise the process of abstraction would entirely fail because of infinite regress. When abstracting, we should rely upon starting points or principles that were not abstracted. These starting points or principles cannot have, thus, empirical or actual grounds. The criteria according to which an abstraction can be made require justification that should be independent of the process of abstraction itself, otherwise we would commit circularity. Individual pure possibilities and their general relations may respectively serve as starting points and principles for any legitimate process of abstraction avoiding any circularity and infinite regress.

Pure numbers and geometrical figures, indeed all objects or entities of *pure* mathematics, are independent of empirical and actual reality and they are not acquired by experience or by abstraction from empirical and actual reality. At least, there are two main reasons for that.

First, the major difference between pure and applied mathematics is simply that, unlike applied mathematics, pure mathematics is independent of any application to empirical or actual reality. Thus, pure mathematicians *may* deal with objects and entities that do not correspond to anything in actual or empirical reality and which do not depend upon anything actual or empirical. There is no way in which such objects or entities were abstracted, directly or indirectly, from actual or empirical ones.

Second, mathematical imagination and mathematical intellect are independent of actual and empirical reality, and pure mathematics deals with mathematical pure possibilities, which have no grounds in actual reality. Mathematical thinking has epistemic access to individual pure possibilities and their general relations, as our imagination and intellect are *free* from actual and empirical reality. At least, our intellect and imagination are not confined to that reality. We are free to think about pure possibilities that are not abstracted from empirical and actual data. Not only are the capabilities of our intellect and imagination innate; equally innate are our acquaintance with and knowledge of some individual pure possibilities and their general relations.[10]

Could a mathematician, born blind, using her intellect and free imagination, abstract the cross-modal pure possibility of a cube from an actual cube of which she has had *only* a tactile experience but, allegedly, no innate concept whatsoever of its pure possibility? No, at least not on the grounds of my current approach. In order to

---

the reader to judge whether my approach is simpler, clearer, and more economical than their "Platonized naturalism", and whether my approach challenges Benacerraf's epistemic problem more convincingly.

[10] Note again that my view is neither empiricist nor rationalist. It adheres to the fact that all we know about actual and empirical reality is by experience and observation, which are irreducible to anything innate, whereas our acquaintance with and knowledge of individual pure possibilities and their general relations are innate, relying upon our innate capabilities of the intellect and the imagination. For the major differences between Platonic or Leibnizean innateness and that according to the panenmentalist approach, see Gilead 2009, pp. 124–126 and 129–135.

abstract from her tactile experience of a cube and to visualize such a cube in her imagination (thus to conceive it as if in another mode of perception), she must have a primary knowledge of the cross-modal individual pure possibility of the cube in question. Furthermore, on this basis, she can imagine how it looks like under other geometries, for instance, a non-Euclidean one. Without such an innate knowledge, she could not know *how* to abstract the cross-modal possibility from tactile data. She had no *criterion* or *model* according to which she could make such an abstraction. Relying upon experience only, she has only the tactile mode of perception of the cube.[11]

Let me give a fascinating example concerning the knowledge of crystalline pure possibilities that was conceived independently of anything actual or empirical and which was not abstracted from empirical or actual data. Some years before the first discovery of an actual quasi-crystal, Alan Mackay, a theoretical crystallographer, discovered quasi-crystalline pure possibilities, which, at the time, were considered to be impossible both on theoretical (purely geometrical) and on empirical grounds (Gilead 2013; 2016). These grounds did not prevent him from surmising, using his imagination, quasi-crystalline pure possibilities and to device computer visual images of them (in re-organizing actual experiments to discover whether diffraction pattern could be obtained). At that time, Mackay had no experience with quasi-crystals or any empirical data about them, whereas on the basis of the official theoretical (purely geometrical) crystallography at the time he could not infer their possibilities let alone existence. Indeed, one cannot infer the possibilities of quasi-crystals on the basis of the theoretical (purely geometrical) classical crystallography or on the basis of empirical data concerning actual classical crystals. Neither could one abstract quasi-crystals from actual classical crystals. Like a born-blind mathematician, such a fully sighted crystallographer could conceive pure possibilities of geometrical structures or figures of which no actual cases were known to him (or to her). This fascinating example also shows how our free imagination and intellect

---

[11] On the grounds of the above discussion, contrary to Gareth Evans (2002), I think that the question of innateness vs. acquired empirical knowledge lies at the heart of Molyneux's problem as well as the problem of cross-modal sensory identification. Referring to Quine's notion of "innate similarity-space", Evans supposed that "according to the most radical empiricist position, an organism has an innate similarity-space defined over *sensations*, and concepts simply result from a partitioning of that space" (op. cit, pp. 327–328). Is this a "radical empiricism" indeed? After all, Quine's holism also acknowledges, in addition to the primary role of experience, some "innate" ingredients of our knowledge, in which there is no *a priori/a posteriori* dichotomy (or *analytic/ synthetic* one). In contrast, the classical empiricist positions, which I challenge in this Chapter, all oppose innateness of any kind. Lawrence and Margolis (2012) consider the way in which Quine associates abstraction with innateness. Quoting Quine, "A standard of similarity is in some sense innate. This point is not against empiricism; it is a commonplace of behavioral psychology" (Quine 1969, p. 123), they comment: "Quine assumes that the fine-grained discriminatory capacities are innately ordered in terms of similarity (an innate 'spacing of qualities'), which he interprets behavioristically. ... Quine's innate similarity metric incorporates a further element of innate generality, but it also facilitates learning, allowing the account to avoid the difficulties that earlier empiricist accounts of abstraction had in capturing the similarity in the input without general representations" (Lawrence and Margolis 2012, p. 11).

can conceive or discover individual pure possibilities and their general relations independently of actual reality and empirical data.

## 6.3 Empirical Answers to Molyneux's Question

One may challenge all my arguments so far as redundant or simply speculative, for there are empirical studies to answer Molyneux's question. Nevertheless, philosophy is always required for understanding scientific findings and explicating their implicit philosophical assumptions and implications.

There are empirical studies concerning vision and active touch (haptics) as the two sensory modalities by which human observers perceive three-dimensional objects and their cross-modal perception as well (for instance, Norman et al. 2004, and Norman et al. 2006). Unfortunately, these studies do not refer to Molyneux's problem or question. In contrast, Pascual-Leone and Hamilton (2001) explicitly refer to Molyneux's problem and the meta-modal organization of the brain.

Cross-modal perceptions have been associated with the intriguing phenomenon of synesthesia (for instance, Fraser et al. 2006), which has been associated with Molyneux's problem (for instance, Ward 2008, pp. 16–18 and 39). If we are allowed to conclude on basis of various studies of synesthesia that newborns are synesthetic and that only a process of maturation enables them later to use their senses specifically (see, for instance, Spector and Maurer 2009), this may indicate that we have an innate capability of cross-modal sensory identification, as our brain actualizes it.[12]

Experimental studies to answer Molyneux's question started with the famous operation of born-cataracts in a boy about the age of 13, and the answer was negative (Cheselden 1727–1728). Nevertheless, there is a neurological view (such as documented for example by Oliver Sacks 1993, mentioning Molyneux, Locke, and Cheselden) according to which after so many years of blindness, the blind persons may lose their capability to recognize shapes even after a successful operation or treatment, as their neural system of sight had no chance to develop normally after such a long period of blindness since birth. A more recent empirical study (Held et al. 2011) shows that "although after restoration of sight, the subjects could distinguish between objects visually as effectively as they would do by touch alone, they were unable to form the connection between object perceived using the two different senses" (Degenaar and Lokhorst 2014), which appears to answer Molyneux's question negatively (as this question is phrased to be about cross-modal identification). Indeed, Held and his colleagues explicitly write:

> Our results suggest that the answer to Molyneux's question is likely negative. The newly sighted subjects did not exhibit an immediate transfer of their tactile shape knowledge to

---

[12] A major difference between normal cross-modally sensory identification and synesthesia is that the latter is involved with hallucinations: while perceiving a color, for example, a synesthete would experience also a hallucination of a sound. In contrast, normal cross-modal sensory identification is not involved with any hallucination.

the visual domain. ... Whatever linkage between vision and touch may pre-exist concomitant exposure of both senses, it is insufficient for reconciling the identity of the separate sensory representations. However, this ability can apparently be acquired after short real-world experiences. (Held et al. 2011, p. 552)

Nevertheless, according to other experiments, because of the neural loss of the capability of normal full-blown sight, this seems not to answer the question, because in fact the operation or treatment did not fully restore the visual mechanism of the subjects. Hence, no wonder that some of them need to continue using their tactile sense to recognize the objects lying in front of them, as their very poor sight is not sufficient for recognizing them. There are some other reasons to consider current empirical studies as not answering Molyneux's question (cf. Schwenkler 2013, pp. 89–91; Cheng 2015). Thus, in such experimental and observational methods, this problem cannot be settled, at least so far, and still it has been left to philosophers.

Note that Held and his co-researchers mention another possibility, suggested by Amir Amedi and his team as well as by other scientists to approach to the question of cross-modal interaction in a way that circumvents the lack of sight. The results of these experiments may demonstrate that:

The rapidity of acquisition suggests that the neuronal substrates responsible for cross-modal interaction might already be in place before they become behaviorally manifest. This appears to be consistent with recent neurophysiological findings documenting neurons that are capable of responding to two or more modalities even in cortical regions devoted mainly to only one modality. Also notable are demonstrations from human brain imaging studies that multimodal responses in primary sensory areas of the cortex can be elicited rapidly during unimodal deprivation, consistent with our findings of a short time course of cross-modal learning. (Held et al. 2011, p. 552)

This may indicate that innateness has to do with our cross-modal cognitive capability as it is actualized in our brain. Indeed, Amedi and his team recently performed some quite new experiments, in which, entirely regardless of Molyneux's problem, they demonstrate that "visual-to-auditory sensory substitutes devices (SSDs) convey visual information via sound, with the primary goal of making visual information accessible to blind and visually impaired individuals" (Levy-Tzedek et al. 2014, p. 1). By means of these devices, the visual cortex is trained "to 'see' again after years of life-long blindness by utilizing the preserved 'visual' task functionality of the occipital cortex of the blind" (op. cit., p. 7).

The reader may remember quite well the following remark by Locke:

A studious blind man, who had mightily beat his head about visible objects, and made use of the explication of his books and friends, to understand those names of light and colours, which often came in his way, bragged one day, that he now understood what scarlet signified. Upon which, his friend demanding what scarlet was? The blind man answered, It was like the sound of a trumpet. Just such an understanding of the name of any other simple idea will he have, who hopes to get it only from a definition, or other words made use of to explain it. (*Essay* II, 4, xi)

As a matter of fact, Amedi and his team have devised an amazing technique by which blind people can use their capability to listen to music in order to "see."

Teaching them a language of sounds that corresponds to visual data makes it possible for them to "see." For these persons, sounds signify figures and colors, as if the "absurd" possibility that Locke surmised becomes a reality.

Tactile experience may also be involved in the experiment with the SSDs technique, which is described thus:

> These devices convey visual information via auditory or tactile input, thus making it possible for people who are blind or visually impaired to acquire information about the world that is not usually accessible through audition or touch. Visual-to-auditory SSDs are a relatively accessible solution, for they usually consist of a simple video camera that provides the visual input, a small computer running the conversion program and stereo headphones that provide the resulting sound patterns to the user. (op. cit., p. 1)

This is even a more fascinating way to approach Molyneux's problem in a sophisticated empirical way, in which two sensory modalities, so different from each other, such as sound and vision, can be unified. The point is that this technique can stimulate the visual cortex of blind persons to render them capable of "seeing." As the experiments made by Amedi and his teams clearly demonstrate, once the visual cortex can be activated, blind person can *learn* how to feel, experience, or receive the audile sensations *as visual ones* (as though they are translating from one sensory language to another).[13] This clearly shows that our *capability* to unite audile or tactile experiences with visual ones, of one and the same object, is innate in us and not acquired by experience or empirical means.

This also shows that the panenmentalist approach that I present in the current Chapter is compatible with some novel empirical findings. Note that Amedi's experiments were performed also with sighted subjects who were blindfolded during the experiments. This demonstrates even more that the conclusions drawn from the experiments are valid to human beings in general and that our capability of sensory cross-modal identification is innate.

## 6.4 Conclusions

My panenmentalist approach to Molyneux's problem rests upon the insight that the newly sighted person has an innate epistemic access to the pure possibilities of the actual sphere and cube, irrespective of any specific geometry and of empirical data. If the new sight of such a person were normal, he or she could, immediately or after some training, distinguish between the sphere and the cube and relate them to his or her tactile previous acquaintance with these objects, while still being blind. This innate capability of cross-modal sensory identification is well understood in terms of a panenmentalist approach and some recent empirical tests as well.

---

[13] The capability to achieve such translations also indicates innateness. Since Chomsky's psycholinguistics, the idea that linguistic competence is innate and not acquired is not a new one. The SSDs technique utilizes a similar innate competence. Furthermore, if indeed babies are born synesthetic, the capability to translate from one sensory modal to the other is innate.

# Chapter 7
# Pure Possibilities and Some Striking Scientific Discoveries

**Abstract** In this Chapter, I demonstrate, on the grounds of panenmentalism, how some of the most fascinating scientific discoveries in chemistry could not have been accomplished without relying on the knowledge of individual pure possibilities and the ways in which they relate to one another (for instance, in theoretical models). The discoveries include the following: Dan Shechtman's discovery of quasicrystals; Linus Pauling's alpha helix; the discovery of F. Sherwood Rowland and Mario J. Molina concerning the destruction of the atmospheric ozone layer; and Neil Bartlett's noble gas compounds. On the grounds of the analysis of these cases, actualism must fail, whereas panenmentalism, at least as a philosophy of science, gains more support.

## 7.1 Introduction

Where everything is possible, no knowledge can exist. To know and understand anything, we must exclude various possibilities to get the valid ones. Predictions, too, require excluding possibilities. "Tomorrow will be either clear or rainy" is not a prediction worth its name. We need to know which one of these two excluding possibilities will be the case; otherwise we have no weather forecast. The same holds for truth. "That proposition is either true or false" provides us with no knowledge of whether the proposition in question is true or not.

Though excluding possibilities is necessary for any scientific progress, no less important is saving possibilities, for excluding some possibilities has turned out to be almost fatal for scientific progress: excluding the possibilities of quasicrystals did not contribute to the recent progress of crystallography, materials science, chemistry, and physics; instead, it hindered it (although excluding these possibilities contributed to the development of classical crystallography at the time). Today, quasicrystals are being considered as actualities although, until his last years, one of the greatest chemists of all times, Linus Pauling, declared them to be impossible. Quite a few great scientific discoverers have been doomed to fight strong opinions about the impossibility of their discoveries.

---

A first version of this chapter was published in *Foundations of Chemistry* 16:2 (2014), pp. 149–163.

A. Gilead, *The Panenmentalist Philosophy of Science*, Synthese Library 424,
https://doi.org/10.1007/978-3-030-41124-4_7

The first step in saving such vital possibilities is to acknowledge them as pure. To exclude the pure possibility of something means, from the outset, that such a thing cannot exist at all. To acknowledge, identify, understand, explain, and predict something, we must first acknowledge its pure possibility. For instance, as long as physicists had neither established final evidence for the actual existence of the Higgs boson nor such evidence for its actual nonexistence, they still had very good reasons to acknowledge its pure possibility on theoretical basis. Theoretical physicists have thought that such possibility *exists* and is not excluded; they knew *a priori* how to *identify* it; they understood and explained *why* it had to exist; they expected to discover its actual existence; and, thus, they *predicted* this discovery.

Logical, mathematical, or theoretical possibilities, for instance, are pure possibilities ("theoretical" refers to any epistemic or scientific field; hence there are physical, chemical, biological pure possibilities and so forth). The philosophical view that does not admit individual pure ("mere") possibilities—altogether or at least as existing independently of actual reality—is called "actualism," whereas the view that does acknowledge such possibilities is called "possibilism." Actualism is generally allowed to use the idea of possible worlds and possible world semantics. Nevertheless, to the best of my knowledge, no actualist theory, including the most recent ones, admits the aforementioned absolutely independent existence of individual pure possibilities *or*, more traditionally, even any existence of them (consult, for instance, Bennett 2005, 2006; Nelson and Zalta 2009; Contessa 2010; Menzel 2011; Woodward 2011; Vetter 2011; Stalnaker 2004, 2012).

Challenging actualism, panenmentalism is a possibilism *de re*, according to which pure possibilities are individual existents, existing independently of actual reality, any possible-worlds conception, and any mind (hence, they are not ideal beings). To the best of my knowledge, panenmentalism differs from any other kind of possibilism. Claiming that, it is not in my intention to argue that it is preferable to the other kinds; I say only that it is a novel alternative to them, by means of which we can clearly and convincingly explain scientific discoveries, such as those that will be discussed below.

The following are the main features in which panenmentalism differs from other kinds of possibilism: First, Panenmentalism is strongly realist about individual pure possibilities, which are thus independent existents rather than mere "beings" or "subsistents." Over this point, panenmentalism disagrees with Meinonigians, Neo-Meinonigians (to begin with Richard Routely [Sylvan]; see Gilead 2009, pp. 23–27, 33–38, 46–47, 83–91, 109–113, and 121) and their followers (such as Graham Priest, Nicholas Griffin, Terence Parsons, and Edward Zalta). Second, it dispenses with the idea of possible worlds, which almost all the possibilists known to me have adopted. Third, panenmentalist pure possibilities are not abstract objects or entities, neither are they potentialities, for abstractions (as abstracted out of actualities or actual reality) and potentialities depend on actualities which are ontologically prior to them, whereas pure possibilities are ontologically prior to and entirely independent of actualities. Four, though using truthful fictions, panenmentalism, acknowledging the full, mind-independent reality of pure possibilities, differs from any kind of fictionalism, especially modal fictionalism (Gilead 2009, pp. 80–83; this difference holds also for Kendall Walton's make-believe theory). Five, as mind-independent, pure possibilities are not concepts, hence panenmentalism is possibilism *de re* and not conceptualism or possibilism *de dicto*.

If some readers may think that the panenmentalist pure possibilities allegedly remind them of Edward Zalta's "possible objects" or "blueprints" (Zalta 1983; McMichael and Zalta 1980) or of Nino Cocchiarella's "possible objects" (Cocchiarella 2007, pp. 26–30; Freund and Cocchiarella 2008), such is not the case at all. First, these possible objects rely heavily on possible-worlds conceptions. Second, according to Cocchiarella's conceptual realism, framed within the context of a naturalistic epistemology, abstract intensional objects "have a mode of being dependent upon the evolution of culture and consciousness" (Cocchiarella 2007, p. 14), whereas panenmentalist pure possibilities are entirely independent of such evolution and of its naturalistic context as well and are *a priori* accessible (as I have argued above). Third, following Meinonigians and Neo-Meinonigians, both Zalta and Cocchiarela consider "actual" and "exists" as synonyms, while, in their view, possible objects are merely "beings." In contrast, panenmentalism treats both pure possibilities and actualities as existents, though in different senses of the term "existence" (distinguishing between the existence of pure possibilities and that of actualities—the former is spatiotemporally and causally conditioned, while the latter is entirely exempt from these conditions).

Indeed, each of the scientific discoveries considered below is clearly and convincingly explained on the grounds of individual pure possibilities that are real, mind-independent existents, without relying on any possible-worlds conception. Each of these discoveries begins with the ontological discovery of the existence of a pure possibility (or of pure possibilities and their relationality), *a priori* accessible, regardless of any reliance upon actualities or actual reality as a whole. Acknowledging the independent existence, the reality, of these possibilities paved the way to the discovery of their actualities. Other scientists could make such actual discoveries, had they not excluded the independent existence of these pure possibilities. No possible-worlds conception was required for any of these discoveries.

Instead of relying on the idea of possible worlds, panenmentalism refers to one total realm of all possibilities, pure and actual. Panenmentalism allows our knowledge a limited *a priori* access to the domain of pure possibilities, whereas the access of our knowledge to actual, empirical reality is *a posteriori* and must rely upon experience only. Actual reality consists of actualities, which are the outcomes of the actualization *of* individidual pure possibilities. Actualization subjects individual possibilities to spatiotemporal and causal conditions. Each actuality has a particular pure possibility serving as the identity of that actuality. It is the knowledge of that pure possibility-identity that enables us to identify the relevant actuality. As actuality, each possibility is not pure and, hence, it is restricted by spatiotemporal and causal conditions, from which any pure possibility-identity is exempt. For reasons I have detailed elsewhere (recently in Gilead 2009 and below), necessity pertains to the domain of pure possibilities and their relationality (namely, the ways in which they relate to each other): each pure possibility necessarily exists and each such possibility necessarily relates to all the others. In contrast, contingency inescapably pertains to the domain of actualities for there is no necessity about the actualization of pure possibilities. There could be nothing actual, whereas there could not be no pure possibilities. They are timeless entities, which are not invented but subject to discovery. Thus, nothing could avoid the existence of pure possibilities and nothing could generate them. As the actual domain is not subject to necessity, it is subject to contingency, randomness, chances, coincidences, serendipity, and the like. This is compatible with the panenmentalist view that the domain of pure possibilities is *a priori* cognizable, whereas the domain of actualities is *a posteriori* cognizable. In panenmentalism, as in the philosophy of Kant and not as in the views of Saul Kripke and others, only *a priori* truths can be necessary, whereas *a posteriori* truths are contingent.

The access of our knowledge to the domain of pure possibilities is limited, for we are finite and limited creatures. Yet our imagination and intellect are sufficient for an adequate *a priori* access to that domain, quite independent of experience and actual reality.

If not by empirical means, how can we have access to pure possibilities, mathematical or otherwise? How can we know them, become acquainted with them, if not by means of actual experience and empirical observations?

Each individual pure possibility is different from any other individual pure possibility, and no two individual pure possibilities can be identical, otherwise "two" pure possibilities would have been one possibility instead of two. In case of actualities, which are actual possibilities, two or more of them might be identical (though numerically different), provided that they did not exist at the same place in the same time, whereas such could not be the case of pure possibilities, which are entirely exempt from spatiotemporal conditions or restrictions. Pure possibilities are different from actual possibilities. We can grasp counterfactual possibilities, even though we know of no similar possibilities, actual or pure. While observing actualities, we can think or imagine that things *could* be different, that there are counterfactual *possibilities*; while considering some pure possibilities, we can think also about quite different pure possibilities, each of which is *different from* the other and, hence, each *relates* to the other. This universal mutual relationality allows our knowledge *a priori* accessibility to a great number of pure possibilities. Thus, our imagination and intellect are not confined to actual reality; they have access to new pure possibilities simply because they are different from the already familiar possibilities, pure or actual. Thus, we are capable of relating and referring to, imagining, thinking about, and understanding possibilities that are entirely different from the possibilities, pure or actual, with which we are already familiar.

Although it is a moot point to what extent mathematics is not empirical but, instead, is independent of actual reality, I have some good reasons to think that pure mathematics consists of mathematical pure possibilities and their relationality and, thus, its discoveries are *a priori* accessible. Mathematical imagination, calculation, and inference have been never confined to actual reality and many most imaginable, even fantastic, mathematical discoveries have been independent of actual reality and empirical facts (Atiyah 1995). It is sufficient to think about Euclidean geometry to demonstrate such independence: for instance, unlike its representation as an actual dot, a point has a position but no dimensions (according to the first definition in Book I of Euclid's *Elements*); it cannot be measured and yet it exists in the Euclidean space (which is a purely possible domain in which geometrical pure possibilities are set side by side). A line (according to the second definition in that book) is "length without breadth," whereas any actual drawn line must have some breadth, however small. Pure geometrical lines and points are not subject to our observation; only their manifestations, phenomena, or representations in actual space are subject so. Pure geometrical points and lines are thus "ideal," but this does not make them abstractions from actual reality (by "ideal" I do not mean a dependence on our mind, as I will explain below). They are neither idealizations of empirical facts, for, if they were, they would be idealized according to some ideal standard or paradigm, which, in turn, must be entirely independent of empirical facts and which, hence, would have been purely possible.

There is another philosophical way to consider mathematical structures as *a priori* cognizable. Except for wrongly considering Euclidean geometry as the only possible and necessary mathematics, there is still force in Kant's conception of mathematics, in the *Critique of Pure Reason*, and of the epistemological status of its sentences ("judgments") as synthetic *a priori* ones. According to Kant, any mathematical calculation, numbering, constructing figures, ordering, proofs, and the like are independent of empirical facts, and our knowledge about them is prior to experience; namely, it is *a priori* (to begin with Kant 1998 [1787], pp. 107–108). We are acquainted with mathematical novelties not by experience, observations, or experiments and not because of an analysis of concepts. We are acquainted with such novelties only because of constructions or forms of order, which our reason generates and which provide us with synthetic *a priori* propositions, wholly independent of experience or experiment. Thus, according to Kant, geometrical or arithmetical kinds of order, structure, and construction are *a priori* cognized. Kant limited these to the restrictions of Euclidean mathematics but it should be valid, I think, for any kind of mathematics, which is *a priori*, namely, independent of any experience, and yet is valid for an experience of this or that kind and is indispensable for the scientific knowledge involved in it. However, Kant did not consider those kinds of order, structure, and construction as possibilities, let alone as pure possibilities. This had to wait for another philosophical turn. Furthermore, he mentioned the "ideality" of mathematical entities, namely, their total dependence on our mind (more precisely, the Human Reason) whereas, according to panenmentalism, mathematical pure possibilities are entirely independent of our mind and, hence, we discover, not invent them.

Supports of the idea that pure mathematics refers to *a priori* accessible possibilities, independently of actual, empirical reality, can be found in Hermann Weyl (1929, p. 249), Howard Stein (1988, p. 252), Geoffrey Hellman (1989, p. 6), and others.

It is an actualist fallacy to think that no individual pure possibilities exist or that actualities are all the individual possibilities that exist. Panenmentalism argues to the contrary—that (1) individual pure possibilities exist independently of our mind, on the one hand, and of actualities, on the other; (2) our mind has enough *a priori* access to the domain of pure possibilities, quite independently of experience or of actual, empirical reality; (3) we discover pure possibilities, mathematical and otherwise, as much as we discover new actualities; (4) ignoring or not acknowledging relevant pure possibilities may result in passing by their actualities without noticing, recognizing, or identifying them and, thus, this hinders the progress of science and knowledge. If I am not mistaken, actualism results in fatal fallacies whenever it rejects these four statements.

Intellect and imagination are indispensable tools for scientific progress, and it is in the nature of these tools not to be confined by the actual and even to proceed beyond it. These faculties of the human ingenuity thus rely heavily upon pure possibilities and their relationality.

Since imagination plays an indispensable role in discovering pure possibilities, we should ask ourselves whether this does not leave us with fictions instead of real possibilities. We create fictions, whereas we discover pure possibilities. Fictions may well serve us in discovering possibilities that are independent of our mind. I called such fictions truthful (Gilead 2009, 2010). By means of truthful fictions we capture mind-independent pure possibilities and their relationality. In scientific discoveries, for instance, the fictions involved in thought-experiments and models serve scientists to discover mind-independent pure possibilities and to illuminate some of the deepest secrets of reality.

The first step in acknowledging the existence, let alone the possibility, of something is to acknowledge its pure possibility. Before Shechtman's discovery of an actual quasicrystal, crystallographers, physicists, chemists, or material scientists argued that quasicrystals were impossible, not only because they contradicted empirical classical crystallography and its basic principles or laws, but also because they were incompatible with the classical *geometry* of crystals and its 230 purely possible space groups—such scientists excluded in fact the very, *pure* possibility of any quasicrystal. Independent of Shechtman's great discovery, such mathematical or theoretical pure possibilities were discovered by Alan Mackay and others. On 8 April 1982, when Shechtman for the first time observed the image of an actual quasicrystal, he could not understand and explain *why* such a crystalline structure was possible,[1] for he was entirely unaware of Mackay's discovery (Hargittai 2011a, p. 159; 2011c) but, as he has claimed, he was aware of the "impossibility" of such a structure according to classical crystallography. To understand and explain why such structure is nevertheless possible, Shechtman should have been familiar with the pure possibility of such a structure. Since I have discussed elsewhere (Gilead 2013) the discovery of quasicrystalline pure possibilities and Shechtman's discovery of an actual quasicrystal, I will not discuss now this most instructive case in

---

[1] An enlightening comment by an anonymous reviewer to Springer Nature should be inserted here: "This is why Shechtman was one of the exceptional scientists in science history who did not discard something entirely new just because nobody before had made such an observation (cf. a similar case: Fleming did not throw out the Petri dish in which penicillin destroyed the sample he wanted to investigate) and because it contradicted accepted dogmas."

detail again (for an updated analysis of Shechtman's discovery see Hargittai 2011a, pp. 155–172; 2011c).

István Hargittai has recently published an important and interesting book—*Drive and Curiosity: What Fuels the Passion for Science*—concisely detecting the process of some great scientific discoveries in the twentieth century (Hargittai 2011a). He ascribes these discoveries to two main motives that impel the discoverers—drive and curiosity. His reports of scientific discoveries have a special significance, because he himself is a distinguished scientist and a scientific editor who has conducted interviews and maintained communication with many of the discoverers mentioned in his books. Thus, he is especially familiar with the context of discovery concerning these discoverers. The clear and accessible information as well as the insightful analysis with which he provides the reader are most valuable.

Hargittai does not refer to the role of saving possibilities, to begin with pure possibilities, in fueling the drive and curiosity motivating the scientific discoverers. Nevertheless, I interpret his reports of these cases in light of some panenmentalist ideas concerning the indispensible role of acknowledging and saving pure possibilities in making such discoveries possible.

## 7.2 Linus Pauling's Discovery of the Alpha Helix

Describing Pauling's discovery of the alpha helix as the structure of protein, Hargittai writes:

> Pauling—ever the model builder—sketched a protein chain on a sheet of paper, which he folded while looking for structures that would satisfy the assumptions he had made. He found two. He called one the alpha helix and the other the gamma helix, but he would quickly discard the gamma helix….Pauling found the model so attractive and so sensible that he had little doubt in its correctness….At this time, he had an opportunity to visit the British group in Cambridge, and Max Perutz showed him his excellent diffraction patterns. From the X-ray diagrams it was obvious to Pauling—though not yet to Perutz—that it corresponded to Pauling's alpha helix model. Pauling was excited by what he saw but kept calm and did not say anything to Perutz. (Hargittai 2011a, p. 98)

First, the model under discussion, though depicted in an actual folded sketch, is purely possible: while Pauling constructed it, the model was entirely independent of any actual evidence. The sketch simply depicts the relationality among the chemical pure possibilities of a protein chain; a relationality that Pauling discovered. Note that pure possibilities are individual entities, whereas relationality is general, which taking part in the nature of models. While the model of alpha helix consists of pure possibilities and their relationality, Perutz's diffraction structure was an X-ray image of an *actual* structure. Pauling recognized it as an actuality of (or corresponding to) the model he had discovered and hence he—unlike Perutz—could identify the structure. As Hargittai reports, "[t]he Cambridge X-ray diffraction pattern showed the helical nature, but Perutz did not think about it and thus did not notice it" (Hargittai 2011a, p. 101). In contrast, Pauling had thought about the purely possible alpha helix structure, of which he had had an *a priori* concept, owing to which he later noticed and identified the actual alpha helix structure in Perutz's findings. It is quite interesting and enlightening (especially from a panenmentalist viewpoint) to compare this with Alan Mackay's warning in the 1980s that crystallographers should be aware of the *possibilities* of structures beyond the classical system, "[o]therwise we might encounter them but walk by them without recognizing them" (Hargittai 2010, p. 82). Similarly, Perutz encountered the novel structure of the alpha helix and "walked by it" without recognizing or noticing it, only because, unlike Pauling, he had no access to the purely possible identity of the alpha helix.

This identification was possible only because Pauling already had the purely possible model in mind. Having read Pauling's paper about the alpha helix model, Perutz performed an additional X-ray experiment that gave further evidence, "showing the correctness of Pauling's result, something that Pauling himself had missed" (Hargittai 2011a, p. 104). Such actual evidence might provide the model with confirmation or with completion and correction. Indeed, in scientific discoveries, empirical experiments and observations are as vital as pure possibilities. Though the purely possible model is subject to discovery and, thus, is independent of the discoverer's mind, the first acquaintance with the model may be incomplete or faulty, as human knowledge is never perfect. Nevertheless, the acquaintance with the empirical data can help scientists in correcting and completing the model, which does not diminish its status as purely possible and as *a priori* cognizable.

Pauling's excitement while watching Perutz's X-ray images reminds me very strongly of James D. Watson's excitement in watching the X-ray diffraction image of the DNA produced by Rosalind Franklin, and which was a "crucial experimental evidence" (Hargittai 2011a, p. 42) for the actual existence of the double helix. Watson already had a model in his mind, a purely possible one, by means of which he could identify the image of the DNA actual structure and realize its great significance.

Still, is any such model really independent of actualities, empirical observations, and experiments? In other words, does it really consist of pure possibilities and their relationality? Is it *a priori* cognizable?

Indeed, in panenmentalist terms, the model relates to the relevant pure possibilities, which can be actualized in empirical reality. Even if the properties on which the model is constructed are first taken from actual reality, based upon actual observations or experiments, the model concerns the relationality (for instance, the general structure) of *all* the relevant pure possibilities that are actualized, actualizable, or *predicted* to be actualized in empirical reality. All the relevant pure possibilities and their relationality lie quite far from the actual cases from which the properties of the model were supposedly selected or chosen by the theoretician-discoverer.

Scientific models draw their principles and patterns from the domain of pure or *a priori* cognizable possibilities or, more precisely, from the relationality of these possibilities. The purity and apriority under discussion are entirely compatible with the indispensable demand that all such models, if or when confirmed as actual, should comprise or capture all the known relevant empirical facts that have happened to be actual (and which can serve in correcting the model and adapting it further to actual reality). In themselves, models consist of pure possibilities whose meanings and significance are independent of actual reality and yet are quite actualizable. Such are all the valid and sound mathematical or theoretical models. Indeed, practical considerations play an essential role in deciding which of the purely possible models can serve us conceptually in capturing actualities or in applying pure possibilities to actualities. Yet nothing in this practicality may exclude the purity and apriority of the model of the alpha helix.

In the case of Pauling's discovery, there was no need, especially for such a self-assured and convinced scientist, to make an empirical adjustment or correction. Although, in the beginning, there was a discrepancy between his calculations and the empirical findings of Perutz, eventually "the origin of the discrepancy was understood: it was caused by the alpha helices twisting together into ropes resulting in a change in the experimental data as compared to what it would be for a single chain for which the model had been constructed" (Hargittai 2011a, pp. 99–100). Thus, in the end, "Pauling's alpha helix was confirmed even in this detail. The alpha helix has proved to be a great discovery because it is a conspicuously frequent structural feature of proteins" (ibid., p. 100).

You may, however, still counter-argue that in constructing successful models, we discern and abstract from actualities the important or significant properties to serve well our knowledge, understanding, and predications, while, at least for a time, we have to ignore other, insignificant properties or effects (this may be modified in the future depending on what the model is supposed to describe or to model). My answer would be that to discern and abstract so we need an earlier acquaintance with pure possibilities-identities and their relationality by means of which we can discern, identify, and capture these significant or important properties as pure or as actualized. Possibilities-identities are *a priori* referable and accessible independent of their actualization.

If indeed Pauling was ever the model builder, he knew admiringly how to use the accessibility of his imagination and intellect to the domain of chemical pure possibilities and their relationality, by means of which he could achieve his astonishing discoveries.

At present, computers help scientists to construct models. This does not render the model *a posteriori* or experimental. Like Pauling's folded piece of paper, computer simulations provide the scientists with images for their thought-experiments. As Mark McEvoy points out, "granted that unsurveyable computer-assisted proofs include only *a priori* methodology, they are *a priori* proofs" (McEvoy 2008, p. 386). Similarly, such computer experiments are made in the *image* of thought-experiments. All the more, Pauling's folded paper does not exclude the apriority and the pure possibility of the alpha helix model, for Pauling could *imagine* how the folded pure structure could look like! At least, his methodology was *a priori*. He studied the pure possibilities in a thought-experiment—how should their arrangement look like in various modes of folding. Pauling's thought-experiment was of a theoretical-geometrical nature.

Ironically, Pauling model did not reveal perfect symmetry, but this did not bother him greatly. He "expanded the realm of crystallography toward structures that were not part of classical crystallography" (Hargittai 2011a, p. 101). What he had allowed himself, he did not allow Shechtman later: although Pauling had been considered as a maverick in his time, he, while confronting Shechtman's daring discovery, reacted like a conservative, dogmatic "classical" crystallographer. Shechtman and others saved possibilities and expanded the knowledge of crystallographic possibilities against the strong opposition of Pauling and his followers. These two cases clearly show how saving relevant possibilities, against the scientists' prejudices, is no less indispensable for scientific progress than excluding other possibilities. Pauling's superior knowledge of structural chemistry helped him greatly in restricting the circle of possible models (Hargittai 2011a, p. 102). Following Hargittai's description, some of the great British crystallographers (Bragg, Jr. and Perutz, on the one hand, and Rosalind Franklin, on the other, extensively using X-ray images of the structures of proteins or of the DNA respectively) appear as distinguished experimenters and empirical observers, empiricists in nature and, in my terms, actualists, whereas Pauling, unwittingly, relied also and no less on pure possibilities, with his great capability to restrict them when necessary and to expand their discovery when necessary.

In 1934, Bernal pointed out the possibility of deducing atomic positions from the X-ray diffraction diagrams of a protein (a single pepsin crystal), however, he did not think in terms of models but in more empirically oriented terms. Since Bernal was quite influential, his followers, including Perutz, were not enthusiastic to use models either. Bernal tried to obtain the structure from the actual determination of the atomic coordinates. Hence, he was far from the theoretical, purely possible model of Pauling, which eventually was empirically, actually confirmed in full. Deducing from actualities, in the manner that is unwittingly actualist yet admittedly empiricist, scientists—like detectives—can achieve great results, but not as great as those achieved by scientists such as Pauling.

## 7.3    Árpád Furka's Discovery of a New Field—Combinatorial Chemistry

In 1999, Árpád Furka stated: "Nowadays combinatorial chemistry is an accepted branch of chemical science. Fifteen years ago it was completely unknown" (Furka 1999, p. 22). Hargittai devoted a whole chapter to this great discovery (Hargittai 2011a, pp. 123–137), which begins with Furka's calculating the number of "theoretically possible sequences in peptide families built up from the twenty natural amino acids. Such collections of compounds are named libraries" (Furka 2007, p. 2). Furka characterized these libraries as "the nonexistent peptide libraries" (ibid., p. 3). In my panenmentalist interpretation, this means that though the peptide libraries have been purely possible, they were not known at that time to exist as actualities. Indeed, the "virtual" peptide libraries comprise chemical individual pure possibilities and their general relationality, and the number of such sequences is so huge that thousands years of nonstop experiments were needed, before Furka's discovery, to synthesize them. He discovered a split-mix method that "opened the possibility" for producing peptide mixtures containing millions of components, though it "seemed unacceptable in the conventional drug discovery practice where single compounds were used in pure form" (ibid., p. 4). Moreover, the "real combinatorial methods can be used in multi-step processes and their most important characteristic feature is that make [actually] possible to prepare in a single run all structural derivatives that can be theoretically [namely, in a purely possible fashion] deduced from the structures of the building blocks" (ibid., p. 55). Thus, Furka converted a purely possible field, practically excluded at the time (around 1982), into an accomplished actual fact, which happened to be most useful for vital pharmaceutical purposes.

Furka also succeeded in representing all the possible combinations of amino acid building blocks in the actual synthesized peptides (ibid, p. 57). His discovery is an illuminating demonstration of a most fruitful use of the pure possibilities of synthesized peptides, whose actualities can be observed, thanks to his discovery, in experimental syntheses, exhausting all the relevant pure possibilities as actualized. To select the useful or promising pharmaceutical possibilities, researchers must screen the largest libraries possible (ibid., pp. 150 and 171). This means that saving as many chemical pure possibilities as possible is the first step, after which the step of excluding the futile possibilities can take place. Furka's discovery has enabled researchers to have access to *all* possible peptide combinations as well as to select from them what is pharmaceutically promising and useful. Furka's methods also provide predictive modeling solutions for all aspects of drug discovery research, utilizing computer updated technologies (ibid., p. 176). As I have argued above, such a utilization is quite compatible with the *a priori* accessibility to pure possibilities and their relationality (which is compatible also with Furka's report (ibid., pp. 65–66). Indeed, Furka's combinatorial methods are not experimental or actuality-based; on the contrary, they are *a priori* accessible (judging also from ibid., pp. 155–156), which is entirely compatible with relating to pure possibilities.

In sum, Furka's discovery in chemistry provides a fine example of the way in which saving pure possibilities as well as excluding or screening other pure possibilities are indispensable for scientific discoveries and progress.

## 7.4   F. Sherwood Rowland and Mario J. Molina's Discovery—The Destruction of the Atmospheric Ozone Layer

Reading about the background of the discovery made by Rowland and Molina, Nobel Laureates in chemistry in 1995, I immediately recalled an episode from a novel by a Nobel Laureate in literature, Aleksandr Solzhenitsyn. In Solzhenitsyn's *Cancer Ward*, Oleg Kostoglotov, while receiving treatment for cancer, is immersed in doubt about the use of X-ray radiation for medical purposes and argues about it with his doctor (Solzhenitsyn 2001 [1968], pp. 79–95). He wonders why nobody at the beginning of the use of X-rays suspected that it might cause severe damage to the patients. Indeed, at that time doctors used X-rays with almost no restrictions, even in treating, with tragic cancerous consequences much later, quite minor skin diseases (such as acne—as in the case of the 1984 Nobel laureate in chemistry, Robert B. Merrifield, as documented in Hargittai 2003, p. 217—or such as tinea capitis in Israel in the 1950s). It was a dogmatic exclusion of valid pure possibilities, which, in fact, had been quite prevalent, not only among scientists (though it fits tyrannical regimes or religious fanaticism). In a somewhat similar dogmatic vein, when, in the seventies, James E. Lovelock published his pioneering findings about the ubiquity in the atmosphere of chlorofluorocarbons [CFCs], which have been synthesized by humankind, he did not consider them to be a hazard; rather, he regarded them instead as inert harmless tracers, quite useful for scientific research (Hargittai 2011a, p. 196). Furthermore, having read of Rowland and Molina's warning the public about the great hazard that these compounds posed to the atmosphere's ozone layer, Lovelock accused them of "being prone to panic, saying that Rowland in particular was acting like a missionary" (ibid., p. 205; later, however, Lovelock changed his mind about the dangers to the environment and about such an attitude). Lovelock's opposition to Rowland and Molina's discovery is similar to Pauling's opposition to Shechtman's discovery of quasicrystals. The name of the game is excluding possibilities, first of all pure ones, on dogmatic, even fanatic, grounds. Even though Rowland and Molina's calculations and consideration concerning this problem were so simple, the dogmatic power of excluding possibilities was strong and quite powerful (similar to that of some philosophical exclusions). Though scientists have to exclude possibilities to pave the way toward scientific progress, they should be extremely careful when excluding such possibilities, for they are prone to exclude most significant, even fatal, ones, as in the case of the CFCs' damage to the ozone layer. Pharmaceutical history demonstrates this very well (think, for instance, about the "absolute harmlessness" claimed by the European manufacturers of thalidomide at the beginning of its production!). Excluding some pure possibilities has entailed grave mistakes and fatal dangers, and the dogmatism involved is fanatic no less than religious fanaticism. A dogmatic accusation made against innovators (such as Rowland) that they were acting like missionaries may reflect more on those who cast the blame than on the innovators.

The story of Rowland and Molina's discovery began with a series of thought-experiments and calculations (Hargittai 2011a, p. 200), which means that it began with studying pure possibilities and their relationality. Instead of experiments, observations, and measurements of actualities, they had to rely on calculations and on using their imagination in investigating pure possibilities and their relationality. Intellect and imagination are indispensable tools in searching the domain of pure possibilities. In the absence of empirical means, scientists must rely upon these tools even more heavily. In Hargittai's report we read:

> If Rowland and Molina had had the means to travel to the atmosphere and carry out measurements there, they would have been more easily convinced that their calculations were

correct. But this was not possible at that time so, lacking the ability to verify their findings through experimentations, they could rely only on calculations. First, they had to convince themselves that the CFCs posed an unprecedented danger to the ozone layer. (Hargittai 2011a, p. 201)

Comparing this to Lovelock's assumption that these products were causing no harm, the result of their thought-experiments and calculations was really shocking—their estimate in 1973 of the loss of the ozone layer was about 7–13%. Thus,

> By December 1973, Rowland and Molina knew they had uncovered an environmental problem of global significance. Their initial hesitation about the validity of their findings came partly from the enormity of the effects and partly from the fact that their calculations and considerations were so simple that it seemed surprising that no one before them had come to similar conclusions. Such hesitation is quite characteristic when a researcher makes a discovery, especially when it seems—at least in retrospect—simple. (Hargittai 2011a, p. 202)

The discovery of some pure possibilities is so simple that their discoverers are really surprised that no one before had thought about these possibilities or thought about them in such a way. It was very simple to imagine or think that X-ray radiation may cause great damage to the patients and, quite surprisingly, only following the empirical evidence of such damage, doctors began restricting their use of it to the possible minimum. The idea—that to expose microorganisms to antibiotics unrestrictedly may bring about their immunity and resistance to these medications—is very simple. Nevertheless, this simple possibility was not considered or taken seriously prior to the empirical evidence about the fatal outcomes of the "domestication" of these bacteria owing to their exposure to antibiotics. We are now facing a medical catastrophe because of the immunity of these microorganisms to antibiotics.

Even the acceptance of the simplest pure possibilities may be quite rare. The reason for this is that usually we think and behave like actualists, assuming that only actualities exist and that to consider counterfactual possibilities or possibilities about which, as yet, we have no actual evidence about because of our limited and confined experience, is simply an imaginary play that should not be taken seriously. Such are some prevailing attitudes toward philosophy and philosophers. Nevertheless, philosophy, or referring to pure possibilities and their relationality, plays a significant role in scientific progress. We should not blame only science or scientists for ignoring this; after all, many, if not most of the philosophers nowadays are not possibilists; many of them are, wittingly or unwittingly, actualists.

## 7.5   Kary B. Mullis's Polymerase Chain Reaction

Hargittai describes Mullis's discovery concerning the polymerase chain reaction in 1983 as a "mental exercise" (Hargittai 2011a, p. 207). Owing to the technical actualization of Mullis's discovery, biochemists are capable of reproducing fragments of DNA in an unlimited number of copies (ibid.). This capability serves many vital purposes in the fields of forensic medicine and identification, medical research, diagnosis, and treatment, and many others. It was this mental exercise that initiated Mullis's long journey that led to his being awarded the 1993 Nobel Prize in chemistry. Yet considerably more effort than "Mullis's thought-experiments" was needed to create the polymerase chain reaction technology (ibid., p. 218). The *actual* experiments made by others were essential to create this technology. Yet Mullis had the concept ("merely a brilliant intellectual achievement"—ibid., p, 221), an offspring of his most original thought-experiments, without which the actual discovery could not have been made. In this case, too, we can see to what extent the investigation of the relevant pure possibilities and their relationality is as essential as is their empirical confirmation. Lacking such confirmation, lacking the laboratory actualization of the concept and lacking experimental support, Mullis's submitted paper, reporting his brilliant concept, was rejected by both *Nature* and *Science* (ibid., p. 219). However amazing and novel are the results of thought-experiments or of the study of scientific pure possibilities, if they lack sufficient support by empirical evidence, such prestigious scientific journals are unwilling to publish them.

In the "Epilogue," Hargittai summarizes Mullis's achievement thus:

> Kary Mullis came to this multibillion-dollar-worth discovery by letting his imagination go. There were more knowledgeable scientists who might have made the same discovery, but they did not; Mullis did. He did not feel confined in his thinking, and his mind was well prepared for his jumping ideas. (Hargittai 2011a, p. 299)

Confined thinking, thinking step-by-step with no jumps, may sometimes block scientific discoveries (concerning the discovery of the DNA structure, such appears to be the unfortunate case of Rosalind Franklin and Erwin Chargaff, contrary to that of Watson, which I will discuss below). What confines thinking most is blindness to pure possibilities or to counterfactual ones.

## 7.6   Neil Bartlett's Discovery of "Noble" Gas Compounds

In my chemistry studies at secondary school, I was quite fascinated by the supposedly established fact that noble ("inert") gases cannot react or combine with any other element. This impossibility appeared to me as solid as the most intriguing fact that the periodic table had proven to be rock-solid and its predictions, concerning all the eka-elements, has been entirely confirmed. No wonder that on hearing for the first time of Neil Bartlett's discovery of a noble gas compound, I was quite surprised. By combining xenon and platinum fluoride, Bartlett created the first noble gas compound (Hargittai 2011a, p. 238, citing the plaque on the building at UBC in which the astounding discovery was actually made). What is so fascinating about this discovery? I believe that one of the most surprising and illuminating traits of science is the intellectual revolution in which what was considered to be impossible proved to be possible and actual. These are special cases of saving possibilities and of expanding our knowledge of the possible as well as the actual.

Indeed, this discovery invalidated the assumption—having been sustained from the discovery of the noble gases, at the end of the nineteenth century, until Bartlett's discovery—that noble gases are inert, namely, as if it were impossible, at least practically, to combine them with each other or with any other element to synthesize molecules. Bartlett transformed this impossibility into a possibility. As in the case of quasicrystals, the definition of noble gases had to be changed owing to Bartlett's discovery—today they can no longer be considered as inert.

The randomness and serendipity of many actual discoveries is quite typical of actualities, especially, of their empirical knowledge. Such is, according to panenmentalism, the contingent nature of actualities and their knowledge. Fortunate circumstances or serendipity is also ascribed to Bartlett's discovery (Hargittai 2011a, p. 226); to the discovery of the first quasicrystal by Shechtman (ibid., p. 155); to the findings of Hideki Shirakawa and Hyung Chick Pyun (ibid., pp. 182 and 299), which led to the discovery of conducting polymers; to Bartlett and Lohman's producing of $O_2PtF_6$ (ibid., p. 226: "but it was no accident that Bartlett understood the nature of its bonding"); and, finally, to the observation of Arno Penzias and Robert Wilson that the cosmic microwave radiation amounted to three kelvins (ibid., p. 289). Hargittai details some fortuitous circumstances in which discoveries were made (including Mendeleev's great discovery of the periodic law and order; ibid., p. 227).

The possibility of combining a noble gas with another element was not invented by Bartlett; it had "always" been there, "waiting" for a discovery. Bartlett discovered that

[T]he energy needed for removing an electron from the oxygen molecule was about the same as for removing an electron from a xenon atom. This energy is called the ionization potential… From the similarity of the ionization potentials of molecular oxygen and xenon, it was a short step to the realization that if he succeeded in combining $O_2$ and $PtF_6$, it should be possible to do the same for Xe and $PtF_6$. On paper, this worked fine; now the question was whether he could do the experiment. (Hargittai 2011a, p. 227)

This was a new *pure possibility* that was revealed to Bartlett's mind. It was an *actualist fallacy* to assume that no noble gas could be combined with another element and that any attempt to do so was doomed to failure from the outset. This was an actualist fallacy, for the excluding of such a possibility was based upon actual findings and not upon theoretical considerations. It was an empirical fact, an actuality—no such compound had been found before—either in nature or among human products. Though Bartlett established the pure possibility of such a compound on theoretical grounds, it needed to be established as actual in a well-planned experiment.

As in the case of Shechtman's first observation of a quasicrystal on 8 April 1982, considering the differences between their discoveries, Bartlett, another lonely discoverer, on 23 March 1962, when he performed the experiment, was alone in the laboratory, with no assistant nearby, with nobody to share his great actual discovery (Hargittai 2011a, p. 227–228). Interestingly, both discoveries were of compounds created by human beings and not by nature. Both great discoverers were witnessing something that nobody had seen before. Both discoverers were flooded with doubts facing the "impossible." Bartlett asked himself whether the xenon in his experiment was not pure, maybe there was some oxygen present, may he was just fooling himself (ibid., p. 228). It appears that "[b]oth the urge to share the news about his experiment and the instant doubts are typical in such moments of discovery" (ibid.). After all, he had discovered an actuality that was considered impossible (at least practically) before the moment of the experimental discovery. Now, he had to *acknowledge* this possibility, and such acknowledgment could be achieved only after the possibility had passed all the tests of possible doubts.

This motif of the "lonely discoverer" is quite instructive especially against the background of saving scientific possibilities. Natural scientists are routinely working in scientific teams or communities, which share common ideas and methods. To save the possibilities that these communities have excluded, as in the cases of Bartlett and Shechtman, the discoverer needs to take a "lonely" stance—to distance or dissociate himself or herself from the prejudices and some of the received ideas at the time. This was also the case of Pauling in his time (although even then his exceptionally vast knowledge of structural chemistry strengthened his confidence in his discovery). The "lonely discoverer" adopts some of the possibilities that the relevant scientific community has excluded. He or she expects that eventually this community will acknowledge the discovery. Such loneliness is valid also even for a team of scientists taking part in a great discovery that still awaits acknowledgement by the scientific community.

There is a further intriguing similarity between Bartlett's discovery and Shechtman's. Each of these discoveries opened up a whole new field in chemistry. In Bartlett's case, within quite a short period of time, more and more researchers reported on producing new noble gas compounds with interesting properties and structures (Hargittai 2011a, p. 222). And in the case of Shechtman, the production and discoveries of more and more actual quasicrystals have been reported since his discovery. In both cases, structural chemistry has been much enriched owing to these great discoveries. Both cases can teach us an important lesson: once a scientist establishes the discovery of a new possibility as real and vital—although previously it had been excluded or not accepted—within quite a short time, a "flood gate" may open for many similar discoveries. These two cases strikingly demonstrate to what extent excluding possibilities, pure or actual, may obstruct scientific progress and scientific vital discoveries.

Although Bartlett's case is quite similar to that of Shechtman, some other scientists before Bartlett tried in vain to combine noble gases with other elements. They made such attempts because the possibility was not theoretically *a priori* excluded unlike the case with Shechtman. Pauling predicted that since he had found fluorine to be "so electronegative, it could attract away an electron even from the atoms of the noble gases. It was a hypothesis, though, and Pauling wanted experimental confirmation" (Hargittai 2011a, p. 231). In this case, unlike that of the quasicrystals, Pauling did not commit an actualist fallacy and he was open to the pure possibilities of such compounds. But no such a confirmation was found at the time, though the experimenters, Don M. Yost and Albert L. Kaye stated that "[i]t does not follow, of course, that xenon fluoride is incapable of existing" (Yost and Kaye 1933, p. 3892).

Thus, in 1933, despite the failure in actualization, Yost and Kaye did not exclude the pure possibility of xenon fluoride. In contrast, the received view that noble gases were inert was simply based on an actualist fallacy, which did not rely upon the relationality of chemical pure possibilities but only on the contingency of the actual failure to combine such gases with other elements or to find such compounds in nature. Thinking in this fallacious way, other chemists argued that there was no point in trying to perform such experiments, simply because there was no actual evidence for the existence of such compounds in nature neither had anyone succeeded in synthesizing them. Contrary to Pauling, Yost and Kaye, all those chemists committed an actualist fallacy. Indeed, this was disclosed quite soon after Bartlett's success, as more xenon compounds were discovered and the experiments to combine more noble gases with other elements were no longer considered as hopeless (Hargittai 2011a, p. 230). The second difference between Bartlett's and Shechtman's discoveries is that Shechtman did not expect to encounter such an "impossible" crystal, whereas Bartlett expected the possibility that the reaction with the xenon would be actualized.

The discovery of noble gas compounds and the research into their molecular structures relate to various models of molecular geometry and chemical bonding (Hargittai 2011a, p. 235, and Gillespie and Hargittai 2012). Models of molecular geometry certainly rely upon mathematical-chemical pure possibilities and their relationality. The discovery of xenon hexafluoride compound served as an early proof for the validity and usefulness of one of these models that is called the valence shell electron pair repulsion (VSEPR) model and whose discoverer was Ronald Gillespie (Hargittai 2011a, p. 235). This model, too, selects some of the properties of the system that it intends to describe and ignores the rest (Hargittai 2011b, p. 5). Yet the choice is a matter of *a priori* considerations, like the taxonomic choice of the shape (*morphē* in classical Greek, hence morphology) of organisms to organize them in a classification of species and genus. Similarly, the chosen properties or structures function as pure possibilities, better, as the general relationality of the relevant pure possibilities. All the more pure are the possibilities that are chosen to construct, better, to discover a geometrical model. The VSEPR qualitative model has been successful because by means of it chemists have been able to predict properties of systems not yet studied and "on occasion, not yet even existing" (Hargittai 2011b, p. 5). In panenmentalist terms, they not yet *actually* existing, but their pure possibilities do exist and are not excluded. The VSEPR model has succeeded in predicting molecular shapes, geometries, and even structural variations in series of substances (ibid.). Similarly, modeling served Pauling very well in discovering the alpha helix structure of proteins (as with the case of Watson and Crick's discovery of the double-helix structure of nucleic acids, which I will discuss soon below), and modeling inevitably relies upon the study of the relevant pure possibilities and their relationality. The model displays the structural pure possibilities that are actualizable, and the actualization can be confirmed only by empirical findings concerning actualities, namely, by means of empirical observation and experiments.

In his foreword to the chapter devoted to Bartlett, Hargittai reminds the reader that Marcellin Berthelot likened chemistry to art, as both create their objects

(Hargittai 2010, p. 223). As I see this, in art, too, creation begins with the discovery of pure possibilities and their relationality. Think, for instance, of Michelangelo's idea about the uncovering of his sculptures by removing anything superfluous out of the marble. In fact, he selected the marble and uncovered his sculptures out of them in light of the pure "models" he had in mind and sketched on the walls of his hideplace in Florence or on paper. Thus, Michelangelo's creation was a kind of discovery, both of the pure "model" and of the actual sculpture. When chemists create new objects—when, like Bartlett, they synthesize molecules that were not known to exist before—they in fact contribute to the discovery of purely possible ways in which pure chemical possibilities relate one to the other. Thus, creation in chemistry, too, implies discovery of pure possibilities and their relationality, on which these creations supervene, wittingly or unwittingly as far as their discoverers are concerned. In the case of Bartlett, the discovery of the pure possibility of the noble gas compound preceded the creation of the actual compound and paved the way for it. Chemists, beginning with Mendeleev, have endeavored to discover all chemical pure and actualizable possibilities as well as all the ways in which they relate to each other (which is their relationality) to enable compounds. As a result, they have endeavored to discover all possible chemical compounds, whether they are created by nature or by human beings. Either way, any chemical reaction is an actualization of the relationality between some chemical pure possibilities-identities. Bartlett helped greatly in expanding our knowledge of this relationality and in exhibiting its actualization.

## 7.7  James Watson and Francis Crick's Discovery of the DNA Structure

The story of this discovery is so intricate that I prefer to quote from Hargittai's concise and clear description of it:

> Modeling was also decisive in the discovery … of the double-helix structure of nucleic acids by James D. Watson and Francis Crick. In addition to physical modeling, Watson and Crick utilized Rosalind Franklin's X-ray diffraction information and Erwin Chargaff's data … It has been suggested that Franklin was close to the solution, but, lacking modeling, the results of her DNA structure would have emerged in steps rather than in one big splash as did Watson's and Crick's.…The data Chargaff collected scattered quite a bit, and the pattern did not emerge unambiguously as we think of it today. Yet Chargaff did notice it and was brave enough to annunciate it. Alas, he stopped short of asking the crucial question of Why? If the regularity was a real phenomenon, there must have been a reason for it; it must have occurred as a consequence of something. Today we know that it was base pairing between the two strands in DNA, and Chargaff might or might not have arrived at the concept of the double helix had he attempted to model what he had observed. … Watson and Crick did not carry out experiments. That did not mean though to rely on model building alone; on the contrary, they relied on experiment and the best one at that, except that it was not their own experiment. Even then, Watson's and Crick's initial model necessitated a large amount of experimental work in Maurice Wilkins's laboratory before the original model could be considered established unambiguously. (Hargittai 2011b, pp. 7–8)

Let me put that in panenmentalist terms. As in the case of Shechtman's discovery, the "crucial question of Why" has to do with individual pure possibilities. Individual pure possibilities provide the identities of actualities, and the relationality of pure possibilities provides us with the explanation and understanding of actualities or actual regularities. Consisting of the relevant pure possibilities and their relationality, scientific models enable scientists to identify phenomena or data, to understand and explain them, and to predict about them as well as about yet unknown or yet not existing phenomena. Models can explain empirical regularities, for the models explicate their reason (like Hargittai above, I use 'reason' by purpose, for it is not a cause, which is an actual factor), and the necessary connections and relations that a model displays are between antecedents and consequences instead of between causes and effects. Reasons pertain to the domain of pure possibilities, whereas causes pertain to the domain of actualities.

In the intricate discovery of the DNA structure, the physical model that Watson and Crick constructed was a physical image, a depiction of the theoretical model, which is the purely possible model that they discovered. With no empirical, actual data, the model would remain merely possible but, lacking it, Rosalind Franklin and Erwin Chargaff did not succeed in discovering the DNA structure, though they contributed decisively to the scientific progress leading to the actual discovery. Lacking such a model, they did not notice the DNA structure, at least not clearly or unambiguously enough. In such a case, too, scientific progress is achievable only when standing upon two indispensable legs—purely possible systems or models, on the one hand, and empirical, actual data, on the other. It is an actualist fallacy to ignore the former. Pure possibilities without experimental studies and actual data are impractical whereas such studies and data without pure possibilities are blind, not noticing the identity and significance of the findings. To identify the DNA, a purely possible model was indispensable.

## 7.8   Leo Szilard's Idea of a Nuclear Chain Reaction

In 1933, while walking along Southampton Row in London, a surprising idea occurred to Leo Szilard. It was the discovery of the *concept* of a nuclear chain reaction (Hargittai 2011a, pp. 244–245). According to Szilard's version of this discovery, that idea was as follows:

> … if we could find an element which is split by neutrons and which would emit *two* neutrons when it absorbed *one* neutron, such an element, if assembled in sufficiently large mass, could sustain a nuclear chain reaction. I didn't see at the moment just how one would go about finding such an element or what experiments would be needed, but the idea never left me. (Lanouette and Silard 1992, p. 133)

At the time, there was no empirical evidence for such an idea; it was simply a discovery of a pure possibility, lacking any empirical evidence by actual experiments. The only relevant established fact was James Chadwick's discovery of the actual existence of neutrons (in 1932; the concept, namely, the pure possibility of neutrons, had been discovered by Ernest Rutherford in 1920). Chemists were familiar with chemical chain reactions, and the concept had been known to them, but there is no evidence that Szilard knew anything about it (Hargittai 2011a, p. 245). Not surprisingly, the very possibility of a nuclear chain reaction was excluded by experts at that time: "When the concept was pronounced for the first time, it was not only shocking, but many experts absolutely refused to consider it *as a possibility*" (Hargittai 2011a, p. 246; my italics, A. G.). It is one of the striking cases that scientific and technological innovations could be avoided or hindered simply because the relevant pure possibilities have been excluded.

Like Pauling's opposition in the case of Shechtman's discovery of quasicrystals, Szilard had to face the angry opposition of one of the most eminent scientists of the time, the Nobel laureate for chemistry 1908, and the father of nuclear science (or, at least, the father of the nuclear atom model)—Ernest Rutherford. As quoted in a report published in *Nature* of September 16, 1933, Rutherford argued in a lecture that "One timely word of warning was issued to those who look for sources of power in atomic transmutations—such expectations are the merest moonshine" (Hargittai 2011a, p. 246). In other words, according to Rutherford, such expectations were groundless, based on a worthless idea with no grounds at all in actual reality. What at that time Rutherford considered as impossible was considered to be a real possibility in Szilard's mind some months later. And the story continues:

> Szilard was puzzled by Rutherford's statement because he did not think anyone could know what someone else might invent. On June 4, 1934, Szilard visited Rutherford, but when he told the great scientist about his idea of a nuclear chain reaction and that he had already patented the concept [on March 12, 1934], Rutherford became very upset and threw Szilard out of his office. Rutherford's intuitive foresight was legendary, but on this occasion he was wrong—very wrong. In fact, this misjudgment was so much in contrast with Rutherford's usual intuition.... (Hargittai 2011a, p. 246).

Again, the power of dogmatism and, primarily, actualism can be so great, that an eminent scientist, such as Rutherford, not only excluded the very possibility of a nuclear chain reaction, just the idea of it made him mad. Nevertheless, acknowledging pure possibilities as valid is not enough for scientific discovery; empirical evidence concerning actualities is equally indispensable. What that brilliant concept left out was the missing indispensable piece of information—what is the element or what are the elements that would be suitable to use for a nuclear chain reaction (Hargittai 2011a, p. 247). It took scientists years to perform enough experiments to find such an element. No brilliant thought-experiment and no most creative idea or concept can take the place of empirical knowledge and actual experience to find such an element. The discovery of the pure possibility of atomic chain reaction was made by Szilard alone; he had adequate access to it; the actualization and the empirical confirmations were left to others.

# 7.9 George Gamow's Big Bang Model

The beginning of this model was a mathematical version of it. Gamow used all the available information concerning nuclear reactions, but he added to it "many reasonable assumptions" to simplify the calculations (Hargittai 2011a, p. 286). The purely mathematical aspect of such a model is enough to indicate its *a priori* accessibility and to establish the view that it consists of theoretical individual pure possibilities and their relationality. As expected, this purely possible model was not readily accepted at the beginning. Even the name "Big Bang" served the opponents of the model—first of all the distinguished astronomer, Frederick Hoyle—to make fun of it. The received view, in contrast, supported the so-called steady-state model, whose main proponent was Hoyle. It was Hoyle who ironically coined the name "Big Bang" (Hargittai 2011a, p. 287).

One of the empirical, actual anchors of the model was the estimate of the high relative abundance of helium in the universe, which strengthened the credibility of the model's results, because the distribution of the light elements was known from entirely independent sources (Hargittai 2011a, pp. 286–287). Another, more crucial, empirical or actual anchor to confirm the model was the necessary condition that if the model was correct, there must be "a remnant heat ... which should have stayed around even billions of years after the moment of the Big Bang" (ibid., p. 288). This crucial anchor further confirmed the model, for it was Gamow's prediction, according to the model, that the temperature of the universe is seven kelvins. From two parameters—the age of the universe and the average density of matter in the universe—he estimated this temperature, corresponding to the cosmic microwave radiation (ibid., p. 289). At this stage, the serendipity or the contingency of actual events enters the dramatic scientific scene: in 1964, Arno Penzias and Robert Wilson, who "were not concerned with models about the origin of the universe," reported on "their serendipitous observation of the cosmic microwave radiation amounting to three kelvins," which "was a stunning confirmation of the Big Bang model" (ibid.). As long as the empirical data indicate a temperature that is clearly above zero, these data confirm the model: in the universe there is a remnant heat of the Big Bang. Hence, three kelvins should be more than enough to serve as a confirmation of the model: "it was definitely established that there is a remnant heat in the universe, giving final and absolutely convincing evidence for the Big Bang model" (ibid., p. 291). It is a mark of a genuine scientist that even if he or she was an ardent opponent of some theory or model, when finally convinced by the evidence, he or she helps to add more evidence to support it. So did Frederick Hoyle. When finally convinced about the model, he contributed evidence to acknowledge it and to support its reception.

Nevertheless, empirical findings or actualities do not provide us with an explanation. This is the function of the purely possible model. First, Penzias and Wilson convinced themselves of the soundness of their measurements (which, unlike calculations, are based on empirical observations), and then they looked for an explanation. They did not connect this explanation with the Big Bang model. At the same time, other scientists studied possible models of the origin of the universe (Hargittai 2011a, p. 290; this is another example of the coincidence that has to do with the contingency of actual matters such as events). Penzias and Wilson, the discoverers of cosmic microwave background radiation, found a cosmological interpretation of their empirical observation in the help of some of these scientists.

Hargittai demonstrates that Gamow brought together seemingly disparate facts, observations, and theories and, from them, he reached a far-reaching conclusion; where he sensed gaps, he augmented the missing links with intuition and imagination (Hargittai 2010, p. 293). In panenmentalist terms, when necessary or only possible, Gamow ventured to proceed beyond the domain of actualities and empirical data to that of pure possibilities and their relationality. Insatiable curiosity attracted him to quite different areas of interest, whether actual or purely possible, but without an access, owing to his imagination and intuition, to the domain of individual pure possibilities and their relationality, he could not discover the Big Bang model.

## 7.10    Frances H. Arnold and a Novel Way to Open up Chemical and Evolutional Possibilities

Frances H. Arnold, a Nobel laureate for Chemistry in 2018, whose scientific contributions are not discussed in Hargittai's book, has pioneered methods of directed evolution to create useful enzymes and other biological systems. Despite being created by human engineering, these systems are not programmed by means of human ingenuity and its computerized auxiliaries, but are empirical products as a result of human manipulation and experiments. What in fact determines the new chemical possibilities is nature, which proceeds contingently or randomly. This fine example shows to what extent human ingenuity, intellect, and mind in general are limited or confined. We can manipulate chemical possibilities for our benefit much beyond what is familiar and given to us, much beyond the actual, but we are incapable of programming, planning, and logically determining the range of such possibilities and their nature. We have to wait and see what nature will say.

Natural (in this case, chemical and biological) possibilities are much more ample than logical possibilities, and the limitation of nature's possibilities lies greatly beyond logics. Arnold's achievements provide a fine example in demonstrating this. The natural possibilities lie beyond the limitation of our logical thinking, programming, as well as our prediction or anticipation.

For this reason, the Periodic Table, which reveals the basic chemical possibilities, is irreducible to quantum calculus or quantum logic, however bizarre and different from classical logic. Moreover, it is much beyond the quantum possibilities. The same holds true for mathematics. As a discovery of the human intellect, mathematical possibilities do not exhaust natural possibilities, and physics is thus irreducible to mathematics or to logic. Descartes was entirely wrong about that and so are all the rationalists following him in this matter. Physics is irreducible to pure mathematics, let alone to formal logic. The great achievement of Frances H. Arnold demonstrates that empiricism may overcome (or rule out) rationalism, especially when human scientific manipulations of natural possibilities are under consideration.

Furthermore, Arnold's great success demonstrates how far evolution goes, for it embraces not only biology but also chemistry (and, of course, biochemistry). She has clearly succeeded in demonstrating that there is not only evolution of organisms but also of molecules! Hence, evolutionary possibilities are not confined by human logic, mathematics, or computer science; they are more far-reaching and there are many more of them.

This refers to what can be called the "irrationality" of nature. In Frances Arnold's words:

> Most protein reengineering efforts have been by so-called rational design. The filtering effect of scientific publication (successes get published, failures mostly do not) might lead one to believe that we can, with reasonable probability of success, identify and modify the amino acids responsible for key properties such as an enzyme's substrate preference, stability, or activity in a non-natural environment. In reality, we are far from being able to do this reliably. This is true even for the relatively small number of enzymes for which considerable structural and mechanistic data are available. Admitting ineptitude in rational design, however, frees us to consider other approaches which are hardly irrational. An alternative and highly effective design strategy can be found by looking to the processes by which all these proteins came about in the first place. (Arnold 1998, p. 125)

Admitting incompetence or a lack of rational design points to the limitation of scientific rationality. Randomness plays an irreducible role in nature. And the universality of evolution, perhaps as valid to the whole of nature (we may be justified in ascribing evolution also to mathematics or logic), is compatible with such randomness or irrationality. Thus, Arnold mentions the "contingent nature of evolution" (Arnold 2001, p. 256).

There are two paths in scientific explanation:

> Broadly speaking, one can identify two philosophies: [1] either existing biocatalysts can be fine-tuned by rational redesign, or [2] combinatorial techniques can be used to search for useful functionality in libraries generated at random and improved by suitable selection methods. (Arnold 2001, p. 253)

And the case appears to be that randomness prevails in this matter. The opportunities, best, possibilities, that randomness opens for us are greater in number and scope than those that rational redesign opens up for us. Randomness is compatible with diversification (op. cit., p. 254), whereas rational redesign goes with reduction and unification, which end with the reduction and closing of possibilities.

This does not mean that rationality or intelligibility plays no role at all in this new field of development and investigation. Thus, Arnold states:

> But the number of possible protein sequences inevitably dwarfs any existing or even conceivable technology for searching it experimentally. So one must make intelligent choices about what and how to search. This may be where 'rational' design will be crucial: identify-

ing the most likely places to search combinatorially for desired functions....But the capabilities of rational design, particularly computational techniques and *de novo* design, are expanding too. (op. cit., p. 257)

On these grounds, expanding the chemical possibilities, as Arnold and others believe, as a result of "a marriage of ingenuity and evolution will expand the scope of protein function well beyond Mother Nature's designs" (Brustad and Arnold 2011, p. 201). This does not change the following:

> The complex and subtle interplay of interactions that dictates fold and function presents a daunting obstacle to rational protein design. Evolution is Nature's solution to the design problem. Scientists have learned how to implement evolutionary strategies to engineer new proteins, exploiting natural protein scaffolds as starting points for breeding improved versions. A not-surprising testament to the power of natural selection is that the most reliable approach to optimizing protein function is by iterative rounds of mutagenesis and screening or selection, i.e. directed evolution. (ibid.)

Nevertheless, the main achievement of Arnold's innovative view is to utilize the unbridgeable gap between the actual and the possible in nature. Her technique of manipulating the actual structures of enzymes, for example, to "create" new versions of them, whose capability is much beyond what has been actually achieved in nature, before human intervention, is nothing but the utilization of that gap, or the fact that the purely possible is wider and more comprehensive than the actual. In this way, she inadvertently maintains the panenmentalist imperative—save as many and various possibilities as possible (Gilead 1999, 2014a, b, c, and his other panenmentalist publications in the philosophy of science). She and her teams have greatly succeeded in saving chemical and evolutional possibilities, in advancing scientific knowledge and practice much beyond the limitations of the actual in chemistry and biochemistry. Moreover, she and her collaborators have broadened the scope of evolution. In their scientific contributions, they have implemented evolutional possibilities in the field of chemistry and biochemistry. For natural evolution and selection, it has to take many years to implement possibilities, while in laboratories it can take hours or days. This means that Arnold and her collaborators have changed the implementation of temporal possibilities, which have never before been actualized in nature.

The pure possibilities of natural evolution is thus much wider and greatly more comprehensive than evolutional actualities or facts. Arnold and her collaborators call this "promotion of evolvability" or increasing a protein's evolvability" (Bloom et al. 2006, p. 5869 and p. 5873). They write: "We discuss the implications of our work for understanding natural protein evolution and designing better protein engineering strategies" (ibid.).[2]

---

[2] Cf.: "Evolution is Nature's solution to the design problem. Scientists have learned how to implement evolutionary strategies to engineer new proteins, exploiting natural protein scaffolds as starting points for breeding improved versions. A not-surprising testament to the power of natural selection is that the most reliable approach to optimizing protein function is by iterative rounds of mutagenesis and screening or selection, i.e. directed evolution" (Brustad and Arnold 2011, p. 1); and "Generating a function where it does not already exist, not even at a low level, however, is a

Frances H. Arnold has contributed greatly to reveal this for our scientific theoretical interests as well as practical ones.

Saving scientific possibilities in a panenmentalist sense can be found in various scientific fields (Gilead 2014a, b, c). A most striking example is the pioneering field entitled "General Crystallography" (Gilead 2017, 2018). J. D. Bernal, the great mentor, and Alan Mackay, perhaps his most noted follower, contributed greatly to establish this most important and interesting scientific field. Crystals take part in what is generally called structures in science, and such structures pertain to the very general field of order, which is, first of all, a metaphysical concept. Thus, structures comprise not only crystals, which are physical, material, biological, or mathematical structures, they also comprise information, computing, communication, and the like. Structures bear meaning and significance; they have to do much with deciphering of various codes and communication. They are thus sharable by very different scientific fields. The discovery of quasicrystals, which has dramatically changed the definition of crystallography, is a fine example of saving crystallographic possibilities in chemistry and material science, and it has much to do with the theoretic discoveries of Alan Mackay and other eminent scientists in General Crystallography.

What Bernal and Mackay did for General Crystallography, Frances Arnold has done for evolution and chemistry. Each of these eminent scientists contributed, metaphysically-panenmentalisticly speaking, to the great achievement of saving scientific possibilities, without which no scientific progress is achievable.

Against this background, I find the following quite illuminating:

Arnold is leading a very interdisciplinary research, at the cross-border between chemistry, biology and engineering.... her team has developed methods to imitate natural evolution and creation, but on a much shorter time scale, and is now exploiting these methods for the synthesis of new compounds. Indeed, nature has evolved highly complex proteins and enzymes, but despite the current advances in biology and our ability to read any sequence of DNA, instructions on how to compose DNA are still missing. How could we artificially obtain molecules and proteins that would perform better than those resulting from thousands of years of evolution? Arnold's answer to this question is simple: let's produce thousands of molecules. Thanks to the automated machines used for chemical synthesis, it is now possible to fabricate many variations of a chemical sequence in a very short time. Such a powerful yet simple approach *opens up many possibilities* but is also terrifying since there cannot be any prediction of what will really work.... Along with optimised proteins comes the *metaphysical question of how to create novelty?* Nature creates novelty everyday but we are not able to "evolve a cat into a dog". "Can we create a wolf-bird?" she asks.... For example, ... some bacteria are now able to eat our plastic waste. To enable synthetic creation, she points out that enzymes, just like a person having a job and hobbies, are able to do a lot of other things than just the enzymatic reaction they are initially designed for. Therefore, it could be that a protein can catalyze a reaction where one of the reactants is slightly different than the "usual" one. Based on this, she addresses the following questions:

---

bigger challenge that will benefit from the use of new technologies that improve the design process or that provide new chemical functionalities not present in natural systems....Computational approaches to protein design are also building on our growing understanding of protein structure and function. Entire protein folds and proteins that catalyze reactions not present in Nature have been designed *in silico* and constructed" (op. cit., p. 2). "Not present in a natural system" is equivalent with "not actual" or "not actually exist," which may remind one of a panenmentalist idea.

Silicon is the second most common element on earth. It is also very similar to carbon. Yet, there are only carbon-based organisms—why are there no silicon-based ones? Since we use Si-C bonds in many artificial chemical and electrical products, *why are there no Si-C bonds in nature?* Can we produce a Si-C bond using natural proteins? Could an iron-carbon binding enzyme create a silicon-carbon bond? Similarly, could a hydrogen-carbon binding enzyme create a boron-carbon bond? And yes, *it is possible.* Using the optimisation algorithm described earlier, her team found that a particular protein, Rmacyty is able to make such a bond, and can even be improved for high production. (Taz 2018; italics are mine)

In light of panenmentalism, this sounds especially interesting. As actual, nature (if we do not include in it humans and their actions and products, which, in fact, take part in nature) does not exhaust all the possibilities that are open for her. Humans can open many more possibilities, first of all by thinking of them and taking them seriously. In contrast, in many cases we dogmatically exclude possibilities. Such an exclusion is necessary for scientific progress (if everything is possible, there is no place for knowledge). But when we exclude possibilities on the basis of our ignorance, limitation of imagination, prejudices, or dogmatic considerations (which certainly prevail also among scientists), we then exclude significant possibilities which are indispensable for scientific progress. Remember Linus Pauling's prejudice that there are no quasicrystals, there are only quasi-scientists (such as the daring Dan Shechtman…). Such a dogmatic exclusion may stop scientific progress. The same holds true for interdisciplinary achievements, which quite usually raise suspicions within many scientists. Bernal, Mackay, and Arnold are fine examples of great scientists who dare to reveal new possibilities and who have been aware of the importance of interdisciplinary searches in order to reveal them.


## 7.11  Conclusion

The analysis of these striking scientific discoveries clearly shows that relying upon pure possibilities and their relationality has been indispensable to achieve these discoveries. If acknowledging and saving pure possibilities in such discoveries is indeed indispensable for the scientific progress, whereas excluding them hinders such progress, actualism is inescapably invalid and panenmentalism gains more support. Scientific progress is achievable only when standing upon two indispensable legs—purely possible systems or models, on the one hand, and empirical, actual data, on the other. It is an actualist fallacy to ignore or exclude the former. To rephrase a Kantian idea for a panenmentalist purpose, pure possibilities without experimental studies and actual data are practically empty, whereas such studies and data without pure possibilities are blind, for without the latter scientists cannot notice, recognize, or identify some of the most vital or significant facts. Furthermore, pure possibilities and their relationality are indispensable for understanding and explaining such facts and their possibilities. Empirical studies and study of the relevant pure possibilities together with their relationality are equally essential for scientific progress. Excluding pure possibilities, on actualist or other grounds, has been shown to hinder scientific progress.

# Chapter 8
# The Philosophical Significance of Alan Mackay's Theoretical Discovery of Quasicrystals

**Abstract** Dan Shechtman was the first to discover an actual quasicrystal (on April 8, 1982). As early as 1981, about 1 year before Shechtman's discovery of an actual quasicrystal, Alan L. Mackay discussed, in a seminal paper, the first steps for the expansion of crystallography toward its modern phase. In this phase, new possibilities of structures and order (such as the structures of fivefold symmetry) for crystals have been discovered. Medieval Islamic artists as well as Albrecht Dürer, Johannes Kepler, Roger Penrose, Mackay himself, and other pioneer crystallographers raised important contributions to the theoretical discovery of pure crystalline possibilities long before or independently of the discovery of their actual existence. Shechtman, however, was not the first to discover the individual pure possibility of this novel structure (the theoretical discovery), which had been excluded from the range of the possibilities of crystals (as it had been fixed by both theoretical and empirical means at the beginning of the twentieth century). Penrose and Mackay, in particular, had contributed to the discovery of the individual pure possibilities of quasicrystals, which are merely structural, and, like purely mathematical entities, they do not exist spatiotemporally and causally, whereas actual quasicrystals exist only spatiotemporally and causally. The individual pure possibilities of quasicrystals do not depend on their actualities, and without these possibilities, those actualities would have been theoretically groundless, meaningless, and could not be correctly identified, if at all. Hence, Mackay's contribution to the meaning and theoretical basis of the discovery of actual quasicrystals is indispensable.

In this Chapter, I discuss further the philosophical significance of Mackay's theoretical discovery and his contribution to the expansion of pure geometrical crystallography, biological crystallography, and generalized crystallography.

Unlike actual discoveries, which are of actualities, theoretical discoveries, such as purely mathematical or purely physical ones, are of pure possibilities, namely, individual possibilities that do not yield to spatiotemporal and causal conditions. A fine example of a theoretical discovery is that of the pure possibility of the Higgs boson much before its actual discovery. Some years before the discovery of the actual

The first versions of this chapter were published in *Structural Chemistry* 28:1 (2017), pp. 249–256 and *Foundations of Chemistry* (2018).

Higgs boson at CERN, the 2004 Wolf Prize in physics was awarded to Robert Brout, François Englert, and Peter Higgs "for pioneering work that has led to the insight of mass generation … in the world of sub-atomic particles."[1] This has been considered as an outstanding *discovery*, and the Wolf Foundation's announcement further states: "The discovery of Brout, Englert and Higgs was essential to the proof … that the theory with massive gauge particles is well defined; and subsequent calculations in that theory, verified experimentally, culminating in the discovery of the massive W and Z particles" (ibid). We can easily understand what a discovery of an actual particle or entity is. Nevertheless, such was not the case of the Higgs particle until July 2012: until then this boson was not discovered or detected as an actual particle (though in 2011 there were already some tentative empirical indications of its actual existence). Thus, the press release of 4 July 2012 at CERN site announces: "… the ATLAS and CMS experiments presented their latest preliminary results in the search for the long sought Higgs particle. Both experiments observe a new particle in the mass region around 125–126 GeV".[2] Even more interesting is the press release of 8 October 1013, announcing the award of the Nobel Prize in Physics for 2013 to **François Englert** and **Peter W. Higgs** (but, unfortunately, not to Robert Brout because of his death in 2011, as there is no posthumous Nobel Prize): "for the theoretical discovery of a mechanism that contributes to our understanding of the origin of mass of subatomic particles, and which recently was confirmed through the discovery of the predicted fundamental particle, by the ATLAS and CMS experiments at CERN's Large Hadron Collider"[3] (my italics. A. G.). Nobel Prize appears not to be awarded for theoretical discoveries unless they turn out to be actual, namely, empirically proven. In the light of panenmentalism and especially the applications of this metaphysics to natural science, the theoretical discovery of the Higgs boson was of its pure possibility, whereas CERN officially announced, in 4 July 2012, the discovery of the actual Boson particle. Without the theoretical discovery of the pure possibility of the Higgs boson, the actual Higgs boson could not be identified or discovered at all.

On these grounds, I believe that in addition to Dan Shechtman, whose courageous actual discovery challenging the antagonism of a huge camp of chemists (led by the Pope of Chemistry, Linus Pauling), certainly merits this prize, Mackay has also deserved it for his theoretical discovery, which clearly anticipates the actual one. In contrast, in 1984, Levine and Steinhardt (1984) contributed to the theoretical understanding of quasicrystals (whose name was given by them) *after* the discovery of an actual quasicrystal by Shechtman.

Equally, the theoretical discovery of the pure possibility of the omega-minus particle by Yuval Ne'eman and Murray Gell Mann (independently), based on the purely mathematical SU(3) symmetry group, was achieved 2 years (in 1962) *before* the discovery of the *actual* omega-minus. This discovery would have been impos-

---

[1] URL: http://www.wolffund.org.il/full.asp?id=17 (2004)

[2] http://press.web.cern.ch/press/PressReleases/Releases2012/PR17.12E.html

[3] http://www.nobelprize.org/nobel_prizes/physics/laureates/2013/press.html

sible without the former theoretical discovery. Like applied mathematics, following the antecedent pure mathematics, the pure physical possibilities, which a theory discovers, precede the discovery of physical actualities and enable one to identify them.

The mathematics of the nineteenth century provided crystallographers with what, before Dan Shechtman's discovery, had been wrongly believed to be the complete geometrical theory of the symmetries of *all possible* crystals—in my terms, the allegedly complete geometry of all pure possibilities of crystals (230 crystallographic space groups in number), which happened to be also empirically, actual discoveries.

Hence, the experimental verification of these possibilities, initiated by Max von Laue in 1912, resulted in what seemed to be the complete scheme exhausting, "closing", also all the *actual* possibilities of the atomic order of crystalline solids. Unfortunately, "this elegant scheme for a longtime had a constraining effect on crystallographers somewhat analogous to the effect of Euclid's scheme on geometers. It became the paradigm. Important features of real materials were called 'defects' and materials that did not fit the scheme dismissed as 'disorder'" (Mackay 2002, p. 5; cf. Gilead 2013). A similar fate was that of "chaotic" phenomena, until the advent of chaology ("chaos theory") made it possible to detect in them other kinds of order, subject to special kinds of geometries, nonlinear ones.

Against this background, Shechtman has completed a revolution, a paradigmatic shift, which has a philosophical significance, especially in the philosophy of science, not only concerning actual possibilities of the atomic order of crystalline solids but reflecting, even more, the pure, mathematical or theoretical, possibilities of such order. Thanks to scientists like him, we are entitled to refer today to new geometries for new materials.

Similarly, until the middle of the nineteenth century, scientists, including great mathematicians and philosophers, believed that the only possible geometry was the Euclidean one and that Aristotelian logic was a complete, final logic, and, too, the only *possible* logic (Gilead 2013). Such, also, were the firm beliefs of Immanuel Kant (who died in 1804). Such beliefs excluded various scientific pure possibilities. Since the middle of the nineteenth century, humanity has become acquainted with other kinds of mathematics and logic. Thanks to Shechtman and others, we know today that many other scientists were wrong in believing that since the end of nineteenth century materials science possessed the complete geometrical theory of the symmetries of all possible crystals (empirically verified in using X-ray examinations since 1912 until 8 April 1982!). Opening further pure possibilities for scientific discoveries is indispensable for the progress of science.

Without diminishing Shechtman's empirical revolution even slightly, Istvan Hargittai, most convincingly, ascribes to J. Desmond Bernal and, even more so, to Alan L. Mackay, the first steps in doubting the crystallographic dogma concerning the exhaustion of the crystalline possibilities and the exclusion of others:

> Mackay was very alert to structures that fell beyond the seemingly perfect system of classical crystallography. He followed in John Desmond Bernal's footsteps in his attempts of broadening the scope of crystallography, which was called generalized crystallography, and

which in essence was the science of structures. Mackay produced his simulated diffraction experiment and issued a warning to other scientists in his lectures and papers that *there might be* structures of fivefold symmetry that *could be missed* because the *dogma* on their non-existence had been ingrained so strongly in our science. (Hargittai 2010, p. 81, referring to Mackay 1997; my italics. A. G.)

Roger Penrose had made a theoretical discovery (Mackay 1997/8). It was a mathematical solution to the problem of covering a space with pentagons. Johan Kepler in 1611 and Muslim artists in Spain of the Middle Ages, had earlier solved similar problems. Each of these discoveries was a purely mathematical, theoretical one or, at least, the theoretical discovery preceded the actual one.

At the time of the actual discovery, Shechtman knew nothing about Mackay's warning, but, undoubtedly, he, too, did not *practically* exclude the pure possibility of fivefold symmetry pertaining to possible new crystals. "There might be structures" clearly suggest *pure possibilities* of structures that *can be actualized*, while the pure possibilities under discussion should be recognized *first* as actualizable. The main point is the warning against the exclusion of such pure possibilities to begin with. Having accepted that, one could recognize actual quasicrystals and analyze such phenomena as intelligible and explicable.

The following passage from Hargittai's paper shows this quite nicely:

Not only did Mackay complain about the restrictiveness of classical crystallography, he urged everybody who would listen to be vigilant and stay aware of the *possibility* of structures outside the classical system. *Otherwise we might encounter them but walk by them without recognizing them.* Beyond issuing warnings, Mackay even produced a simulated electron diffraction pattern of one of the several Penrose tilings…. It was a *forerunner* of Dan Shechtman's *actual* diffraction pattern of a quasicrystal, alas, Shechtman was not aware of Mackay's *prescient* simulation at the time of his discovery. Mackay gave two presentations on fivefold symmetry in Budapest in September 1982 in which he once again pointed out the *possibility* of the existence of so-called non-crystallographic structures. In his utmost defiance of classical crystallography, Mackay even produced—computationally—pentagonal snowflakes citing Kepler's 1611 work, *De Nive Sexangula* when his paper referred to "De Nive Pentangula."

Unbeknownst to Mackay, by the time of his Budapest lectures, Dan Shechtman … had performed his experiment leading to his seminal discovery of quasicrystals. … It would have been wonderful had Shechtman known about Mackay's simulated experiment, but he did not and neither did he know about Mackay's warning. He was slightly familiar with the Penrose pattern, but did not make the connection between his work and the pattern until long after his discovery. (Hargittai 2010, p. 82; my italics. A. G.)

The paper referred to above by Mackay was published in *Soviet Physics Crystallographica* 26 (1981). Internationally, this was a rather obscure periodical and it seems that those for whom Mackay's suggestion might have presented interest, might have not been aware of it. A somewhat similar fate greeted Shechtman's first publication in 1984 (even in taking into consideration its shortcomings). Did these two great discoverers, Mackay as the discoverer of new crystalline pure possibilities and Shechtman of a new actual one, still have some doubts about their novel discoveries and thus tried to hide them in some way from the vast scientific public till more evidence could be added? I do not know. One has to ask them.

I translate the above-cited passage into a metaphysical, panenmentalist language as follows: Mackay discovered the *pure possibility* of a structure that would be named, by Dov Levine and Paul Steinhardt (1984), "quasicrystal", and his contribution joined the discovery of the mathematical pure possibility of an ordered non-periodic (or "quasi-periodic") structure, searched for by Johannes Kepler and Albrecht Dürer and had been discovered earlier by some Muslim mathematicians in the Middle Ages (following the golden ratio, a mathematical discovery by the Ancient Greeks of a most interesting and fruitful irrational number; were it rational, the structure under discussion could not be non-periodic). Much later, the pure possibility was mathematically developed by Penrose and, finally, three-dimensionally enlarged and computationally simulated as an electron diffraction by Mackay, Levine, and Steinhardt to render it *possible* for an *actual* crystallographic discovery. And this could be considered as a prediction and a theoretical foundation as well of such a discovery, to the extent that Mackay's simulation is concerned. As a result, the 2010 Oliver E. Buckley Prize of the American Physical Society for an outstanding *theoretical* contribution to condensed matter physics was awarded to Mackay, Levine, and Steinhardt, "for pioneering contributions to the *theory* of quasicrystals, including the *prediction* of their diffraction pattern" (Mackay 2002; my italics. A. G.). It should be emphasized that Mackay was the first to produce an optical transformation of that pattern, which "exhibits local tenfold symmetry and repeats the shape of the quasi-lattice cells which gave rise to it" (Mackay 1997, p. 612).

All this has to do with the *pure possibility* (as captured by means of theoretical constructions) of a quasicrystal, whose *actuality* was first discovered by Shechtman. This means that—until Shectman's discovery—the existence of such a crystal was considered, in fact, as purely possible by Mackay and his followers. Until then it was known only as a pure possibility, if possible at all, as the great majority of materials scientists, almost all of them, considered it as simply impossible. More importantly, the pure possibilities of quasicrystals were accessible to some scientists independently or before Shechtman's formidable discovery of an actual quasicrystal.

Pictures may save us many words. Two simple pictures *show* what I have just *told* above. Each of these pictures, though quite similar, is independent of the other, as there is no any causal connection between them. The first one is Mackay's optical transformation of the diffraction pattern of an atomic crystalline order that "exhibits local tenfold symmetry" (Mackay 1997, p. 612). This is the optical transformation of the pattern in figure 4 on p. 611, intended "to model a *possible* atomic structure" (my italics. A. G.). Mackay mentions that the pattern "has local fivefold axes and thus represents a structure outside the formation of classical crystallography and *might* be designated as a quasi-lattice" (op. cit., p. 609; my italics. A. G.). At that time, this pattern could be considered only as a *pure possibility*. The second one is the image of a tenfold symmetry of the atomic order of an *actual* quasicrystal that Shechtman recorded by electron diffraction on 8 April 1982.

Compare Mackay's figure (op. cit., on the left) with that by Shecthman (on the right):

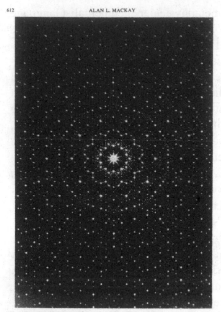 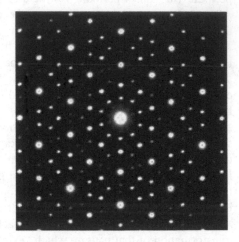

Fig. 5. The optical transform of the pattern of fig. 4. The annular objects show in the circular strong and weak modulations of the transform. The pattern itself exhibits local ten-fold symmetry and repeats the shape of the quasi-lattice cells which gave rise to it.

What were considered before Shechtman's great discovery as impossibilities or, at the best (in the case of Mackay), as pure possibilities and possible predictions, have been discovered as empirical facts, namely, actualities. Mackay's figure implies that "Alan Mackay … [was] better prepared for Shechtman's discovery than Shechtman was" (Hargittai 2010, p. 85), as Shechtman wrote down three question marks in his notebook to express his great surprise and perplexity concerning the tenfold crystalline symmetry detected in his figure, whereas, in the same year, Mackay independently referred to such a symmetry as a possibility that crystallographers should be prepared to face in actual reality.

Hargittai emphasizes—in my philosophical, panenmentalist translation—that with no awareness of the possibility, namely, the *pure* possibility, of crystalline structure outside the classical crystalline system, materials scientists might encounter quasicrystals "but walk by them without recognizing them." Indeed, in my panenmentalist publications, I have emphasized, repeatedly, that knowledge (tacit or explicit) or awareness of pure possibilities is indispensable to recognize or identify, understand and explain any empirical fact or event. In contrast, excluding some pure possibilities can result in ignoring some of the most meaningful and significant phenomena and in rendering any great discovery impossible. Had Shechtman *conclusively* (and not merely feeling perplexed, initially, or provisionally) thought like Pauling and many others and *decided* that "there is no such animal" (as Shectman actually wrote in his notebook in Hebrew אין חיה כזו to express his surprise and perplexity (Hargittai 2011a, b, c, d), he would have excluded the pure possibility of a quasicrystal and he would not have been the discoverer of its actuality. With no knowledge of the pure possibility under discussion, Shechtman could not explain his discovery, and in order to achieve that he needed the mathematical and other purely theoretical aids of other scientists. Indeed, Denis Gratias, a mathematical crystallographer, helped to devise an improved model for interpreting Shechtman's experimental observation (Shechtman et al. 1984).

Note that Levine and Steinhardt, who were informed of Shectman's amazing discovery of 1982, were the first to *name* this discovery; the title of their paper was "Quasicrystals: A new class of ordered structures" (Levine and Steinhardt 1984). Again, this does not diminish Shectman's great achievement even slightly; it shows, instead, in what way scientific novelties are achieved and received. As Hargittai puts it, "The quasicrystal story shows the complex interrelationship of various branches of sciences and other human activities. There were scientists like Alan Mackay, who were better prepared for Shechtman's discovery than Shechtman was, but, to Shechtman's credit, once he made his discovery, he grew to handling it in a seasoned way" (Hargittai 2010, p. 85).

Panenmentalism ascribes priority and independence to pure possibilities, many of which are indispensable for scientific progress. There is an indispensable *a priori* part in any serious philosophical thinking and in scientific thinking as well (especially concerning its purely mathematical and theoretical aspects). When preconceptions were concerned, "people were continually trying to see *a priori* forms in Nature and this sometimes obscured their vision of what was actually there" (Hargittai 1997; though in Mackay's paper, referring to the Penrose tiling and the experimental observation of quasicrystals, Mackay reminds the reader of Einstein's comment that "the human mind has first to construct forms, independently, before we can find them in nature" [Mackay 1997, p. 307]). Nevertheless, this is not valid for the pure possibilities, *a priori* cognizable, that are indispensable for scientific progress and to which Mackay raised a significant contribution in the novel discovery of the pure possibilities of quasicrystals.

Though some of Mackay's philosophical ideas express an empiricist-actualist view, quite common in the British and American philosophies and which panenmentalism opposes, not all of his philosophical ideas in "Lucretius or the philosophy of chemistry" (Mackay 1997), for instance, are like that. Let us have a closer look at them. Mackay acknowledges that mathematics "appears to be a logical system existing in the *noösphere* independently of human beings, but discovered by them, spun out of a few premises" (op. cit., p. 306). "Noumental" means "to be an (ideal) object of the intellect," whereas "phenomenal" means "to be an object of the senses or of empirical knowledge." This is a well-known Platonic distinction and, undoubtedly, Mackay refers to it at this point. Such a view would be considered in philosophy of mathematics today as "platonist" (to be distinguished from Platonism which is a realist view of transcendent, noumenal—noömental—Ideas, which Plato adopted); and I am not sure whether Mackay would like to be assigned such a label. Like a genuine platonist, Mackay considers the Mandelbrot set, for instance, as a discovery instead of an invention (op. cit., p. 306). As he states, "There is a close, but perhaps not exact, correspondence between mathematical relationships and physical measurements, which is extremely useful in following through the consequences of hypotheses" (ibid.), though there is a "cleavage between mathematics and physics" (op. cit., p. 307). This is compatible with his statement, "Mathematics is simply the investigation of structures in the *noösphere* but with applied mathematics one returns to the real world with suggestions as to how real matter may behave" (ibid.). Instead of *noösphere*, I would prefer "the realm of pure possibilities." Translating this into panenmentalist terms, instead of "mathematics" I would mention "*pure* mathematics," to be distinguished from *applied* mathematics. Finally, instead of "the real world" I would mention "actual reality." As I see them, pure possibilities are not simply concepts in our mind but they exist independently of it and are accessible to and discoverable by it.

Despite his strong reservations of Platonism and Pythagoreanism, whose influence is still with us, and despite blaming them for regarding "the world of mathematics as this ideal world, where knowledge *a priori* form whole numbers" (Mackay 1990, p. 3), Mackay concedes that "in a sense, various mathematical objects [such as Mandelbrot's set] do exist in the *noösphere*, quite apart from their materializations. … Thus, the Platonic view of the existence of a library of ideal objects is not altogether perverse" (op. cit., p. 4). Translating again into panenmentalist language, instead of the "library of ideal objects" I would mention "the realm of pure possibilities," and instead of "materialization"—"actualization." Having said that, note that in almost any other aspect, panenmentalism differs from Platonism. First of all, actualities are inseparable from their pure possibilities, as the former are *included*, as spatiotemporally and causally conditioned *parts*, in the latter. Pure possibilities are not transcendent; unlike the Platonic ideas, they are immanent.

Mackay further states, "The most recent history of fivefold symmetry may be compared to the escape from the idealistic Platonic influence toward a more realistic computational attitude of the Babylonians" (op. cit., p. 4.). Yes indeed, but it should be compared, first of all, to the escape from the Euclidean geometry toward non-Euclidean geometries, consisting of quite different geometrical pure possibilities and their relationality (namely, the sum of the relations of each pure possibilities with the rest of them). At his point, Mackay makes a fascinating comparison: "The Greeks [more precisely, the Pythagoreans] were shocked to find that the diagonal of a square could not be represented as the ratio of two integers. Our own scientific culture has been similarly shocked by the recent discovery of Shechtman … that extended assemblies of atoms could have fivefold symmetry as evidenced experimentally by their diffraction patterns" (ibid.). First, irrational numbers are mathematical pure possibilities, which the Pythagoreans refused to accept and thus attempted to exclude them. Of course, secondly, the fivefold symmetry as evidenced experimentally is essential for modern crystallography, but Shechtman's actual discovery could not be intelligible and explainable without the theoretical, "pure" contributions with which Mackay and others have provided crystallography. Scientists may be dogmatic about pure possibilities no less than about actualities, and open-mindedness, such as Shechtman's and Mackay's, is indispensable for good science in both of its aspects—as actual and as pure or theoretical.

In the following paragraph, Mackay actually mentions opening scientific possibilities contrary to restricting or closing them:

> No theory is secure as is the theory of the [230] space groups. This gave some crystallographers the illusion of acquiring *absolute* knowledge *a priori*. The iconoclastic aim of overthrowing this model theory seemed quite unrealistic but *this has always been the possibility* of outflanking the restrictions ... (Mackay 1990, pp. 7–8; my italics. A. G.)

Yes indeed, this has always been the possibility, although almost all of the crystallographers before Shechtman's discovery, and some after it, refused to accept it or ignored it. *A priori*, theoretical considerations, concerning pure possibilities and their relationality, side by side with *a posteriori* ones, considering known actualities, restricted their view and understanding. What should have been considered as possible in crystallography was considered as impossible.

Mackay emphasizes a major point, relevant to my discussion of mathematical pure possibilities:

> At the earliest stage Euclid's *Elements* (including the books 14–15, added later) was a guide to the geometry of space and in particular it provided the five regular platonic solids as ideal objects which might be seen in imperfect form in the real world. Today there is an infinitely greater range of *possible* abstract forms some of which may be *realized* in practice. Einstein wrote: "The human mind has first to construct forms, independently, before we can find them in things." This is part of the role of mathematics. (Mackay 1997, p. 307; my italics. A. G.)

This fits perfectly with some of my ideas concerning pure possibilities and the difference—what Mackay names "cleavage"—between pure possibilities and their actualities. As I have argued above, mathematical pure possibilities are not abstractions but independent "constructed forms," which the human mind discovers. Instead of "realized in practice" I would term it "actualized in empirical reality." Of course, non-Euclidean geometries have enabled us to expand greatly our acquaintance with mathematical pure possibilities, many of which have proven to be most useful for modern physics. Yet, mathematical pure possibilities are not sufficient, and scientists need also philosophical or metaphysical pure possibilities, for they tacitly assume philosophical and metaphysical assumptions or postulates such as those dealing with existence, order, and causality.

Returning to chemistry, Mackay states, "Philosophy of chemistry began when Lucretius began to organize what facts he had into a coherent recognizable building" (Mackay 1997, p. 307). Nevertheless, as I see it, neither Lucretius nor modern chemists could manage without relying *also* on physical and chemical *pure* ("theoretical") possibilities. Such organizing possibilities, including their connections and functions (or generally relations, in a panenmentalist term, "relationality"), pertain to the most important aspect of Mendeleev's periodic table of the elements. This has been clearly recognizable in the indispensable role of eka-elements from Mendeleev's time until the present day. As chemical pure possibilities, eka-elements have been *predictions* of actual elements, some of which have been actually discovered later (or will possibly be discovered in the future), in a way similar to that of the pure possibilities of quasicrystals becoming predictions of their actualities. This is beautifully demonstrated in the case of Mackay's pioneering theoretical contribution to the discovery of the pure, theoretical possibilities of quasicrystals.

Indeed, Mackay refers to the organization of the elements into the periodic table and to Mendeleev's important predictions (Mackay 1997, p. 307). Undoubtedly, the table summarizes all the chemical empirical findings and the chemical pure possibilities of the elements that so far have not been actually, empirically discovered or synthesized. The way in which the elements are organized, the taxonomic system according to which they are organized, is *a priori*, like other taxonomic systems. Similarly, as Mackay mentions, "there were many insights like that of Democritus. For instance, chemistry is characterized by structure and by change" (ibid.). These insights, like Mendeleev's taxonomic ideas, did not rely upon experience; instead, they theoretically preceded it and were independent of it and of empirical facts in general (such is the case of Democritus, one of the post-Eleatic philosophers who challenged Eleatic philosophy). In these lines, Mackay pays special attention to the role of arrangement and *postulations* in chemistry as follows:

> Before Jacobus Henricus van 't Hoff, who *postulated* (1874) the tetrahedral carbon atom and that chemical compounds had actual structures in three-dimensional space, there was only a steady accumulation of facts which formed a random heap with some fragments of regularity rather than an organized edifice. The philosophy of chemistry began when Lucretius began to organize what facts he had into a coherent recognizable building. (ibid; my italics. A. G.)

First, this reminds us of what Mackay says, on the previous page, about mathematical postulates, which are independent of any experience and, logically or philosophically, precede it. Secondly, as *organizing* the facts, Lucretius relied upon a relationality of organizing pure possibilities. The same holds for the structures and the organized edifice to which van 't Hoff's *postulate* referred. These were *a priori* structures and edifice that today are subject to chemical observations, experience, and experiments. Today we are familiar with them as actual; but such was not the case from the beginning, neither of chemistry, nor of the philosophy of chemistry. There must be some part of the chemical knowledge that is *a priori*, relying upon chemical and physical *pure* possibilities and their relationality, independently of experience or empirical facts.

Since the "philosophy of chemistry … involves N-dimensional geometry" (op. cit., p. 308), and since such a geometry cannot rely upon empirical facts but it is a relational system of pure possibilities, the *a priori* part of the philosophy of chemistry should not be dismissed. This aspect of Mackay's philosophy of chemistry (and of physics as well) is compatible with another aspect, the one that inescapably relies upon experience or experiment as follows: "The world of quantum mechanics is quite unfamiliar to people who live on the scale of human beings and could not be deduced by reflection, but only by extremely subtle experiment" (ibid.).

We should pay attention to the following:

> The recognition of hierarchic structures in quasi-crystals has broadened the horizons of what is *possible*. Many questions arise about the origin of life and the evolution of the genetic code, the operation of the brain, etc. but all these are to be answered by the methods of science rather than from the ruminations of academic philosophy. Serious philosophical questions arise at the level of chemistry. (op. cit., p. 310; my italics. A. G.)

Judging from the history of the revolution of the modern crystallography, in which Mackay and others have played a significant pioneering role, the horizons of the possibilities opening for crystallography could not be restricted by known empirical facts. Mackay was among the first modern crystallographers to show this on a purely theoretical basis! Philosophical ruminations are not restricted to academic philosophers, for major philosophical questions bother natural scientists no less than they bother children; we are guided by various tacit or implicit philosophical assumptions about existence, knowledge, possibilities, order, lawfulness, structures, symmetry, causality, and so on. Indeed, serious philosophical questions arise at the level of chemistry, but these questions must also be treated and, in the first place, by means of thought and reflection, referring not only to scientific facts but also, and sometimes no less, to pure possibilities that are indispensable for scientific progress and predictions. At least, this is the way that panenmentalism views these matters.

I have sympathy for Mackay's blaming post-modernists, among others, for the "confusion between the real world and the noumenal world," a confusion which "is still with us" (Mackay 2002). Panenmentalism, keeping the *distinction* between actual reality and the realm of pure possibilities quite sharp, is involved with a harsh critique of post-modernism.

Finally, no matter what is the way or the temporal order by which we are acquainted with pure possibilities, owing to experience or without it, their independence of actualities and empirical reality is intact. In any event, these pure possibilities have been *a priori* accessible, entirely independently of empirical, actual reality. Indeed, such is clearly the case of Mackay's crystalline theoretical discovery, which was independent of Shechtman's actual great discovery of 8 April 1982 (see more about this in Hargittai's enlightening publications [30–35]). When Mackay published his seminal paper of 1982, the publically known crystalline facts were overwhelmingly against his suggestion to expand the realm of crystalline possibilities beyond the restriction of the accepted 230 crystallographic possibilities of space groups, and all he could suggest then to materials scientists referred, at that time, only to pure possibilities.

Alan Mackay's "Crystal Symmetry" (Mackay 1976) is a seminal work for understanding his leading project to expand the theoretical realm of crystalline possibilities within the new field of "generalized crystallography". The same holds true for his "De nive quinquangula: On the pentagonal snowflake" Mackay has represented a daring, new project of developing.

> a unified way of dealing with spatial structures (and their changes, growth, evolution and transformations) which will cover the range from diamond, naphthalene and iron, to recognizably living objects (such as bacteriophages)....concepts of cellular automata, having their origins in computer theory, and now being elaborated for the purposes of developmental biology, provide a suggestive ideology for recasting crystallographic ideas so that they be carried forward. (Mackay 1981, p. 495)

Against this background, he mentions "von Neumann's theory of self-reproducing automata, which was later seen to be realized in the genetic code" (ibid.). Such a model deals "with information rather than energy" (ibid.).

Generalized crystallography is a project that has attempted to unify all the disciplines in which crystals play a role. Crystals play crucial roles not only in chemistry and in material sciences but are also vital for biochemistry, biology, bioinformatics, structural and molecular biology, life sciences, genetics, virology, medicine, mathematics, information theory, nanotechnology, and other disciplines. Can we construct, then, a generalized crystallography that unifies crystalline possibilities of all these disciplines, first of all as theories? Mackay and other crystallographers have asked themselves this most important question, the answer to which reveals that chemical structures or forms of order, embodied as crystals, have roots and branches as well in other scientific disciplines, which appear to be quite different or far from chemistry. Thus, the significance of these structures, shapes, or forms of order, as reflecting internal order, lies much beyond the boundaries of chemistry (see Hargittai 2010, 2017, and the other papers included in this issue dedicated to Mackay's 90 birthday). Moreover, Mackay's papers (Mackay 1976, 1981) clarify more deeply his crucial theoretical discovery concerning quasi-crystalline possibilities.

In my discussion above, in this Chapter, concerning Mackay's scientific contribution, I simply mentioned but did not analyze this paper, which deserves special consideration. Now I have the opportunity to discuss this contribution, too, in a philosophical light. I owe much to Mackay (1982, 1990, 1997, 1997/8, 2002; Lord et al. 2006) and especially to Istvan Hargittai (1992, 1997, 2010, 2011a, b, c, and 2017; Hargittai and Hargittai 2000, 2009) for my understanding of the revolutionary discovery of new unorthodox crystalline possibilities.

Philosophers, including philosophers of science, are quite familiar with the notion of possible worlds ("maximally consistent sets of statements" or "maximal, global states of affairs or ways in which the actual world might have been"), which philosophical semantics and modal logics usually employ with various benefits. Not ignoring these benefits, I still find this notion quite problematic and controversial, whenever ontological and metaphysical topics are concerned. Thus, I have attempted to replace the problematic, even quite obscure and, indeed, controversial, notion "possible worlds" with a simpler and clearer one—*individual* pure possibilities. Assuming that each individual actual entity—each actuality—is different and distinct from any other actuality, as each actuality has an individual pure possibility that exclusively belongs to it, renders it possible from the outset (namely, a priori or prior to any actual circumstances), and individuates or distinguishes it, ontologically and epistemologically, from any other actuality. Hence, thanks to our knowledge of individual pure possibilities and their relations, we are capable of recognizing, identifying, and understanding actualities, namely individual actual things or phenomena, whereas denial or ignorance of such possibilities results in not recognizing, in entirely missing, such actualities even when we encounter them. The knowledge of individual pure possibilities might prepare us to expect the existence of their actualities and not to ignore it when we encounter these actualities in empirical experiments or observation. The denial of the theoretical, pure possibilities of quasicrystals led eminent chemists, such as Pauling, to deny the very existence of actual quasicrystals even while watching clear images of them.

Penrose, Mackay, and others had contributed to the discovery of the pure possibilities of quasicrystals well before the discovery of actual quasicrystals. These pure possibilities are mathematical–structural and, like purely mathematical entities, they do not exist spatiotemporally and causally, whereas actual quasicrystals exist only spatiotemporally and causally. The pure possibilities of quasicrystals do not depend on their actualities, and without them, these actualities would have been theoretically groundless, meaningless, and could not be correctly identified if at all.

Contrary to actualities, each individual pure possibility is independent of any spatiotemporal and causal conditions and of any actual circumstances. When you refer to any actual individual entity regardless of any of its spatiotemporal and causal properties or conditions, you refer, in fact, to the individual pure possibility of that entity. This actual entity is, in fact, the restricted part of this pure possibility from which this part is inseparable. The restriction is subject to spatiotemporal and causal conditions.

This Chapter is devoted to a treatment of special kind of individual pure possibilities—crystalline pure possibilities, which, like any other individual pure possibilities, are discoverable not empirically but by theoretical means, mathematical and otherwise. As we shall see below (and see, furthermore, Gilead 2013), long before the discovery of actual quasicrystals by means of empirical experiments and observations, the individual pure possibilities of such quasicrystals were discovered by mathematicians and artists as well as by theoretical crystallographers such as Mackay and others. As we shall see, of special interest is Johannes Kepler's discovery, according to Mackay's interpretation as well as others',[4] of unorthodox crystalline possibilities considered long before the discovery of actual quasicrystals in the twentieth century.

Another kind of interesting chemical individual pure possibilities are the eka-elements, namely, the predicted chemical elements in the periodic table well before their discovery or production as actual elements (Gilead 2016). Independently of actual discoveries, the periodic table is the maximal attempt to exhaust all the chemical pure possibilities (including their actual parts) concerning the elements and their possible reactions. For further examples of chemical pure possibilities, consider my papers (Gilead 2014a, c).

In all these examples, the term "pure possibility" might be partly replaceable by the terms "theoretical possibility" or "predicted possibility." However, such terms do not convey the special significance that panenmentalism ascribes to individual pure possibilities of any kind, namely, their ontologically independent status, as panenmentalism is realist about these possibilities. This means that panenmentalism considers them as mind-independent and, thus, as discoverable and not as inventible. Our intellect and imagination, and not empirical observations and experiments, discover and reveal such possibilities for us. Such discoveries are one of the main

---

[4]Cf. Hargittai 2017, p. 8: "The search for extended structures with five-fold symmetry had been going on for centuries and involved excellent minds, such as Johannes Kepler and Albrecht Dürer. Roger Penrose came up with such a pattern in two dimensions and Mackay crucially extended it to the third dimension, and urged experimentalists to be on the lookout for such structures."

functions of scientific theories. Individual pure possibilities are subject to truth, which is different from empirical or actual truth, but this major issue is clearly beyond the scope of this Chapter. For the time being, think of purely mathematical truths, which refer to mathematical objects or entities, which are mathematical pure possibilities and which, as such possibilities, are independent of actual or empirical facts.

Do not let the term "pure possibilities" confuse you. Every reader is familiar with the concept of pure mathematics, which deals not with physical or material, actual phenomena or objects but with the mathematical pure possibilities, which are numbers and geometrical figures, and especially with their relations. While actualities are discovered by empirical means—experience, observations, and experiments—pure possibilities are only theoretically discoverable, by means of our intellect and imagination alone and not by empirical means. Such is the nature, for example, of pure mathematics—its objects and their relations are not discoverable by empirical means. In contrast, applied mathematics deals with actual phenomena and objects and it draws from empirical sources of knowledge.

This very distinction is valid also for crystallography: There is, first, a geometrical crystallography (a term Mackay uses in Mackay 1981, p. 518; cf. Mislow and Rickart 1976, and Hargittai and Hargittai 2009, and discussing the application of mathematics, for instance, in matters of chirality and symmetry, to the description of nature), in which theoretical discoveries of crystalline pure possibilities are made (for instance the Penrose pattern (Mackay 1981, p. 519; cf. Mackay 1976, p. 497), which, in panenmentalist terms, is a geometrically crystallographic pure possibility); second, there is an applied geometrical crystallography, which is applied to actual, material crystals (whether these material substances are physical, chemical, or biological). Thus, mentioning "semiregular quasilattice" patterns, Mackay explains that "these patterns arise from *purely* mathematical considerations, but even if they do not correspond to physical structure, they may provide interesting suggestions as to some ways in which a biological crystallography may develop" (Mackay 1981, p. 519; my italics, A. G.; cf. Mackay 1981, p. 522: the significance of the pattern "shows, perhaps, one step from classical crystallographic structures towards the biological").

Translating this to panenmentalist terms, I suggest that this is precisely one of the important functions of discovering new crystalline pure possibilities and their relationality as well. Relationality exhausts all the relevant relations that these possibilities have, and, hence, unlike the individual possibilities as such, relationality is general or universal. When Mackay mentions "various attempts to extend the conventional compass of symmetry theory" (ibid.), this can be translated into the panenmentalist term of saving pure possibilities instead of excluding them ("There is no such animal"! was Shechtman's first response while observing an actual quasicrystal). Saving crystalline pure possibilities means extending the compass of symmetry theory in particular and theoretical crystallography in general.

Considering mathematical pure possibilities, think especially of the pure geometry of crystals or crystalline structures. At the end of the nineteenth century and until the second half of the twentieth century, the orthodox or "classical" crystallographers were familiar only with the 32 crystallographic points groups and the 230 space groups. These were all the crystalline possibilities that crystallographic pure geometry assumed. Empirical findings by means of X-ray procedures applied these possibilities to actual crystals observed in the laboratory (cf. "the experimental procedures of X-ray diffraction" Mackay 1981, p. 519). Owing to its deep interest in saving pure possibilities, a panenmentalist, like a theoretical crystallographer, had to ask himself or herself whether the 32 crystallographic points groups and the 230 space groups really exhausted the pure crystalline possibilities. Today we know for sure that such is not the case, and we know this thanks to brave and unorthodox, and surely undogmatic, scientists such as Bernal and Mackay.

As soon as the first paragraph of Mackay's paper, Mackay 1981, the text appears to prepare itself to be fully translated ("like a glove to a hand") into philosophical panenmentalist terms. According to these terms, we have to distinguish between individual pure possibilities (of the same kind of purely mathematical figures or numbers) and their actualities (which, like the empirical solar system mentioned below, are subject to spatiotemporal and causal conditions). Let us read this first paragraph with some minor omissions:

> Johannes Kepler was mistaken in this book *Mysterium Cosmographicum* (1596) when he suggested that the five regular polyhedra were the "spherical harmonics" which were the key to the structure of the Solar System. Unfortunately, he never knew that they were, in fact, the eigenfunctions of Schrödinger's equation and the solution to the mystery of the microcosmos at the atomic level. He just knew that they must be the answer to something. Kepler was, however, one of the first to explain the visible forms of matter in terms of the spatial arrangement of their component atoms.... he attributed the hexagonal symmetry of the snowflake to the close packing of spherical atoms. (Mackay 1981, p. 517)[5]

Now, the following is my philosophical, panenmentalist translation of this paragraph (with some omissions) as follows:

> *Johannes Kepler was mistaken...when he suggested that the* **pure possibilities** *of five regular polyhedra as well as their relationality were the "spherical harmonics" which were the key to the* **actual** *structure of the Solar System. Unfortunately, he never knew [but which is known to us now] that these* **pure possibilities***, whose general relations serve as a model, were of the functions of Schrödinger's equation and the solution to the mystery of the microcosmos at the atomic level. He just knew that these* **pure possibilities** *were of some actualities (or as a model of them). Kepler was, however, one of the first to explain the* **actual***, visible forms of matter in terms of the* **purely possible** *spatial arrangement of their component atoms....he attributed the hexagonal symmetry of the snowflake to the close packing [or tiling]of spherical atoms.*

While pure possibilities are individual, not general, possibilities—models, however pure, are general. Note that pure possibilities are not possible worlds, a fashionable

---

[5] Cf. Mackay 1976, p. 497: "We must ask as many have asked since Kepler what the rules which lead to the formation of a snowflake are. We are just beginning to see how the rules for the growth of a tree are written in the genetic code. Is there any resemblance between these two extremes of complexity? Does it make any sense to look back and ask where the program for providing a snowflake may be stored?"

philosophical term that my philosophy—panenmentalism—entirely dispenses with because of the controversial nature and obscurity of this term. Instead, I use individual pure possibilities and their relations. As general, models are not individual pure possibilities. The model consists of the general relationality of such possibilities. Recall that relationality consists of all the possible relations that these possibilities maintain with regard to one another.

No two pure possibilities can be identical and the law of the identity of the indiscernibles is certainly valid for pure possibilities (Gilead 2005a). The simple reason for this is that because individual pure possibilities do not exist in space and time, we are not allowed to consider apparently two "identical" pure possibilities as discernible on the grounds of spatiotemporal positions. It is impossible to spatiotemporally locate pure possibilities, as they cannot occupy any space or time. Because no two individual pure possibilities can be identical, they are different *from* each other and, hence, they must *relate to* each other. A model refers to the general relationality of individual pure possibilities that have something in common, sharing some properties. Thus, a model is general.

A structure consisting of fivefold symmetry composes an individual pure possibility, say, of an actual snowflake. All the pure possibilities of such a kind maintain a relationality that can serve as a model of various interesting, non-classical, actual structures, which was the theoretical way that Mackay and some other open-minded, non-orthodox crystallographers, have taken. Thus, the general relationality of a structure of fivefold symmetry can serve as a model for various phenomena—snowflakes, quanta, quasicrystals, and the like.

Moreover, as early as 1979, Mackay referred both to *experimental* and theoretical evidence for the existence of such stable structures of gold atoms (Mackay 1981, p. 517). May we say then that Dan Shechtman was not the first to observe an actual structure of fivefold symmetry? Did not Kepler, according to Mackay's analysis, observed a snowflake, which had such a structure? This has been quite probable for other cases, as quasicrystals, mostly unknown until the second half of the twentieth century, have existed for many years, and there was a chance, however small, that somebody recognized these extraordinary crystals even long before Shechtman. In any event, Mackay's claim of 1981—"a very small crystal has properties which may be very different from those of a large crystal" (Mackay 1981, p. 518)—holds prophetically true for the 1982 discovery by Shechtman of a tiny quasicrystal that could be observed only through an electron microscope. Furthermore, the theoretical discovery of the pure possibilities of quasicrystals happened long before and it *could* have happened even longer before. Think of an analogy—could somebody discover, much before the middle of the nineteenth century, the possibility that the Euclidean parallel postulate could not be valid? Yes, this was not impossible, because such a theoretical discovery has depended only on personal imagination and intellect, not upon historical or cultural circumstances. Skeptical and imaginative or creative enough minds could have discovered this possibility much before the middle of the nineteenth century. The same holds true for the theoretical discovery of quasicrystalline pure possibilities. Mackay himself made such an analogy in explicitly writing that "the formalism of *International Tables of Crystallography* is similarly restrictive" to the formalism of Euclidean geometry (Mackay 1976, p. 495).

Thinking about the breaking of such a restriction, take, for example, the Penrose pattern to which, as early as 1974, Roger Penrose was the principal contributor but not the only one (Mackay 1981, p. 519). In 1975, 7 years before Shechtman' s actual discovery, Mackay incorporated "local fivefold axes into sphere packings by the hierarchic packing of pentagons in the plane and icosahedra in space, came upon essentially the same pattern.... However, very many people from Albrecht Dürer to A. V. Shubnikov, have examined the intriguing properties of packing of pentagons" (Mackay 1981, p. 520). And, of course, long before these theoretical discoveries, in the Middle Age, Islamic decorations portrayed symmetric units repeating in an infinite non-periodic pattern (Mackay 1981, p. 520). The same holds true for Kepler's discovery, according to Mackay, of the pattern of fivefold symmetry of snowflakes, which he attempted to use as a model to explain other phenomena. Note that the symmetry of all these patterns is fivefold, which is that of quasicrystals. Relying upon Istvan Hargittai's strictly erudite studies, Shechtman was not aware of any of these theoretical discoveries as being relevant to his own discovery, and thus in April 1982 he was bewilderment and surprised when witnessing an actual quasi-crystal of fivefold symmetry through an electron microscope. He could not understand or explain how this crystal was possible according to a known crystalline pure geometry in particular and theoretical crystallography in general. The source for identifying, understanding, and explaining actual crystals are crystalline pure possibilities and their relationality. Without being familiar with them, the identification, understanding, and explanation of the structures of actual crystals is simply impossible or at least very dubious. In Mackay's terms—the significance of the pattern discussed in his paper (Mackay 1981) "gives an example of a pattern of the type which might well be encountered and which might go unrecognized if unexpected" (Mackay 1981, p. 522).

A most important issue discussed in Mackay's seminal paper (Mackay 1981) has to do with crystalline complexity (and Mackay discusses stages of complexity, for instance in Mackay 1981, p. 518). Classical crystals have no fivefold symmetry and their components are absolutely identical, whereas the components of modern or non-classical crystals are quasi-identical and quasi-equivalent (Mackay 1981, p. 517). Now, the more complex crystals are, the more they tend to be subject to this kind of symmetry, which is valid for quasi-identical and quasi-equivalent components (Mackay 1981, p. 518). The more individual are the components, the more complex is the crystal. No wonder, then, that such structures suit crystals related to organisms (ibid.), such as adenovirus, which are more complex than inert matter (in Mackay's words, an "example, which represents the next stage of complexity, is that of the adenovirus particle" [ibid.]). This suits very well the *individual* nature of pure possibilities and the complexity of their relationality. The more complex a physical, chemical, or biological entity is, the more apparent is its individuality, and the component particles in biological structures are "only quasi-equivalent and some may indeed also be chemically somewhat different from others" (Mackay 1981, p. 518).

Having mentioned "some of the features … in which biological structures begin to differ from purely chemical ones" (ibid.), when Mackay points out "the development of a new biological crystallography" (ibid.), we have a glimpse of what Mackay calls "generalized crystallography." According to this crystallography, any physical, chemical, or biological entity is not subject to disorder or to real chaos; it is rather structurally organized or ordered and, in fact, not chaotic. The apparent disorder in the physical, chemical, or biological world is what meets the eye; it is not there when we examine the phenomena more deeply and meticulously. The order in question is embodied as structures or crystals. Crystals are inseparable from material entities and they consist not only of solid materials but also of fluid ones, provided that crystalline structures are not limited to the possibilities of classical crystallography. Moreover, there is also evidence for the existence of gaseous phases of crystals (for instance, Poole 2004, p. 1231: "particular types of solids are *quasicrystals* and gaseous crystals" and there is a gaseous crystalline state [Poole 2004, p. 508]). As early as 1928, Eddington considered the possibility of gaseous crystals (Edington 1928, p. 369). Magma is fluid but it also contains solid and gaseous crystals. In fact, we find crystals (or, better, structural regularities (Hargittai 2010, p. 485)[6] in each of the four states of matter: solid, liquid, gas, and plasma (Hargittai 1992, p. 1013).[7] Moreover, all living substances are made of structures relating to crystals—protein molecules, DNA molecules, viruses, bacteria, and the like all are made of such structures. This kind of order entails *relationality*, in fact a general relationality, which is in the nature of individual pure possibilities and, hence, of their actualities. As any actuality is of an individual pure possibility, which has relationality, equally there is no actuality that is not subject to some kind of order, whether we are familiar with it or not. The more organized or ordered a substance is, the more individual it is and its components are never identical to each other. At

---

[6] "Liquid structures … cannot be characterized by any of the 230 three-dimensional space groups and yet it is unacceptable to consider them as possessing no symmetry whatsoever. Bernal noted presciently that the major structural distinction between liquids and crystalline solids is the absence of long-range order in the former…. A generalized description should also characterize liquid structures and colloids, as well as the structures of amorphous substances. It should also account for the greater variations in their physical properties as compared with those of the crystalline solids. Bernal's ideas have greatly encouraged further studies in this field which is usually called generalized crystallography" (Hargittai 2010, p. 485).

[7] The electron diffraction pattern originating from a gaseous or a plasma sample is the result of intramolecular interferences and to only negligible extent to intermolecular ones, if any, in conrast to solid samples and to a lesser extent to liquid ones. Stating that all living substances are made of crystals (or, better, molecular structures), I mean to say that when wet proteins (which have the structure of crystals) are investigated, they are closer to the living matter, because they better approximate their existence in aquaeous solution, than the "dry" crystals. Crystallization in its *classical* meaning leads living matter to death. I own this comment to the kindness of Istvan Hargittai. But perhaps it will be more helpful to adopt Mackay (and Hargittai's following of him) that „crystalography" should be replaced by „structural chemistry" (see Hargittai 2017, p. 9; cf. Hargittai 2010, p. 81: in essence, generalized crystallography is the science of structures, crystalline and otherwise).

this point, too, generalized crystallography has a very significant and interesting novelty to suggest, especially under the light of panenmentalism.

Actual reality as a whole is the actualization of individual pure possibilities and their relationality as well. Hence, actual reality must be organized, ordered, or structured. Crystallographers have expanded their discoveries of structures to information theory (Mackay 2002, pp. 215–216: "The relation of DNA to protein, the dialectical relation between description and referent, was the key discovery of the last century and has meant that crystallography is also now concerned with information, informational structures, growth and form and morphogenesis generally, involving dynamic as well as static structures in space"). Hence, crystalline structures hold also for information.

The panenmentalist or philosophical significance of generalized crystallography is that the order that individual pure possibilities and their relationality entails shows that all actualities, as actualities of these possibilities, are subject to structural order. Generalized crystallography is, then, of such structures, which have the aforementioned relationality. Yet, we have to distinguish between generalized crystallography in this sense and purely geometrical crystallography: the structures treated by purely geometrical crystallography are spatial (Mackay 1981, p. 518), better spatially possible, whereas generalized crystallography is not confined to spatial structures; it is valid for any kind of order and organization.

To demonstrate how generalized crystallography works in an amazing way, I refer the reader to a fascinating lecture by the physicist, Freeman J. Dyson (Dyson 2002), attempting to suggest a possibility of proving the Riemann Hypothesis by relying upon insights concerning quasi-crystalline possibilities (cf. Lapidus 2008). Dyson explains:

> Quasi crystals were one of the great unexpected discoveries of recent years. They were discovered in two ways, as real physical objects ... and as abstract mathematical structures in Euclidean geometry. They were unexpected, both in physics and in mathematics.... Quasicrystals can exist in spaces of one, two or three dimensions. From the point of view of physics, the three-dimensional quasi-crystals are the most interesting, since they inhabit our three-dimensional world and can be studied experimentally. From the point of view of a mathematician, one-dimensional quasi-crystals are more interesting than two-dimensional quasi-crystals because they exist in far greater variety, and the two-dimensional are more interesting than the three-dimensional for the same reason. ...a quasicrystal is a pure point distribution that has a pure point spectrum. (Dyson 2002, p. 11)

Such insights are along the lines of the creative ideas of Roger Penrose, John Desmond Bernal, and Alan Mackay. To translate the main points of this passage into panenmentalist terms, there are two kinds of discovery concerning quasicrystals: that of crystalline pure possibilities and that of crystalline actualities. Both kinds of discovery were unexpected. The purely mathematical discovery of the three-dimensional pure possibilities of quasicrystals is of possibilities that are actualizable in physical three-dimensional space. As for the pure possibilities of two-dimensional quasicrystals, they were discovered by Penrose, "before the general concept of quasi-crystal was invented" (Dyson 2002, p. 12; I suggest using a more accurate verb which is compatible with Dyson's lecture as a whole—

"discovered" instead of "invented"). The discovery of purely mathematical crystal-line possibilities appears to Dyson to be highly relevant to the long-expected proof of the Riemann Hypothesis. This mostly unexpected extension of crystalline pure possibilities is certainly a new discovery pertaining to generalized crystallography, and Dyson declares: "I take the liberty of generalizing the definition of quasicrystal so that $Z$ will qualify.... $Z$ is generalized quasicrystal if and only if the Riemann Hypothesis is true" (Dyson 2002, p. 16). I wonder whether Mackay has thought about such an expansion but, in light of the arguments in the current Chapter, it is certainly possible and suits his published insights. In fact, in his lecture, Dyson discusses pure structures, which is the main interest of generalized crystallography along the lines of Mackay's approach to it.

As for panenmentalism, there are two more important points to be referred in this lecture: (1) Dyson mentions the fact that there are more crystalline one-dimensional possibilities and two-dimensional ones than three-dimensional possibilities, which, in panenmentalist terms, implies that the realm of crystalline pure possibilities is wider or more comprehensive than that of crystalline actualities; (2) There is a hope that young mathematicians will prove the Riemann Hypothesis, as the "history of mathematics is the history of horrendously difficult problems being solved by young people too ignorant to know that they were impossible" (Dyson 2002, p. 17). Indeed, in panenmentalist terms, a necessary condition for discovering new pure possibilities is not to consider them as impossible and excluded in advance. Scientific preju-dices, to which young scientists may be less enslaved, have prevented elder scientists, such as Dyson, from the discovering and saving of new vital possibilities. In an analogous vein, Mackay is one of the crystallographers, if not the principal of them, who call the attention of crystallographers not to exclude new, however unex-pected, crystalline possibilities at least on a theoretical basis.

Quite recently, Julyan H. E. Cartwright and Mackay have continued in advancing generalized crystallography, relating information to forms and structures, much beyond classical crystallography (Cartwright and Mackay 2012). Considering their paper from a panenmentalist viewpoint, they appears to remain mainly on the level of actualities and do not continue to dig more deeply toward the fundamental basis consisting of individual pure possibilities at the level of the logic and mathematics of quantum computation, in which basic information, form, and structure are involved. Cartwright and Mackay describe their project in the following words:

> We argue for a convergence of crystallography, materials science and biology that will come about through asking materials questions about biology and biological questions about materials, illuminated by considerations of information. The complex structures now being studied in biology and produced in nanotechnology have outstripped the framework of classical crystallography, and a variety of organizing concepts are now taking shape into a *more modern and dynamic science of structure, form and function.... The fundamental level is that of atoms.* As smaller and smaller groups of atoms are used for their physical properties, quantum effects become important; *already we see quantum computation taking shape.* Concepts move towards those in life with the emergence of specifically informa-tional structures.... *We must integrate unifying concepts from dynamical systems and infor-mation theory to form a coherent language and science of shape and structure beyond crystals.* To this end, we discuss the idea of categorizing structures based on information

according to the algorithmic complexity of their assembly. (Cartwright and Mackay 2012, p. 2807; the italics are mine)

I wonder, why does the fundamental level consist of atoms rather than of sub-atomic particles and, more importantly, of quantum individual pure possibilities and their relationality? Quantum computation, I think, is a better idea to start with for the benefit of generalized crystallography and the "science of shape and structure beyond crystals". Following David Deutsch's ideas of qubits (quantum bits), as the basic or minimal amounts of information in physics (Deutsch 2004)—though not in his terms but, instead, in panenmentalist terms—the fundamental level of reality should not be of atoms but of computational pure possibilities (such as "the five regular polyhedral" that Kepler could never know that "they were, in fact, the eigen-functions of Schrödinger's equation" as Mackay states in Mackay 1981, p. 517). Cartwright and Mackay postulate that "underlying all the structures are constraints of time and space" (2012, p. 2087) but, in panenmentalist terms, constraints of time and space underlying the *actualities* of all the structures, not the structures as individual pure possibilities or their relationality. If qubits consist of mathematical-logical pure possibilities concerning quantum computation, not only in artificial products but also in the entirely natural ones (as Deutsch argues), individual pure possibilities and their general relationality serve as the fundamental substratum for the material-actual reality and not vice versa. However, the main point is that generalized crystallography or the "science of shape and structure" may gain support from such quantum theories that pertain to the realm of the investigation into information, order, forms, structures, and, eventually, crystals. This investigation should not ignore the scientific significance of metaphysical view that is realist about individual pure possibilities and their general relations. Quantum possibilities or states provide us with information, which is becoming increasingly complex in macroscopic systems, all of which are actualized in matter, inorganic or biological, and hence are subject also to the study of chemistry and physics.

Crystals are made of atoms, structurally ordered, and atoms are made of sub-atomic particles subject to and actualizing quantum possibilities. Because these individual pure possibilities and their relationality are informative, meaningful, and intelligible, sub-atomic particles, atoms, and crystals are informative, meaningful, and intelligible. Think of a DNA molecule, which is a crystal consisting of atoms whose sub-atomic particles actualize quantum information in qubits, which are the minimal amounts of information. DNA is a crystal that actualizes intelligible and most meaningful information. This is one examples of many other crystals, also actualizing individual pure possibilities and their relationality, thanks to which these crystals are informative, meaningful, and intelligible. Such a general (or generalized) view embraces the whole of nature, including all kinds of matter, organisms, and artifacts, all of which are actualities of individual pure possibilities and all are intelligible, meaningful, and informative. If computation pertains to the substratum of nature as a whole, like other mathematical-logical entities, this substratum consists of individual pure possibilities and their *relationality*.

Mackay's description of the "paradigm shift" or revolution from classical to modern crystallography (which, in his terms, is simply the expansion of crystallography) suits well the meaningful order of individual pure possibilities and their relationality, even though Mackay has not been familiar with panenmentalist metaphysics and has appeared to side with materialist approaches. The discovery of individual pure possibilities and their relationality is entirely independent of the discovery of their actualities but preceded it. Such was precisely the case with Mackay's theoretical discoveries of crystalline pure possibilities and their relationality, which were entirely independent of the later discoveries of their actualities, such as Shechtman's discovery of an actual quasicrystal. Metaphysical possibilities are independent of anything actual and precede it. Indeed, scientists have some tacit or implicit philosophical assumptions of which they are not at all aware and which are really about individual pure possibilities or their relationality that can serve as models. In any of these examples, the ontological and epistemological priority of individual pure possibilities and their relationality is well secured.

# Chapter 9
# Shechtman's Three Question Marks: Possibility, Impossibility, and Quasicrystals

**Abstract** The revolutionary discovery of actual quasicrystals, thanks to Dan Shechtman's stamina, is a golden opportunity to analyze once again the role that individual pure ("theoretical") possibilities and saving them plays in scientific progress. Some theoreticians, primarily Alan Mackay, contributed to saving individual pure possibilities of quasicrystalline structures and to opening materials science for them. My analysis rests upon panenmentalism, which I introduced in 1999, quite independently of any familiarity with modern crystallography, and which deals with saving individual pure possibilities as indispensably contributing, inter alia, to our knowledge and sciences.

The Press Release of the Royal Swedish Academy of Science on 5 October 2011, concerning Dan Shechtman's nomination for the Nobel Prize in chemistry "for the discovery of quasicrystals," describes his achievement in the following words:

> On the morning of 8 April 1982, an image, counter to the laws of nature, appeared in Dan Shechtman's electron microscope. In all solid matter, atoms were believed to be packed inside crystals in symmetrical patterns that were repeated periodically over and over again. For scientists, this repetition was required in order to obtain a crystal.
>
> Shechtman's image, however, showed that the atoms in his crystal were packed in a pattern that could not be repeated. Such a pattern was considered just as *impossible* … . His discovery was extremely controversial. In the course of defending his findings, he was asked to leave his research group. However, his battle eventually forced scientists to *reconsider their conception of the very nature of matter*.[1]

Indeed, Linus Pauling, a famous celebrated Noble laureate in chemistry, ridiculed Shechtman's interpretation of that experiment and observation as a quasi-science and simply dismissed it as impossible. Many well-established scientists, who followed him at the time, considered quasicrystals as impossible. In other words, they absolutely excluded such a possibility from the outset.

---

A first version of this chapter was published in *Foundations of Chemistry* 15 (2013), pp. 209–224.

---

[1] See http://www.nobelprize.org/nobel_prizes/chemistry/laureates/2011/press.html My italics. A. G. More precisely, the image was counter to the *received* crystallographic laws *at that time*.

© Springer Nature Switzerland AG 2020                                                  161
A. Gilead, *The Panenmentalist Philosophy of Science*, Synthese Library 424,
https://doi.org/10.1007/978-3-030-41124-4_9

Discovering the first quasicrystal, in the morning of 8 April 1982, Dan Shechtman wrote in his notebook, documenting his findings, "10 Fold ???".[2] These three question marks beautifully indicate the great surprise of this great open-minded discoverer—he could not believe his own eyes, for, according to everything he knew, crystallography simply excluded such a possibility: There could be no such crystals. Instead of clinging to the firm dogma, universally accepted at that time, he chose to remain skeptical: A surprised, yet not dogmatic, scientist. At that time, as far as crystals were concerned, tenfold or fivefold symmetry was considered impossible; it was a "forbidden symmetry … in the world of crystals" (Hargittai 1992, xiii). Rotating the inspected crystal, Shectman realized that, unlike the diffraction pattern in the image he saw, the crystal did not have tenfold symmetry but instead had fivefold symmetry; at that time, this was also assumed to be impossible.

Until Shechtman's discovery (first published as Shechtman, Dan et al. 1984), crystallographers and materials scientists relied upon a very simple proof that two parallel fourfold or sixfold axes of rotation generate translational symmetry, whereas two parallel fivefold axes of rotation clearly cannot coexist (Lidin 2011, p. 2). After all, three-dimensional bodies with fivefold symmetry could not be packed with the full utilization of the available space, and fivefoldedness was "excluded from the possible operations when crystal structures are built" (Hargittai 1992, p. xiii). Such is the impossibility of covering a surface with same-size regular pentagons without gaps or overlaps (ibid.).

Shechtman and his supporters showed that the old definition of crystallinity was not adequate to cover a new class of ordered solids and, as a consequence, the definition of "crystal," given by the International Union of Crystallography, was changed (Lidin 2011, p. 2). Such a change of definition is not merely semantic; it is a conceptual change that has philosophical or metaphysical major significance, as it was the "conception of the very nature of matter" that has been dramatically changed. More specifically, Shechtman had opened new possibilities for crystallinity, materials science, the science of structures (and order) in particular and for chemistry and physics in general. He has enlarged and enriched our knowledge about the possibilities of the atomic order in the solid matter. He opened the eyes of materials scientists to realize what before his discovery they had ignored or dismissed, even at the level of mere possibilities. He had changed the way in which scientists think which crystals *may* exist and which *actually* exist. The change under discussion is, thus, both epistemological (namely, concerning our knowledge) and ontological (concerning existence or reality), which means that it has metaphysical significance. "Matter," "structure," and "order" are not only physical terms; they are also

---

[2] www.quasi.iastate.edu/discovery.html. Cited in the site of Royal Swedish Academy of Science. See Lidin (2011, 3). Shechtman reported that he was so surprised to see such unexpected findings that he told himself: "there is no such creature" (in Hebrew: אין חיה כזו). See Hargittai and Hargittai (2000, p. 159); compare Hagrittai (2011a). Indeed, "what we call today quasicrystals used to be considered impossible until Shechtman's experiment" (ibid., pp. 745–746). For more recent publications by Hargittai about Shechtman's discovery see Hargittai 2011b, c.

philosophical concepts. As fundamental, primary concepts, they have metaphysical significance.

Modality is a special branch of philosophy, treating concepts such as "existence," "possibility," "impossibility," "necessity," and "contingency." The change under discussion is also a modal change, for it opens up new possibilities for sciences, possibilities that were explicitly excluded in the history of science.

This is not only a good lesson in humility of which scientists are demanded (Lidin, ibid.) and in scientific dogmatism and blindness; it is a major lesson about the indispensable role that metaphysical assumptions, usually tacit or implicit, play in scientific capability or incapability to grasp new discoveries. Especially, it points out clearly to what extent modal assumptions may determine scientific progress. Owing to Shechtman and his followers, the most prolific possibilities of non-periodic crystalline structures are open for scientific discoveries, and we have no way to predict what new discoveries the future is holding for us. Of course, science cannot do without restrictions, but some of these restrictions happened to be revealed as dogmatic, as obstructing scientific knowledge instead of contributing to its progress. Some restrictions or exclusion of some possibilities are necessary for that progress, for if everything is possible or possibly true, no scientific progress is possible. Nevertheless, saving possibilities is no less indispensable for this progress, and Shechtman's case clearly demonstrates this![3]

Linus Pauling and his followers argued that non-periodic crystal structures—quasicrystals—are impossible. In other words, those scientists excluded the possibility of any quasicrystal, for this possibility was forbidden by the received views at that time about fundamental crystallographic laws. Excluding such a possibility may have two different senses: (1) The actual possibility of any quasicrystal does not exist; or (2) Even the mere, pure possibility of any quasicrystal does not exist ("there is no such possibility at all;" "no quasicrystal possibly exists. In the first sense (1), those scientists could rely on induction, however problematic inductive inferences and the laws based on them are, whereas in the second sense (2), they could rely upon theoretical, logical, mathematical, or philosophical considerations or on considerations of conceivability (namely, that such a possibility is inconceivable). The second sense has metaphysical significance and implications and it is more important for the discussion of this Chapter.

The exclusion of the pure possibilities of non-periodic crystalline structures has been incompatible with the mathematical possibility of a similar, though two-dimensional, non-periodic tiling which was well known before Shechtman's discovery—Penrose's pentagonal tiling (Penrose 1974, p. 266).[4] Moreover, the golden ratio appears in all manifestations of fivefold symmetry.[5] It is the relation between

---

[3] Hence, "Rather than making the mistake of again being overly restrictive, science now treats exclusive statements about long-range order with caution" (Lidin, ibid.).

[4] In 1982, Alan L. Mackay extended Penrose tiling into three dimensions (Hargittai 1992, p. xv). See Mackay 1982.

[5] As the golden ratio "is inextricably linked to the Fibonacci sequence" (Lidin 2011, p. 6), this may reflect even more clearly its nature as a mathematical pure possibility.

the diagonal and the edge in the *geometrical pure possibility* of a regular pentagon.[6] I mention a geometrical *pure* possibility, for the possibility under discussion is independent of any actuality. Even if no actual regular pentagons had been observed, the pure possibility of a regular pentagon necessarily exists in a pure geometry. Even if there were no actual applications of such a pure possibility, it would necessarily exist as a discoverable pure possibility. As purely mathematical, the pure possibility of the golden ratio is a discovery of mathematics and aesthetics, and it could and would exist with no actual manifestations at all. To say that even the pure possibility of a quasicrystal could not exist is equal to the exclusion of the pure possibility of the golden ratio or of that of pentagonal tiling.

Moreover, in medieval aperiodic mosaics, such as those in the Alhambra Palace in Granada in Spain or in the Darb-i Imam Shrine in Iran, the regular patterns follow mathematical rules but never repeat themselves, just as in the case of quasicrystals.[7] Such purely mathematical rules consist of the way in which mathematical pure possibilities relate one to the other. If mathematicians, physicists, and artists had excluded such pure possibilities, many of their major discoveries and achievements could not have been attained.

I am aware of the fact that not a few philosophers deem numbers and geometrical figures as abstractions, abstracted from empirical, actual reality. If this were indeed the case, figures and numbers could not be considered as pure possibilities, for they were dependent on some actualities or empirical facts, from which they were abstracted. Those who consider geometrical figures as antecedents of numbers are strongly inclined to such a view. According to this view, irrational numbers, such as the golden ratio, were abstracted from the ratio between lines actually drawn upon some plan and which were empirically observed. Nevertheless, this view is wrong, for mathematical imagination, calculation, and inference have been never confined to actual reality and many most imaginable, even fantastic, mathematical discoveries ("inventions") have been independent of actual reality and empirical facts.[8] It is sufficient to think about Euclidean geometry to demonstrate that: Unlike its representation as an actual dot, a point, for instance, has a position but no dimensions (according to the first definition in Book I of Euclid's *Elements*); it cannot be measured and yet exists in the Euclidean space. A line (according to the second definition in that book) is "length without breadth," whereas any actual drawn line must have some breadth, however small. Pure geometrical lines and points are not subject

---

[6] See the Press Release mentioned in first page of Chap. 2: "When scientists describe Shechtman's quasicrystals, they use a concept that comes from mathematics and art: the golden ratio. This number had already caught the interest of mathematicians in Ancient Greece, as it often appeared in geometry. In quasicrystals, for instance, the ratio of various distances between atoms is related to the golden mean" (p. 1).

[7] See ibid. These mosaics "have helped scientists understand what quasicrystals look like at the atomic level" (ibid.). As Penrose's mosaics, which are mathematical purely possible structures, were used to analyze those medieval Islamic mosaics, the role of individual pure possibilities in analyzing and understanding the phenomena of quasicrystals is quite apparent.

[8] See, for instance, Atiyah 1995. Atiyah demonstrates, *inter alia*, how pure mathematics has borne unexpected fruits in "the recent story of quasicrystals" (ibid., pp. 144–145).

to our observation; only their manifestations, phenomena, or representations in actual space are subject so. Pure geometrical points and lines are thus "ideal," but this does not make them abstractions from actual reality. They are neither idealizations of empirical facts, for if they were, they should be idealized according to some ideal standard or paradigm, which, in turn, must be entirely independent of empirical facts and which, hence, would have been pure possibilities.

This independence raises a crucial epistemological question: If not by empirical means, how can we have access to mathematical pure possibilities? How can we know them, become acquainted with them, if not by means of actual experience and empirical observations? The following is a panenmentalist answer to that question. There are no two identical pure possibilities. Each pure possibility is different from any other pure possibility, and no two pure possibilities can be identical, otherwise "two" pure possibilities would have been one possibility instead of two. In case of actualities, which are actual possibilities, two of them might be identical, provided that they did not exist at the same place at the same time, whereas such could not be the case of pure possibilities, which are entirely exempt from spatiotemporal conditions or restrictions. Pure possibilities are different from actual possibilities. We can grasp counterfactual possibilities, even though we know of no similar possibilities, actual or pure. While observing actualities, we can think or imagine that things *could* have been different, that there are counterfactual (better, counteractual) *possibilities*; while considering some pure possibilities, we can think also about quite different pure possibilities, each of which is *different from* the other and, hence, each *relates* to the other. This universal mutual relationality allows us *a priori* accessibility to an infinite number of pure possibilities. Our imagination and intellect are not confined, thus, to actual reality; they have access to new pure possibilities simply because they are different from the already familiar possibilities, one from the other as well as from actual possibilities. Thus, we are capable of relating to, imagining, thinking about and understanding possibilities that are entirely different from the possibilities, whether pure or actual, with which we are already familiar. For instance, since the nineteenth century, 230 crystallographic space groups have been known, but even then, in the nineteenth century, someone could imagine or consider different crystallographic groups as pure possibilities. The same holds for the Euclidean triangle: Even before the middle of the nineteenth century, someone could imagine or consider the pure possibility of a triangle of the sum of whose angles is less or more than $180°$. In any case, pure possibilities are *a priori* accessible and cognizable, altogether independently of empirical, actual reality. This holds especially for pure mathematical possibilities.

There is quite another philosophical way to consider mathematical structures as *a priori* cognizable. Except for wrongly considering Euclidean geometry as the only possible and necessary mathematics, there is still force in Kant's conception of mathematics, in the *Critique of Pure Reason*, and of the epistemological status of its sentences ("judgments") as synthetic *a priori* ones. According to Kant, any mathematical calculation, numbering, constructing geometrical figures, ordering, proofs, and the like are independent of empirical facts, and our knowledge about them is prior to experience; namely, it is *a priori*. We are acquainted with mathematical novelties not by experience, observations, or experiments, on the one hand, and not owing to analysis of concepts, on the other, but only owing to constructions that our reason generates and which provide us with synthetic *a priori* propositions, wholly independent of experience or experiment. According to Kant, geometrical or arithmetical kinds of order, structure, and construction are *a priori* cognized. Kant limited that to the restrictions of Euclidean mathematics but it should be valid, I think, for any kind of mathematics that is *a priori*, namely, independent of any experience and yet is possibly valid for experience of this or that kind and indispensable for knowing it scientifically. However, Kant did not consider those kinds of order, structure, and construction as possibilities, let alone as pure possibilities. This had to wait for another philosophical turn.

Warning his listeners and readers that referring to the possible does not depend on the actual existence of objects in the real world, Hermann Weyl, one of the prominent mathematicians of the twentieth century, who devoted a major effort to studying symmetry, including crystalline symmetry, stated that the roots of the mathematical method lie "in the *a priori* construction of the *possible* in opposition to the *a posteriori* description of what is actually given" (Weyl 1929, p. 249).[9] Nevertheless, Weyl was later considered to be a mathematical empiricist.

Mathematical apriority is valid, too, for the knowledge of the golden ratio, Penrose tiling, non-periodic crystalline structures, and the like. In my terms, they all are mathematical pure possibilities and their relationality (the ways they relate one to other), and pure possibilities are inescapably *a priori* cognizable or accessible to our knowledge, namely, our knowledge of them is entirely independent of empirical reality.[10]

Indeed, if numbers and ratios were abstracted from geometrical figures, and if geometrical figures were abstracted from the observations of empirical data, my stance that quasicrystals and their *intelligible and explainable* discovery depend on geometrical *pure* possibilities and their relationality (especially in the case of the

---

[9] See also: "*Mathematics is the free exploration of structural possibilities, pursued by (more or less) rigorous deductive means*" (Hellman 1989, p. 6. Italics in the original). Hellman is indebted to Howard Stein's view according to which "*The role of a mathematical theory is to explore conceptual possibilities*—to open up the scientific *logos* in general, in the interest of science in general. ... [and] the usefulness of elaborating *in advance*, that is, independently of empirical evidence (and, in this sense, *a priori*), a theory of structures that need not but may prove to have empirical application" (Stein 1988, p. 252; italics in the original). Hellman follows Hilary Putnam somewhat in "interpreting mathematics ... in a modal language, in which a notion of mathematical or logical possibility is taken as primitive" (Hellman 1989, p. 8). He also examines the "holistic, quasi-empirical platonism of Quine, according to which pure mathematics receives its justification through its empirical, scientific applications" (ibid., p. 3). Nevertheless, philosophical defenses of mathematical apriorism still exist, also in cases that mathematics relies upon computer-assisted proofs. See, for instance: "granted that unsurveyable computer-assisted proofs include only *a priori* methodology, they are *a priori* proofs, even if not known *a priori* by any one individual" (McEvoy 2008, p. 386).

[10] No wonder, then, that E. Prince explicitly refers to "*a priori* classification of space groups" (Prince 2004, p. 897) and to "*a priori* predictions of molecular crystal structures" (p. 906). Of course, this does not reduce the indispensable empirical aspect of crystallography.

relevant ratio) might be found groundless. If such was the case, Shechtman's great discovery could be described as follows: Suppose that another crystallographer saw the findings of Shechtman's discovery (as was the actual case with Pauling), and concluded: "Oh, there is nothing new about it, for it is a formation of a twin crystal" (the phenomenon of "twinning" in crystallography). In this case, Shechtman would certainly react: "I am quite familiar with twinning, and there is nothing like it in this case. To prove that my diffraction patterns did not originate from twins, I have generated a series of experiments (Hargittai and Hargittai 2000, pp. 163–164). Undoubtedly, I have discovered a new crystal, against all my expectations as a materials scientist, and it is an empirical fact, which is undeniable, for there is no error in my experiments or observations, and any crystallographer who repeats them must reach the same findings." Indeed, who can deny empirical facts? If they are incompatible with one's theory, with one's crystallography, the latter must be changed. The case *appears* thus to be that no pure possibilities, mathematical or otherwise, are needed in such a case. But this is only an appearance. Some other materials scientists, especially crystallographers, were also familiar with the difference between Shechtman's findings and twinning, and yet they excluded the very possibilities of quasicrystals and considered them as errors in the observation or the experiment, only because they excluded such pure possibilities on the grounds of purely theoretical considerations and the empirical knowledge at that time. The great merit of Shechtman lay not only in his empirical knowledge of the actual reality of solids; his merit was in open-mindedness toward *new possibilities* of crystals. Despite his initial hesitation, skepticism, and great surprise, graphically symbolized by the three question marks in his notebook, Shechtman *did not exclude the pure possibility of a non-periodic structure of solids, the very possibilities of yet-unknown kinds of order or arrangement of the atoms of solids or crystals.* Instead, he practically acknowledged such possibilities. He acknowledged a not-yet known new kind of crystalline order and structure.

It is not enough to conduct observations and experiments that are free from errors and mistakes; the scientist must recognize, identify, and understand what he or she sees or observes. I assume that quite a few crystallographers before Shechtman may have seen such phenomena but they did not realize their meanings or significance; they did not see, identify, and understand that in front of their very eyes there was a new kind of crystal, quite unknown before. They interpreted such phenomena as errors of experiments or observations or as "twinning." Similarly, many scientists before Pasteur saw microorganisms through the microscope, but nobody before him recognized, identified, and interpreted them as microorganisms. Among the crystallographers who looked at quasicrystals before Shechtman, there must have been some who suspected that something was, or looked, different, but they excluded the pure, very possibility of non-periodic structure of solids, the very possibility of quasicrystals. At the time of the discovery (in April 1982) and some years later, Shechtman could not explain his discovery, for he was not yet familiar with its mathematical and theoretical pure possibility, which has been indispensable for identifying, understanding, and explaining the phenomenon he observed.

The mathematics of the nineteenth century provided crystallographers with what, before Shechtman's discovery, had been believed to be the complete geometrical theory of the symmetries of *all possible* crystals[11]—in my terms, the allegedly complete geometry of all pure possibilities of crystals (230 crystallographic space groups in number). Shechtman has completed a paradigmatic, conceptual revolution, which has metaphysical significance, let alone in the philosophy of science, not only concerning actual possibilities of the atomic order of crystalline solids but, even more, the pure, mathematical or theoretical, possibilities of such order. Thanks to scientists like him, we are entitled to refer today to new geometries for new materials.

Similarly, until the middle of the nineteenth century, scientists, including great mathematicians and philosophers, believed that the only possible geometry was the Euclidean one and that Aristotelian logic was a complete, final logic, also the only *possible* logic. Such, also, were the firm beliefs of Immanuel Kant. Such beliefs have excluded various scientific pure possibilities. Since the middle of the nineteenth century, humanity has become acquainted with other kinds of mathematics and logic. Thanks to Shechtman and others, we know today that Linus Pauling and many other scientists were wrong in believing that since the end of nineteenth century materials science had the complete geometrical theory of the symmetries of all possible crystals (empirically verified since 1912 till 8 April 1982!). Opening further pure possibilities for scientific discoveries is indispensable for the progress of science.

Without diminishing Shechtman's revolution even slightly, István Hargittai ascribes to John Desmond Bernal and, even more so, to Alan Mackay, the first steps in doubting the crystallographic dogma concerning the exhaustion of the crystalline possibilities and the exclusion of others:

> The crystallographer Alan Mackay was intrigued by Penrose's pattern. He thought that if he could simulate a diffraction experiment from the Penrose pattern, it *would* be an indication that structures exist that *could* be built in three dimensions, according to the principle of what Penrose did on a sheet of paper. Mackay was very alert to structures that fell beyond the seemingly perfect system of classical crystallography. He followed in John Desmond Bernal's footsteps in his attempts of broadening the scope of crystallography, which was called generalized crystallography, and which in essence was the science of structures. Mackay produced his simulated diffraction experiment and issued a warning to other scientists in his lectures and papers that *there might be* structures of fivefold symmetry that *could*

---

[11] See Eric A. Lord et al. 2006, p. 4. The experimental verification of these possibilities, initiated by Max von Laue in 1912, resulted in what seemed to be the complete scheme exhausting, "closing," all the *actual* possibilities of the atomic order of crystalline solids. Unfortunately, "this elegant scheme for a longtime had a constraining effect on crystallographers somewhat analogous to the effect of Euclid's scheme on geometers. It became the paradigm. Important features of real materials were called 'defects' and materials that did not fit the scheme dismissed as 'disorder'" (ibid., p. 5). A similar fate was that of "chaotic" phenomena until the advent of chaology ("chaos theory") made it possible to detect in them other kinds of order, subject to special kinds of geometries, nonlinear ones.

*be missed* because the *dogma* on their non-existence had been ingrained so strongly in our science. (Hargittai 2010, p. 81. My italics. A. G.)[12]

At the time of the discovery, Shechtman knew nothing about this warning, but, undoubtedly, he, too, did not *practically* exclude the pure possibility of fivefold symmetry pertaining to possible new crystals. "There might be structures" clearly suggest pure possibilities of structures that *can be actualized*, for the possibilities under discussion should be recognized *first* as actualizable. The main point is the warning against the exclusion of such pure possibilities to begin with. Having accepted that, one could recognize actual quasicrystals and analyze such phenomena as intelligible and explicable.

The following citation from Hargittai's paper shows this quite nicely:

Not only did Mackay complain about the restrictiveness of classical crystallography, he urged everybody who would listen to be vigilant and stay aware of the *possibility* of structures outside the classical system. *Otherwise we might encounter them but walk by them without recognizing them.* Beyond issuing warnings, Mackay even produced a simulated electron diffraction pattern of one of the several Penrose tilings .... It was a *forerunner* of Dan Shechtman's *actual* diffraction pattern of a quasicrystal, alas, Shechtman was not aware of Mackay's *prescient* simulation at the time of his discovery. Mackay gave two presentations on fivefold symmetry in Budapest in September 1982 in which he once again pointed out the *possibility* of the existence of so-called non-crystallographic structures. In his utmost defiance of classical crystallography, Mackay even produced—computationally—pentagonal snowflakes citing Kepler's 1611 work, *De Nive Sexangula* when his paper referred to "De Nive Pentangula."[13]

Unbeknownst to Mackay, by the time of his Budapest lectures, Dan Shechtman ... had performed his experiment leading to his seminal discovery of quasicrystals. ... It would have been wonderful had Shechtman known about Mackay's simulated experiment, but he did not and neither did he know about Mackay's warning. He was slightly familiar with the Penrose pattern, but did not make the connection between his work and the pattern until long after his discovery. (Hargittai 2010, p. 82. My italics. A. G.)

Translating this into a metaphysical, panenmentalist language: Mackay discovered the *pure possibility* of a structure that would be named, by Dov Levine and Paul Steinhardt, "quasicrystal," and his contribution joined the discovery of the mathematical pure possibility of an ordered non-periodic (or "quasi-periodic") structure, searched for by Johannes Kepler and Albrecht Dürer and had been discovered earlier by some Muslim mathematicians in the Middle Ages (following the golden ratio, a mathematical discovery by the Ancient Greeks of a most interesting and fruitful irrational number; were it rational, the structure under discussion could not be non-periodic). Much later, the pure possibility was mathematically developed by Penrose and, finally, three-dimensionally enlarged and computationally simulated as an electron diffraction by Mackay, Levine, and Steinhardt to render it *possible* for an *actual* crystallographic discovery. And this could be considered as a prediction of

---

[12] Referring to Mackay 1982.

[13] Published in *Soviet Physics Crystallographica* 26 (1981), pp. 517–522.

such a discovery, to the extent that Mackay's simulation is concerned.[14] All this has to do with the *pure possibility* (as captured by means of theoretical constructions) of a quasicrystal, whose *actuality* was first discovered by Shechtman. This means that—until Shectman's discovery—the existence of such a crystal was considered as purely possible by Mackay and his followers. Until then it was known only as a pure possibility, if possible at all, as the great majority of materials scientists, almost all of them, considered it as simply impossible. More importantly, the pure possibilities of quasicrystals were accessible to some scientists independently or before Shechtman's formidable discovery of an actual quasicrystal.

Pictures may save us many words. The above-represented two simple pictures *show* what I have just *told* above. Each of these pictures, though quite similar, is independent of the other, as there is no any causal connection between them. The first one is Mackay's optical transformation of the diffraction pattern of an atomic crystalline order that "exhibits local tenfold symmetry,"[15] which at that time could be considered only as a *pure possibility*. The second is the image of a tenfold symmetry of the atomic order of an *actual* quasicrystal that Shechtman saw through an electron microscope on 8 April 1982.

Today, we know that "there is such an animal" and many other quasicrystals exist in actual, empirical reality (not only as synthesized in laboratories but some also in nature[16]). What were considered before Shechtman's great discovery as impossibilities or, at the best (in the case of Mackay), as pure possibilities and possible predictions have been discovered as empirical facts, namely, actualities. The first picture implies that "Alan Mackay ... [was] better prepared for Shechtman's discovery than Shechtman was" (Hargittai 2010, p. 85), as Shechtman wrote down three question marks to express his great surprise and skepticism concerning the tenfold crystalline symmetry detected in the second image, whereas, in the same year, Mackay independently referred to such a symmetry as a possibility that crystallographers should be prepared to face in actual reality.

Hargittai emphasizes—in my metaphysical, panenmentalist translation—that with no awareness of the possibility, namely, the *pure* possibility, of crystalline structure outside the classical crystalline system, materials scientists might encounter quasicrystals "but walk by them without recognizing them." Indeed, in my panenmentalist

---

[14] As a result, the 2010 Oliver E. Buckley Prize of the American Physical Society for an outstanding theoretical contribution to condensed matter physics was awarded to Mackay, Levine, and Steinhardt, "for pioneering contributions to the *theory* of quasicrystals, including the *prediction* of their diffraction pattern" (my italics. A. G.). See: http://www.aps.org/programs/honors/prizes/buckley.cfm . It should be emphasized that Mackay was the first to produce an optical transformation of that pattern, which "exhibits local tenfold symmetry and repeats the shape of the quasi-lattice cells which gave rise to it" (Mackay 1982, p. 612).

[15] Mackay (1982), p. 612. This is the optical transformation of the pattern in figure 4 on p. 611, intended "to model a *possible* atomic structure" (my italics. A. G.). Mackay mentions that the pattern "has local fivefold axes and thus represents a structure outside the formation of classical crystallography and *might* be designated as a quasi-lattice" (ibid., p. 609. My italics. A. G.).

[16] See Bindi et al. 2009; Bindi et al. 2011; and Bindi et al. 2012. Paul Steinhardt is one of the co-authors of these three papers.

publications, I emphasize, repeatedly, that knowledge or awareness of pure possibilities is indispensable to recognize or identify, understand and explain any empirical fact or event. In contrast, excluding some pure possibilities can result in ignoring some of the most meaningful and significant phenomena and in rendering any great discovery impossible. Had Shechtman *conclusively* (and not merely skeptically, initially, or provisionally) thought like Pauling and many others and *decided* that "there is no such creature," he would have excluded the pure possibility of a quasicrystal and he would not have been the discoverer of its actuality. Yet, with no knowledge of the pure possibility under discussion, Shechtman could not explain his discovery, and in order to achieve that he needed the mathematical and other purely theoretical aids of other scientists. Indeed, Denis Gratias, a mathematical crystallographer, helped to devise an improved model for interpreting Shechtman's experimental observation (Shechtman et al. 1984). At about the same time, Dov Levine and Paul Steinhardt provided a purely theoretical pattern that resembled Shechtman's experimental pattern. Note that Levine and Steinhardt, who were informed of Shechtman's amazing discovery of 1982, were the first to *name* this discovery; the title of their paper was "Quasicrystals: A new class of ordered structures" (Levine and Steinhardt 1984). Again, this does not diminish Shechtman's great achievement even slightly; it shows, instead, in what way scientific novelties are achieved and received.[17]

Excluding the possibilities of quasicrystals, Pauling and his followers made one of the major mistakes in the chemistry, physics, materials science, and, particularly, in the crystallography of the twentieth century. Pauling's rejection of Shechtman's discovery is a paradigmatic case that excluding possibilities, especially pure ones, may blind eminent scientists to scientific truths. Until Shechtman's discovery of an actual "quasicrystal" (namely, an actual solid with an ordered non-periodic structure), this possibility was either excluded (by the great majority of crystallographers) or known (by a small minority of them) as simply pure, not yet actualized. Shechtman was the first to show that it is not merely a pure possibility but also an actual one. In any event, before his discovery, almost all of the materials scientists, physicists, chemists, and crystallographers considered such a crystal as simply impossible, and they considered it impossible whether actualities or pure possibilities were concerned. This tremendous change, a paradigmatic change, indicates the highest importance of Shechtman's discovery. Hargittai reports that Pauling received from Shechtman a full report of his experiments; he found these experiments fine, "but not the interpretation of the observations" (Hargittai 2010, p. 86). The interpretation under discussion rests upon mathematical and theoretical individual possibilities, which are pure and which the actualities under observation actualize. The debate was over the identification of the observed solids—are they a result of twinning, some experimental error, or simply a heretofore unobserved or unrecognized ordered non-periodic structure of solids, eventually often called quasicrystals?

---

[17] As Hargittai puts it, "The quasicrystal story shows the complex interrelationship of various branches of sciences and other human activities. There were scientists like Alan Mackay, who were better prepared for Shechtman's discovery than Shechtman was, but, to Shechtman's credit, once he made his discovery, he grew to handling it in a seasoned way" (Hargittai 2010, p. 85).

Levine and Steinhardt's purely theoretical simulated diffraction pattern that resembled Shechtman's experimental pattern *appears* as if demonstrating an actualist-empiricist view, opposing panenmentalism, for in this case it *happened* that an actuality (Shechtman's discovery of the actuality) preceded the discovery of its pure possibility. But this is an appearance only, for no matter what is the way or the temporal order by which we are acquainted with pure possibilities, owing to experience or without it, their independence of actualities and empirical reality is intact. In any event, these pure possibilities have been *a priori* accessible, entirely independently of empirical, actual reality. Indeed, such is clearly the case of Mackay's crystalline theoretical discovery, which was independent of Shechtman's actual great discovery of 8 April 1982. When Mackay published his seminal paper of 1982, the publically known crystalline facts were overwhelmingly against his suggestion to expand the realm of crystalline possibilities beyond the restriction of the accepted 230 crystallographic possibilities of space groups, and all he could suggest then to materials scientists referred, at that time, only to pure possibilities.

# Chapter 10
# Eka-Elements as Chemical Pure Possibilities

**Abstract** From Mendeleev's time on, the Periodic Table has been an attempt to exhaust all the chemical possibilities of the elements and their interactions, whether these elements are known as actual or are not known yet as such. These latter elements are called "eka-elements" and there are still many of them in the current state of the Table. There is no guarantee that all of them will be eventually discovered, synthesized, or isolated as actual. As long as the actual existence of eka-elements is predicted, they cannot be considered as actual but only as purely possible. Given that eka-elements are chemical individual pure possibilities, panenmentalism, at least as a philosophy of science, can gain support as well as an important implication.

The periodic table of the chemical elements has been a grand attempt to exhaust all the chemical possibilities concerning all the elements and their possible combinations, whether they are empirically known or not. The table not only exhausts everything, which chemists have known about chemical elements and their reactions, but it also makes possible to predict the existence of elements—called "eka-elements"—which are not known yet as actually existing but still are theoretically, purely possible (the term "pure possibility" will be defined and clarified below). The prediction includes their reactions with other elements. Note that the prediction under discussion is based upon the assumption that the table can exhaust all chemical possibilities whether these possibilities are actual, known as being actual, or simply chemically purely possible. The prediction under discussion is eventually about the actualities of the relevant chemical pure possibilities; hence, as pure possibilities, eka-elements *per se* are not predicted or hypothetical elements; the predication is about their actualization only. To consider some elements as purely possible means that they can be discovered as actual or be synthesized in the future but this need not be the case. In any event, the relations between their possible existence and the rest of the table are systematically necessary. In other words, as long as the table concerns possible elements, whether actual or not, and their possible reactions as well, it reveals the *theoretical necessity* about them, especially the

A first version of this chapter was published in *Foundations of Chemistry* 18: 3 (2016), pp. 183–194.

necessity of the relations among them. Nevertheless, to the extent that actual, empirical reality is concerned, the actualization of the elements and their reactions is left to the contingency of actual, empirical reality. The same holds all the more for the possible discovery, isolation, or synthesizing of the eka-elements as actual: this discovery or creation, too, is contingent. The elements and the relations concerning them are, to begin with, purely possible as well as necessary, analogously to pure mathematical or logical systems whose physical actualizations or applications must be left to contingency. Yet each of these systems reveals the mathematical and logical necessity of their items and the relations among them. Analogously, the periodic table thus reveals not only the chemical possibilities concerning all the elements and their relations, but also the chemical necessity of them all.

Note that eka-elements are not only expected or predicted natural elements; nowadays chemists and physicists refer to some tens of eka-elements that are planned to be *synthesized*, whether they will be synthesized only or also happened to be discovered eventually as natural, too. As such, we may first consider them as discoveries of pure possibilities and not necessarily as discoveries of natural actualities. Having born in mind such purely possible or theoretical syntheses, chemists and physicists try to actualize them in actual syntheses performed in laboratories and nuclear reactors. *If* such eka-elements were thus to be synthesized in the future, it may be possible to predict their actual existence in nature, too. Whether they eventually happen to exist actually in nature or to remain artificial products only, we already have enough knowledge about their chemical *pure* possibilities. In the case of artificial, synthesized elements, it becomes even clearer that eka-elements should be considered as pure possibilities independently of any actualization, whether natural or artificial, of them.[1] Furthermore, in such cases, it should be clear that at least eka-elements are not natural kinds.

According to Eric Scerri, *A Tale of Seven Elements*, which appeared in June 2013, "The most recently synthesized element was number 117, which was announced in 2010. This means that for the first time, and perhaps for the last time, the periodic table is absolutely complete with not one single gap along any of the rows. This feature will disappear just as soon as element 119 is synthesized since this will signal the start of extremely long period of fifty elements, most of which will never see the light of day" (Scerri 2013a, p. 210, endnote 10; cf. Scerri 2013b). Indeed, on August 2010 Vanderbilt University announced the discovery by a Vandrbilt physicist, Joe Hamilton, and his colleagues from Russia and the USA of "a new super-heavy element [117, which] sheds light on the basic organization of matter and strengthens the likelihood that still more massive elements may form an

---

[1]As radically possibilist and, hence, as anti-actualist, panenmentalism treats pure possibilities as entirely independent of any actualization and actuality. In contrast, Le Poidevin's treatment of the predicted elements as "mere possibilities," which are simply *combinations or re-combinations* of physical actualities, is clearly actualist, physicalist, and ontologically reductionist (Le Poidevin 2005, suggesting "a form of reductionism about *possibilia*" [p. 129]). Thus, my panenmentalist view of the eka-elements as chemical individual pure possibilities, on which chemical actualities supervene, is radically different from his treatment of such "mere possibilities."

'island of stability'—a cluster of stable super-heavy elements that could form novel materials with exotic and as-yet-unimagined scientific and practical applications."[2] Element 117 is the 26th new element that has been added to the periodic table since 1940. The studies of this element test theoretical predictions that elements beyond 112 could have "unexpected positions in the periodic table of elements" (ibid.). The discovery was reported in a paper (Hamilton et al. 2010). The present name of this element is ununpentium (Uup). In the list of the officially acknowledged elements known as actual in 2005, also appear elements no. 116 (ununhexium; Uuh) and no. 118 (ununoctium; Uuo). Element no. 117 (ununseptium; Uus) appeared in the list of 2011.

Very recently (on 30 December 2015), the International Union of Pure and Applied Chemistry (IUPAC) officially announced the verification of the discoveries of elements with atomic numbers 113, 115, 117 and 118. These discoveries complete the seventh row of the periodic table.[3]

Relying on data and theoretical considerations alike, the late Israeli nuclear physicist and chemist of the Hebrew University, Amnon Marinov, and others predicted and discovered the actual existence of the eka element—eka-Th—of the atomic number 122 (Marinov et al. 2010). Its existence was first predicted by the Nobel laureate G. T. Seaborg. Although, so far, some physicists have failed to authenticate Marinov's evidence as to the actual existence of this element (Türler and Pershina 2013, p. 1238),[4] its purely possible existence was established on theoretical grounds (Seaborg 1968, pp. 96, 104, 106–108), and there is still hope that further exact experiments will authenticate Marinov's evidence. Furthermore, as early as 1971, more than 20 years *before* the officially confirmed actual discovery, Marinov and others announced the discovery of the eka-element   eka-Hg—of the atomic number 112 (Marinov et al. 1971),[5] which was also predicted by Seaborg

---

[2] See:   http://www.vanderbilt.edu/magazines/vanderbilt-magazine/2010/08/super-heavy-element-117/

[3] See: http://www.iupac.org/news/news-detail/article/discovery-and-assignment-of-elements-with-atomic-numbers-113-115-117-and-118.html

[4] Moreover, "the claimed observation [Marinov's] of extremely long-lived, high spin, super-or hyperdeformed isomeric states in neutron deficient heavy nuclei, … which was used as an argument to explain the observation of long-lived $^{292}122$, could not be observed in independent experiments" (ibid.). However, see also Dellinger 2010, p. 1290: "Summarizing, we could not confirm the findings of Marinov et al. … using AMS on thorianite samples, nor could we find a superheavy element with A = 292 in commercially available $ThO_2$ powder. Considering the importance of Marinov's findings for the stability of the heaviest nuclides, we plan to remeasure the material used by Marinov et al. … with the AMS method described in this paper."

[5] For a similar experiment, made by the GSI physicsts in 1990, "detected fission fragments that probably came from element 112" see Emsley 2011, p. 144. This team "considered the results not to be fully convincing, although this work is revealing because the tentative results seem to support Marinov's 1971 claim" (ibid.; cf. p. 588, mentioning "a similar, but not exact, experiment" with regard to eka-element 122). According to Emsley, "the cold fusion method was developed in 1973–4 by Yuri Oganessian and Alexander Demin …, although the first person to use it was Amnon Marinov in 1971 (see copernicium)" (ibid., p. 75). Though mentioning the fact that Marinov was not officially declared as the first discoverer or producer of this element (eka-Hg,

(Seaborg 1968, p. 94–95, to which Marinov et al. 1971, p. 464 refers).[6] Like those of other highly innovative scientists, Marinov's discoveries met with harsh doubts and even ridicule, which was at first the same fate, for example, of the great discovery of quasicrystals by Dan Shechtman. All such doubts and ridicules actually based on rejection or exclusion of pure possibilities (and of experiments and observations as to their actualities as well) which should have been treated quite differently by the scientific community had it not occasionally been so dogmatic and narrow-minded. There are some similarities between Shechtman's discovery and Marinov's great discoveries.[7]

The whole very idea of superheavy elements and the "islands of stability"—that is, stable superheavy nucleus isotopes—within them, consists of expanding, saving, or opening up the chemical pure possibilities that the periodic table can or should comprise. The very ideas, the pure possibilities, of such islands of stability were at first—at least until the discoveries of Marinov and others—considered as almost

---

Z = 112), which hence Emsely names "element of doubt" (ibid., p. 144), in a concluding table of the elements discovered and synthesized in the 1900s, Emsley considers Marinov as the *first* discoverer (or producer) of copernicium in 1971, whereas Hofmann and his colleagues are mentioned in the same table as the producers of this element, 25 years later, in 1996 (ibid., p. 656).

[6] "Elements 104 through 112 should be formed by filling the 6d subshell and should have chemical properties analogous to the elements hafnium (atomic number 72) through mercury (atomic number 80) although these elements formed by the filling of the 6d subshell may exhibit a greater variety of oxidation states than their lighter analogs" (Seaborg 1968, pp. 94–95).

[7] I am grateful to Yona Siderer for drawing my attention to Marinov's amazing discoveries. Rama Marinov-Cohen provided me with vital details about the achievements of her late father. Is this one more tragic scientific story? Some eminent scholars and experts strongly believe that Marinov should be acknowledged as the first discoverer of eka-elements number 112 and 122. Such an acknowledgment would naturally endow him with *two* Nobel prizes for discovering *two* new elements (as in the case of Madame Curie), had he lived to gain this acknowledgment, which he officially did not receive until his death in 2011. There is still some room to seriously doubt the official decision of the scientific establishment according to which the "priority of discovery of … element 112 … was assigned to Hofmann et al. … in … 2009" (Türler and Pershina 2014, p. 1238; this paper does not even mention Marinov's report about the discovery of element 112 and his paper in *Nature* of 1971!). Nevertheless, of a special, even decisive, weight is the following: "There are two independent claims for the discovery of element 112: The claim by Hofmann et al. from 1996 and the older claim from 1971 by Marinov et al. This Comment will not challenge the experimental results of Hofmann et al., but it will discuss one aspect of the claimed discovery of element 112 by Marinov et al., as their experiment has never been reproduced in exactly the form in which the original experiment has been carried out. The reasons for this deficiency may not be found in the field of science, but possibly in radioprotection restrictions for researchers who want to carry out such an experiment. However, such is not a sufficient reason to exclude the original claim from all considerations of the responsible international authorities, who have to settle such priority questions. It may be in agreement with scientific traditions, that when the responsible international committees do not feel to be able to come to a positive decision on the '1971' claim, they could keep the priority problem unsettled for the time being" (Brandt 2005, p. 170). Are we entitled to say that for Marinov and for unbiased science as well, scientific justice was not carried out in his case? Scientific dogmatism, narrow-mindedness, and politics gained the upper hand in some tragic cases, which, like the case of Shechtman until the death of Pauling, have been strongly linked with the dogmatic exclusion or denial of vital scientific pure possibilities and of the experiments and observations concerning their actualities.

actually impossible. These chemical pure possibilities have been discovered by means of theoretical considerations, using scientific imagination, even scientific speculation, and based on original meticulous reasoning and calculations. In contrast, actual discoveries—discoveries of actualities—can be made only by means of empirical observations, experience, or experiments. Sometimes, because of the lack of knowledge concerning the theoretical grounds and explanation for the discovered entities or, better, because there was no epistemic access then to the pure possibilities that these entities actualize, the actualities under discussion were not well understood and even unexplainable. Without access to the relevant pure possibilities, no established identification of their actualities is possible. Such was *at first* the case of the discovery of the quasicrystal by Shechtman and such was *at first* the case of the discovery of element 112 by Marinov. Marinov could not explain the stability of the discovered element, a stability that could not be explained or understood by the theoretical knowledge at the time. Only after the connections between isomers and superheavy elements had been established, could these strange or unexpected discovered entities be understood and explained. In a similar vein, Shechtman could not at first explain his discovery, which was quite incompatible with the crystallography accepted at the time. Only after Steinhardt and Levine (whom Alan Mackay had preceded, as was Penrose's mathematical solution for the relevant problem of tiling) had laid down the theoretical grounds—better, the discovery of the relevant pure possibilities and their relationality—to identify, explain, and understand the existence of quasicrystals (the name that Steinhardt and Levine first dubbed!), could Shechtman and other crystallographers understand and explain his great discovery.[8]

In any event, the theoretical expansion of the periodic table beyond uranium, including the superheavy eka-elements and the pure possibilities of the islands of stability in them, consists of saving and expanding *chemical pure possibilities*. Marinov attempted to save the pure possibilities of *stable natural* super-heavy elements—112 and 122—and to demonstrate their actual existence as well. His attempts met two kinds of objections and doubts—regarding the pure possibilities of such elements as well as his actual findings.

Nevertheless, Scerri writes: "Countless revisions of the table followed, but all of them had holes—until now. With element 117, the periodic table is complete for the first time" (Scerri 2013b, p. 70). But this need not be the case. Rows can be always be added to the table to include more and more chemical possibilities. In this sense, the periodic table cannot be considered as absolutely complete, and Scerri acknowledges that whether there is an end to the periodic table or not is a mute point (Scerri 2013b, p. 73). Otherwise, the fantasy (or "dream") of Kant and other philosophers that well-established science (or the sciences of Human Reason) can be complete, this fantasy would be reborn, though since the middle of the nineteenth century, scientists have realized that such cannot be the case and that there is no end to scientific progress and no exhaustion of scientific possibilities, not only in theoretical

---

[8] See: Gilead 2013 and the relevant references, especially Istvan Hargittai's publications concerning the discovery of quasicrystals.

physics but also in logics and mathematics. Perhaps, the dream of the completion of the table, the dream that Scerri rightly ascribes to Mendeleev, is quite similar to that obsolete dream or fantasy.

According to Scerri, "even as Mendeleev's creation has filled up and scored its successes, [the discovery of new super-heavy elements] may have begun to lose its explanatory and predictive power" (ibid), for the chemical periodicity, which has so far endowed the table with its predictive power, may be lost. But perhaps not, as Scerri mentions that it is not yet clear whether the principle that elements in the same column in the periodic table behave similarly remains valid for superheavy atoms, which leaves such an option possible. Be that as it may, I counter-argue, in any event, the chemical merit of discovering new superheavy eka-elements is certainly reserved to the extent that *chemical pure possibilities* (a distinction that Scerri seems to ignore) and their predicted actualities are concerned. As for chemical periodicity, time will tell, as so far we have known only a part, even a little part, of the elements populating the new rows. Thus, Scerri's doubts are at least premature. Time will tell, whether he is right or wrong about that, but, surely, for the time being, we know about a very small part of these possible super-heavy elements, so it is simply premature to tell anything well-established about their periodicity. This situation is clearly different from the one that Mendeleev and many others had to face—until the recent developments, the eka-elements in any row and column have been quite few in number in comparison with the actual elements, and the periodicity was valid for them and for much many elements known then as actual. Such is not yet the case with the discovery of the new superheavy eka-elements.

In the event that Scerri's doubts and criticism being accepted, still the story of the eka-elements until what he calls the "completion of all the gaps in the periodic table" is that of the discovery of chemical *pure* possibilities. This holds true even if in the future super-heavy eka-elements are discovered as being entirely unstable and thus will not gain any chemical significance but remain as actual physical possibilities only. Indeed, "super-heavy elements also tend to be very unstable, decaying into lighter elements in a fraction of a second" (Scerri 2013b, p. 73), but thanks to the discoveries of Marinov and others,[9] there are also long-lived isomeric states in some of their atoms (in addition to some other atoms which are not super-heavy and which are isomeric). Scientists search for more and more evidences for such states and for "islands of stability" in them. Until 2012, 14 new long-living super-heavy elements were discovered by the Proton-21 laboratory and one discovered by Amnon Marinov (Kostyghin 2012). Such stability has various far-reach chemical implications. Hence, we should doubt whether the alleged instability excludes or will exclude super-heavy elements from being included in the periodic table as chemically significant and as subject to periodicity and prediction. In this case, too, only time will tell and there are well-established theoretical grounds that these elements will have a chemical significance and hence should be appropriately included

---

[9] For instance, Marinov, Eshhar, and Kolb 1987; Marinov et al. 2003; Jones 2002; Herzberg et al. 2006; Türler and Pershina 2013; and Moody 2014.

in the periodic table. Though the predictions of Marinov and others, which state that some super-heavy elements might be unusually stable, faced grave doubts (Brumfiel 2008), since then the perspective has become much more optimistic. As Matthias Schädel and Dawn Shaughnessey claim, enlightening us in a beautifully figurative language,

> …the term 'superheavy elements' was first coined for elements on a remote 'island of stability' around atomic number 114 … . At that time this island of stability was believed to be surrounded by a 'sea of instability'. By now … this sea has drained off and sandbanks and rocky footpaths, paved with cobblestones of shell-stabilized *deformed* nuclei, are connecting the region of shell-stabilized *spherical* nuclei around element 114 to our known world. (Schädel and Shaughnessey 2014, p. x)

The isomeric states in some atoms of the super-heavy elements have some deep panenmentalist implications. According to panenmentalism, no two individual pure possibilities can be identical, otherwise they would have been one and the same possibility. The law of the identity of indiscernibles is strictly valid for individual pure possibilities, as they are not subject to spatiotemporal conditions, hence it is impossible to distinguish between two individual pure possibilities by "locating" them in space and time. Based on this principle, no possibilities of two atoms, as purely possible or as actual, of any element can be identical. As an individual pure possibility determines the identity of the actuality of this possibility, the law of the identity of indiscernibles is derivatively valid for actualities, too. This does not mean that isomeric states are valid for all atoms, but it means that it can be valid for some of them. Physical pure possibilities are more similar to one another than are chemical pure possibilities, and biological pure possibilities are even less similar to each other than are both these kinds of pure possibilities. The reason is that a higher grade of complexity entails a higher grade of individuation. On these grounds, from a chemical perspective, atoms should be more diverse than and different from a physical perspective.[10]

---

[10] "The same and not the same" and the question of identity are perfectly applied to individual pure possibilities and to their actualities. The Nobel laureate of 1981 in chemistry, Roald Hoffmann, has interestingly referred to the question of identity, similarity, and difference in chemistry (Hoffmann 1995; see especially pp. 26–31, concerning the problem of identity in chemistry, the periodic table, complexity, diversification, and isomerism). From a panenmentalist perspective, there are *relational* similarities between individual pure possibilities on the basis that no two of them can be identical but each *relates* to all the others: because each pure possibility is necessarily *different from* any other pure possibility, each pure possibility necessarily *relates to* all the others. On the basis of this mutual universal relationality of differences, *similarities* necessarily appear. From a chemical perspective, there are surely two identical molecules of water, whereas the "number of different isotopomers of hemoglobin is astronomical" (Hoffmann 1995, p. 34); hence, "the chances of two tiny hemoglobin molecules … being exactly the same, in every isotopic detail, are very, very small!" (ibid., p. 35; in this context, Hoffmann mentions isotopomer abundance and Henning Hopf's "individualization of compounds"). Hoffmann's concluding answer to the question "Are there two identical molecules?" is thus: "No, for a really large molecule, probably there are no two identical molecules in that Burmese cat" (ibid.), though chemically or biologically such differences may or do not really matter. Nevertheless, from a *metaphysical* (at least panenmentalist) perspective, no two water molecules can be identical, despite their many physical and chemical

Scerri's concern—"the question of special relativity's effect strikes at the very heart of chemistry as a discipline. If the periodic law does lose its power, then chemistry will be in a sense more reliant on physics, whereas a periodic law that holds up would help the field maintain a certain level of independence" (Scerri 2013b, p. 73)—should be taken very seriously. Nevertheless, as early as 7 December 2004, Kolb and Marinov reported about their analysis on grounds of the relativistic calculations of the electronic structure of element 112 (eka-Hg) of the assumption that this element 112 actually behaves like Hg in the chemical separation process.[11] This is one example of the possibility that relativistic calculations concerning the structure of the atoms of the super-heavy elements need not compromise chemical independence, periodicity, and predictability.

Scerri's comment on the chemical properties of element 112 is of a great interest: "If element 112 truly behaves like a metal, it will bind to gold. If it is more like the noble gas radon, it will tend to deposit on the ice. To date, different laboratories have obtained different results, and the situation is still far from settled" (Scerri 2013b, p. 73). Again, this is an empirical question and time will tell about settling it, but, the opening up or saving of quite new possibilities, some of which are unexpected and surprising, reveals the vitality and abundance of chemistry and its independence as well. Finally, all of Scerri's concerns as to the impact of relativistic and quantum-mechanical calculations on the independence of chemistry should not compromise its predictability and periodicity, for we should not ignore the fact that chemistry mainly deals with *macroscopic* phenomena which are not affect by the considerations concerning quantum states which, in definition, are microscopic. In summary, the discoveries of Marinov and others concerning eka-superheavy-elements do not compromise in any way the independence, periodicity, and predictability of chemistry; on the contrary—they support all these vital properties of chemistry as they have been familiar to us since the time of Mendeleev until the present.

On 17 November 2011, the Research Magazine of the University of Oslo, *Appolon*, announced the world of Jon Petter Omtvedt's (professor of nuclear chemistry at the University of Oslo) plan or "hope" to extend the periodic table with elements 119 and 120.[12] Omtvedt explained that he and his colleagues were "working right at the cutting edge of what is experimentally possible. In order to study the heaviest elements, we have to stretch the current technology to its utmost and even a little further" (ibid.). The goal was to induce a titanium atom to fuse together with a berkelium atom. As titanium has an atomic number of 22, and berkelium has an atomic number of 97—together, these two atoms have a total of 119 protons; i.e. exactly the right number to create an atom of element 119. Omtvedt declared then:

---

similarities. Panenmentalistically speaking, the law of the identity of indiscernibles is valid for pure possibilities and, derivatively, for their actualities as well. From such a metaphysical viewpoint, the differences between any two entities *really matter*. I would like to thank Robin Hendry for drawing my attention to Hoffmann's book.

[11] See: http://arxiv.org/pdf/nucl-ex/0412010v1.pdf

[12] See: http://www.apollon.uio.no/english/articles/2011/element1.html by Yngave Vogt; last modified on 1 Feb 2012.

"One of the biggest and most exciting questions is to find out how heavy an element we are capable of creating. Even though it is extremely difficult to create elements 119 and 120, we do not believe that these elements will be the end of the periodic table" (ibid.).

Before the announcement, in October 2010, of these planed experiments, a Finnish chemist, Pekka Pyykkö, at the University of Helsinki, had announced about a mapping out of an extended periodic table with 54 predicted, eka-elements. He "has used a highly accurate computational model to predict electronic structures and therefore the periodic table positions of elements up to proton number 172—far beyond the limit of elements that scientists can currently synthesize" (according to James Hodge's report in the Chemistry World blog). Pyykkö published his prediction in a paper (Pyykkö 2011; cf. the comment in Scerri 2013b, p. 73).

As Hodge notes,

The extra 54 super heavy elements predicted by Pyykkö may exist under extreme conditions with very short lifetimes owing to radioactive decay, but have not yet been synthesized. Pyykkö says that the value of the work is in showing "how the rules of quantum mechanics and relativity function in determining chemical properties." He gives the example of the potentially record-high oxidation states his work predicts.

Hodge adds that as an expert in electronic structure theory, Peter Schwerdtfeger at Massey University in Auckland, New Zealand, comments: "Pyykkö has used relativistic calculations to go beyond the known elements into unknown territory." Hodged cited Pyykkö's conclusion: "It is hard to say how far experimentalists will get during this century, maybe close to 130, if not more. Although the experimental results may never appear, the basic physics of the problem are sound."

One may criticize such and other predictions as "science fictions," nevertheless, such criticism has been refuted in the history of science: for instance, Rutherford's rejection of Szilard's prediction concerning the possibility of physical chain reactions (Gilead 2014c; Rutherford characterized "such expectations" as "the merest moonshine"), and there is the similar fate in which the possibility of chemical chain reactions had been also rejected at first (ibid.). Pauling's rejection of the pure possibility of quasicrystals and all the more so of their actual existence is another famous example (Gilead 2013). Pauling rejected the existence of quasicrystals as impossible and of their so-called discoverers as quasi-scientists, but as everybody knows by now this is not regarded as one of his merits, to say the very least. These are but a few examples of which there are many others. In the long history of modern science, not a few scientists who have dared to think independently, originally, undogmatically, and "outside the box" have had to face harsh doubts and even ridicule. Another point: science fiction may not be scientifically insignificant or useless. Take, for example, Jules Verne's discoveries of the pure possibilities of the Internet or of lunar journeys (Gilead 2010). It is the imagination of scientists that has grasped the pure possibilities of some of the most important discoveries of science, and scientific truthful fictions have played major roles in scientific progress (Gilead 2009). The discovery of pure possibilities is vital for any scientific progress. To reject pure possibilities on dogmatic grounds or on actualist considerations has led to some of the most erroneous and misleading steps taken by scientists, however distinguished.

To be fair to Scerri, in fact he does not ignore the option of discovering new chemical possibilities concerning eka-elements, for he notes:

> With element 121, a wholly new block would start, at least in principle, which would involve orbitals never encountered before: the g orbitals. As before, the new orbital types add *new possibilities* for the electrons and thus lengthen the periodicity, raising the number of columns. This block of elements would broaden the table to as many as 50 columns (although chemists have already devised more compact ways of arranging such an expanded table). (Scerri 2013b, p. 70; italics are mine)

Nevertheless, he ignores the status of these elements as *pure* possibilities, which may open new horizons for chemistry in general and for the periodicity and predictability that the periodic table may further achieve. As he rightly points out, there is no conclusive evidence regarding chemical reactions involved with super-heavy elements. Time will tell about that, too.

To return to my panenmentalist analysis, while experimental results have to do with actualities only, *pure, theoretical* physics, or "the basic physics of the problem" has to do with purely physical individual possibilities and their relations (in a word, relationality). Purely physical individual possibilities are more basic than purely chemical individual possibilities[13] (while purely mathematical possibilities are more basic than purely physical possibilities), but the gist of the matter is that they are individual pure possibilities that can be discovered as actualities but not necessarily so, for their actualization depends on physical and chemical circumstances and technologies, which, unlike pure mathematics, pure physics, and pure chemistry, are simply contingent. In contrast, pure mathematics, pure or theoretical physics, and pure chemistry are concerned with necessary connections and relations. Today, pure mathematics embraces also computational models, as such models help mathematicians to solve highly theoretical problems and, as models, they consist of individual pure possibilities and their general relations—their relationality—or constructions. Such models may help greatly in discovering yet-unknown *pure* possibilities and also in help predicting their actualities. All the more theoretical or pure, in my terms, is the use of "relativistic calculations to go beyond the known elements into unknown territory," as Peter Schwerdtfeger nicely put it (see above). In my panenmentalist terms, it is the use of relativistic calculations to go beyond the known actual elements into actually unknown territory of novel purely chemical possibilities unknown so far as actualities. The Dirac-Fock calculations solve the Hartree-Fock or Dirac-Fock equations in the algebraic approximation. Such calculations are inevitably valid for pure, theoretical possibilities and their relationality. Hence, Pyykkö's explicit use of these calculations certainly concern pure possibilities. In the philosophy of mathematics and science, the applicability, such as that which Dirac notably made or "manipulated," of pure mathematics to actual-physical, empirical reality is

---

[13] Note that this by no means renders chemical pure possibilities reducible to physical ones. Panenmentalism avoids reductions as much as possible, for it adheres to pluralism (which should not lead to relativism) and to the abundant nature of pure possibilities.

a major question. Some thinkers and scientists, such as Eugene Wigner, mentioning the "unreasonable effectiveness of mathematics in natural sciences," have believed it to remain a great mystery. Indeed, on what grounds did Dirac dare to think that a purely mathematical manipulation might result in discovery of a new sub-atomic particle? Mark Steiner, in his *The Applicability of Mathematics as a Philosophical Problem* discusses this question in extenso. My panenmentalist approach to it is quite different—as each actuality is an actualization of an individual pure possibility, and as no actuality can escape having some mathematical properties (such as numbers and geometrical structures), the actualization or applicability to nature of mathematical entities and their relations is valid and sound. Nevertheless, our accessibility to mathematical actualities or applications rests only upon empirical means, whereas our accessibility to the purely mathematical possibilities that these actualities actualize, may rest solely upon our imagination and intellect.

Pyykkö explicitly writes: "the chemistry of the not yet made superheavy elements must rest on a purely theoretical basis. A list of some possible molecules was given … for the range E121 to E164." He also mentions "a systematic mapping of possible new species" and that "a combination of ab initio calculations with the isoelectronic principle and chemical intuition is a useful way to predict new species… For obvious reasons, much of the molecular chemistry of the superheavy elements is based on studies of hypothetical model systems" (Pyykkö 2012). Except for the term "hypothetical," all these statements are perfectly compatible with my panenmentalist view as I apply it to the periodic table as an achievement of pure chemistry.

By "pure chemistry," I mean precisely what the periodic table has so wonderfully achieved. First of all, the table as a whole refers to all the chemical pure possibilities concerning chemical elements, their reactions, properties, and necessity (in my panenmentalist view, the necessary is a kind of, a part, of the possible, while the actual also takes part in the possible, as all actualities are actual possibilities). According to panenmentalism, pure individual possibilities are also identities. Thus, the pure possibility of the element of the atomic number 1, hydrogen, is the identity of each actual atom of hydrogen. In other words, thanks to our accessibility to such a pure possibility-identity, we can identify an actuality of hydrogen as such. The same holds true for any of the elements and eka-elements—when or if we encounter it as actual, we can identify it only because of our prior acquaintance with its pure possibility-identity. Otherwise, we could encounter an actuality of an element without recognizing or identifying it at all. Without a prior acquaintance with its pure possibility as an element or as an eka-element and without the periodic table as a whole, we actually could not discover, isolate, or synthesize an actuality of an element that so far has not been empirically known at all.

Before concluding, let me answer two questions.

1. Why do I not call the elements in the periodic table "abstract entities"?

As I have argued above, elements are, first of all, chemical pure possibilities. As pure possibilities, they are *individual* existents, existing independently of any actualization and of our mind as well and which are discoverable, not inventible. As individual existents, they are specific and concrete, bearing specific properties, and, in this sense they should not be considered as abstract, universal or general, entities. In contrast, abstract entities are general and not specific or concrete.

The individuality, specificity, and concreteness of the element as a chemical pure possibility are not compromised by the shared locus or topos of the element—this locus at the periodic table is shared by the individual pure possibilities of each of the atoms of this element, bearing the same atomic number and thus the same locus.

Furthermore, to the extent that chemistry (and not physics) is concerned, elements are irreducible fundamental entities or existents. No abstract entity can be considered as an irreducible fundamental entity or existent.

2. What is the difference between atoms and elements?

An element is the chemical pure possibility-identity of an atom, which is the actuality of this element. Yet, as actuality, any atom is also an actualization of its physical pure possibility.

In conclusion, as long as the actual existence of eka-elements is predicted, they cannot be considered as actual but *only* as purely possible.

# Chapter 11
# Quantum Pure Possibilities and Macroscopic Actual Reality

**Abstract** This Chapter treats quantum pure possibilities as individuals, existing independently of any observer or mind. These pure possibilities are also absolutely independent of any metaphysical or logical view that endorses the notion of possible worlds. In my view, the relationship between quantum possibilities and classical physical reality is not between reality as such, as it is in itself, and its phenomena. It is rather between fundamental or primary reality, consisting of quantum individual pure possibilities, on the one hand, and its actualization in what classical physics has discovered so far, on the other. As individual pure possibilities, quantum entities must be, at least ontologically, distinct and different from one another, regardless of the epistemological standing of quantum physics. Hence, quantum metaphysics is committed to the principle of the identity of indiscernibles. I also analyze the two-slit experiment, interference, and entanglement in the light of my approach.

Some of Yakir Aharonov's ideas, concerning quantum mechanics, are purely philosophical. His novel ideas of retro-causation, in which the future determines the present, those of free choice and free will, and those of the final state of the universe are gaining support and clarification by independent and antecedent ideas that philosophers had suggested. Aharonov's independent conception of a temporal direction from the future can be paralleled to quite a similar idea in Kant's moral philosophy in the light of a variation on the theme of this idea (which is my original contribution in interpreting and elaborating on Kant, introduced in 1985 to treat this major idea in Kant's philosophy). I called this variation "teleological time." As for the existence of free will, in the light of a panenmentalist, realist approach about individual pure possibilities, rejecting actualist prejudices against their existence, alternative possibilities are always open for our free choice, even in a deterministic reality. Analogous to the fact that there is no illusion of pain, there is no illusion of free will, and the existence of free will must be compatible with the new physics that Aharonov surmises. These are striking examples of interesting meeting points of philosophy and physics.

© Springer Nature Switzerland AG 2020                                      185
A. Gilead, *The Panenmentalist Philosophy of Science*, Synthese Library 424,
https://doi.org/10.1007/978-3-030-41124-4_11

## 11.1  Introduction

The idea that quantum reality (or "quantum realm") has to do with some kind of possibilities is not a new one. Quantum probabilities certainly have to do with possibilities. Some physicists and philosophers of physics speak of such possibilities as potentialities, propensities, tendencies, or dispositions.

The first scientist-philosopher who considered quantum reality as consisting of possibilities was Werner Heisenberg. Nevertheless, he did not distinguish between pure possibilities and Aristotelian potentialities (or "tendencies"):

> The probability function combines objective and subjective elements. It contains statements about possibilities or better tendencies ('potentia' in Aristotelian philosophy), and these statements are completely objective, they do not depend on any observer; and it contains statements about our knowledge of the system, which of course are subjective in so far as they may be different for different observers. In ideal cases the subjective element in the probability function may be practically negligible as compared with the objective one. The physicists then speak of a 'pure case'. (Heisenberg 1959, p. 53)[1]

The present Chapter differs from any of the previous views in discussing quantum pure possibilities as individuals, existing independently of any observer or mind and which are real as much as actual entities (actualities) are. These pure possibilities are absolutely independent of actualities, whereas potentialities (or propensities, tendencies, and dispositions) depend on actualities from which they drive their possibility and identity. For instance, the seed is a potential plant, and the seed drives its nature and identity from the actual plant. Hence, existing independently of and prior to the physical-actual, pure possibilities are not potentialities, Aristotelian or otherwise. In my view, however, individual pure possibilities, not potentialities, maintain "fundamental ontic status." Such is the case because individual pure possibilities are primary or fundamental entities whose existence does not depend upon any spatiotemporal and causal condition or restriction and upon anything actual, and because individual pure possibilities determine or fix the identities of all individual actual entities, that is, actualities. The existence of actualities, in contrast, depends upon spatiotemporal and causal conditions or restrictions.

---

[1] Cf.: "The probability function does—unlike the common procedure in Newtonian mechanics—not describe a certain event but, at least during the process of observation, a whole ensemble of possible events. The observation itself … selects of all possible events the actual one that has taken place. … Therefore, the transition from the 'possible' to the 'actual' takes place during the act of observation. … the word 'happens' can apply only to the … physical, not the psychical act of observation, and we may say that the transition from the 'possible' to the 'actual' takes place as soon as the interaction of the object with the measuring device, and thereby with the rest of the world, has come into play; it is not connected with the act of registration of the result by the mind of the observer" (op. cit., p. 54). Citing a part of this paragraph, Dorato 2011 replaces "possible" by "dispositional" (or "possibilities" by "dispositions" or "propensities"). Cf. Suárez 2007, pp. 423. According to the modal interpretation of Lombardi et al. 2011, propensities belong to the realm of possibility and they produce definite effects on actual reality even if they never become actual. In their view, probabilities measure ontological propensities, which embody a possibilist non-actualist possibility. Cf. Lombardi and Castagnino 2008, pp. 391 ff.

Primary, fundamental, or basic entities are those that are absolutely irreducible to other entities and without which nothing could exist. Anything that exists is possible, otherwise it could not have existed. Anything that exists is thus an existing possibility, actual or purely possible. Given that my main interest in the modal term "possibility" or "possibilities" is ontological and that I defend a realist approach to individual pure possibilities, I refer to possibilities as substantive modal entities or objects (to be distinguished from modal properties) rather than to modal operators or quantifiers. Given that, what makes an existent possible from the outset is the individual pure possibility of that existent. What makes an existent possible from the outset is what distinguishes it, ontologically and epistemologically, from any other entity, purely possible or actually possible.

To remind you, the individual pure possibilities discussed in this book are mainly mind-independent and hence they are quite different from the "pure possibilities" that some phenomenologists, to begin with Edmond Husserl, have assumed and which are clearly mind-dependent or ideal entities. "Pure", in my terms, indicates an exemption from any spatiotemporal and causal property, condition, or restriction, whereas "actual" implies yielding to such properties, conditions, or restrictions. Furthermore, "pure" means exemption from, and ontological and epistemological independence of, anything actual.

Even though individual pure possibilities are non-actual, this does not mean that they are not actualizable. Furthermore, an individual pure possibility is a non-actual or unactualized possibility also to the extent that something must be left of any individual pure possibility that remains pure and not actualized despite its actualization as an actuality. The reason for this is that any actuality could have changed or been different and yet remains one and the same entity, and, thus, no actuality can exhaust all that pertains to its individual pure possibility. No actuality can completely, entirely actualize its individual pure possibility, which, thus, necessarily remains pure, at least to some extent. Any actuality is simply the spatiotemporally restricted part of its individual pure possibility, which is, thus, always more comprehensive or "larger" than its actuality. Hence, the distinction (which is not a separation) between individual pure possibilities and their actualities, whenever such exist in fact, is always maintained.

As I will show below, quantum possibilities are individual pure possibilities that are actualized as physical entities, first as subatomic particles and, at the end, as parts of macrophysical objects, which are subject to classical physical theory, a theory of classical physical objects, which are not quanta. Individual pure possibilities are entirely independent of actual, classical physical reality, of any mind, and of any notion of possible worlds. Quantum pure possibilities are discoverable by our imagination and intellect (as in pure mathematics or pure physics), not by empirical means. Only actualities, physical entities, are discoverable by empirical means. Finally, individual quantum pure possibilities are like pure mathematical objects or entities, which are also individual pure possibilities, existing quite independently of actual, physical reality and of any mind. Hence, the entities of pure mathematics are *discoverable* by means of our intellect and imagination. This can be proved quite briefly: there is a necessity about mathematical pure entities and their relations; they are never contingent, whereas invention is usually subject to contingency and the arbitrary will of the inventor. Hence, mathematical pure objects or entities are discoverable and not inventible. Generally speaking, individual pure possibilities are discoverable, not inventible. They share this quality with mathematical pure possibilities.

Physical reality is vitally dependent upon its mathematical individual pure possibilities and their universal relationality. Pure mathematics, as its name indicates, is independent of any applicability to anything actual-physical, whereas anything actual-physical is vitally dependent upon the applications of pure mathematics. Any physical entity, any actuality, is quantitative and must have quantitative and other arithmetical and geometrical properties. These properties are actualizations of the general or universal relationality of mathematical individual pure possibilities. Without the applications of pure mathematics, quantities and other arithmetical and geometrical properties make no sense. But the reverse is excluded, as pure mathematics is entirely independent of actual properties and of being applied to the actual-physical reality. Even if actual quantities and, hence, the actual-physical reality as a whole did not exist, pure mathematics, as comprising mathematical pure possibilities, would have existed as it is. Pure geometrical or arithmetical entities have all the meaning, sense, and significance they need without relying upon any actual quantities. The reverse, however, is simply impossible.

Furthermore, actual-physical entities are also dependent upon *physical* individual *pure* possibilities, for these possibilities determine and fix the identities of these actual entities. For instance, any actual particle depends for its identity upon its individual pure possibility, otherwise it would not have been possible from the outset. Physical individual pure possibilities are discovered by means of pure, theoretical physics, which our intellect and imagination consturct. Thus, the pure possibilities of atoms and of subatomic particles (such as the Higgs boson, the omega-minus, and the like) were discovered, and necessarily so, independently and, in many cases, before the discovery of the actual particles. The theoretical discovery is of an individual pure possibility that determines and fixes the identity of the actual particle. Actual physical reality is thus dependent upon the pure possibilities that a purely physical theory discovers. Whenever such a discovery is correct and complete, physical-actual reality is as this theory truly describes it, based upon the purely possible grounds of this reality. There are individual pure possibilities that theoretical physics or chemistry discovered in the past and yet these possibilities (for instance, aether, caloric, and phlogiston) happened to remain unactualized and the theories that discovered them proved to be empirically wrong. The contrary is the case of physical individual pure possibilities that have been actualized and empirically confirmed as actualities. In each of these examples, the relevant individual pure possibilities exist independently of the actual-physical reality and of empirical observations. Nevertheless, only experiments and empirical observations can tell whether we conceive the possibilities that happen to be actualized and whether the relevant physical theory is empirically valid.

## 11.2   Quantum Possibilities and Reality

Returning now to quantum mechanics, my view differs considerably also from Abner Shimony's metaphysics of quantum mechanics (Shimony 1978, following Werner Heisenberg), whose definition of potentiality as "a modality that is somehow intermediate between actuality and mere logical possibility" (Shimony 1999, p. 6) is quite obscure. Somewhat less obscure is the following comment by J. S. Bell, referring to an interference of particles mechanics:

> There are places on the screen that no electron can reach, when two holes are open, which electrons do reach when either hole alone is open. Although each electron passes through one hole or the other (or so we tend to think) it is as if the *mere possibility* of passing through the other hole influences its motion and prevents it from going in certain directions. Here is the first hint of some queerness in the relation between *possibility* and *actuality* in quantum phenomena. (Bell 1987, p. 185; the italics are mine)

Contrary to Bell, in my realist view, the pure possibility of passing of the electron through the other hole *really determines* the actual movement of the electron and prevents it from going in certain directions.

Now, call the purely possible "non-actualized possible," and you may find the following quite relevant to our discussion:

> ... modal interpretations interpret quantum mechanics by slightly changing the standard understanding of the modalities "actuality" and "possibility." ... the terms ... that refer to the non-actualized outcomes are not removed from the state of the device. This procedure of removing the non-actualized possibilities is, however, quite standard in statistical theories. ... In modal interpretations the state is now not updated if a certain state of affairs becomes actual. The non-actualized possibilities are not removed from the description of a system and this state therefore codifies not only what is presently actual but also what was presently possible. These non-actualized possibilities can, as consequence, in principle still affect the course of later events. (Vermaas 1999, pp. 26–27)

Although this modal interpretation of quantum mechanics differs from my approach, it does not contradict it. The "effect," better, the significance of the purely possible has a prominent position in Vermaas's description, which thus shares something of importance with my approach to quantum metaphysics. If the actual is phenomenal, observable, and measurable, quantum mechanics, according to Vermaas's modal interpretation, describes reality including its unobservable or immeasurable states of affairs, which Vermaas calls noumenal (ibid., pp. 209–211), namely, not phenomenal, not perceived by the senses but conceived by the intellect. Nevertheless, in my view, the relationship between quantum possibilities and classical physical reality is not between reality as noumenal, as it is in itself, and its phenomena, as it is perceived by us. It is rather between fundamental or primary reality, consisting of quantum pure possibilities, on the one hand, and its *actualization* in what classical physics has discovered so far, on the other.

A declared possibilist approach to quantum theory is that of Ruth E. Kastner. Claiming that "quantum theory is about possibility" (Kastner 2013, p. 2), she further explains:

> ... there is a well-established body of philosophical literature supporting the view that it is meaningful and useful to talk about possible events, and even to regard them as real. For example, the pioneering work of David Lewis made a strong case for considering possible entities as real... In Lewis' approach, those entities were "possible worlds ... My approach here is somewhat less extravagant: ... I wish to view as physically real the possible quantum events that might be, or might have been, experienced. So, in this approach, those possible events are real, but not actual; they exist, but not in spacetime. The actual event is the one that is experienced and that can be said to exist as a component of spacetime... we can think of the observable portion of reality (the actualized, spacetime located portion) as the "tip of an iceberg," with the unobservable, unactualized, but still real, portion as the submerged part. (Kastner 2013, p. 2)

So far so good, even though I would prefer "possible state" to "possible event" as I consider an event as actual (see below). More problematic is the following: "I thus dissent from the usual identification of 'physical' with 'actual': an entity can be physical without being actual" (ibid). In contrast, as the physical and the actual yield the same spatiotemporal and causal conditions or restrictions, they must be identical. Hence, in my view the actual *is* the physical, although there are physical or quantum individual pure possibilities that, unlike actualities, cannot cause or prevent anything physical-actual. But the major difference between Kastner's view and mine lies in the following:

> Heisenberg ... made the following statement: "Atoms and the elementary particles themselves are not real; they form a world of potentialities or possibilities rather than things of the facts" ... This assertion was based on the fact that quantum systems such as atoms are generally described by quantum states with a list of possible outcomes, and yet only one of those can be realized upon measurement. I think that he was on to something here, except that I would adjust his characterization of quantum systems as follows: they are real, but not actual. In his terms, they are something not quite actual; they are "potentialities" or "possibilities." Thus my proposal is that quantum mechanics instructs us that we need a new metaphysical category: something more real than the merely abstract (or mental), but less concrete than, in Heisenberg's terms, "facts" or observable phenomena. The list of possible outcomes in the theory is just that: a list of possible ways that things could be, where only one actually becomes a "fact." (op. cit., p. 36)

As I have said above, I do not follow Heisenberg's notion of potentialities. Individual pure possibilities are not Aristotelian or otherwise potentialities, for these possibilities are entirely independent of actual reality, whereas Aristotelian or otherwise potentialities are clearly dependent upon it (the seed of a plant is the potentiality of the plant, yet, undoubtedly, the existence of the seed depends upon that of the plant that produced this seed). The major problem is with the semi-status of Kastner-Heisenberg's possibilities: although, unlike concepts or ideas or other mental entities, these possibilities are "real" and not mind-dependent, they are not as real as actualities are. In contrast, in my view, individual pure possibilities are as equally real, though differently, as actualities are.[2] Individual pure possibilities are fully

---

[2] At this point, I agree with Lombardi and Castagnino 2008 and Lombardi et al. 2011 that quantum possibilia are as real as quantum actualities are. Nevertheless, *individual* pure possibilities are quite different from those quantum possibilia in various fundamental respects, especially in the complete independence of individual pure possibilities of actualities, propensities, and properties

particular or specific. My approach is *completely* realist about individual pure possibilities, whereas Kastner appears to have a semi-realist view about quantum possibilities, conflating possibilities with potentialities.[3] Kastner's "possibility as physically real potentiality" (Kastner 2013, pp. 149 ff.), following Heisenberg and Aristotle, is, clearly quite different from my realist approach to individual pure possibilities. The status of Kastner's approach as a kind of possibilism or realism about possibilities is not clear and decisive enough.

Aristotle was not a possibilist. On the contrary, he was an actualist all the way. It is not by chance that the Aristotelian unmoved mover is an *actus purus*. This pure form that has nothing potential or material in it indicates the supremacy of form over matter, *energia* over *potentia*, in Aristotle's philosophy. It is the form (as *energia*) that fixes and determines matter (as *potentia*), not the other way round. If Kastner wishes herself to be a full-blown or a determined possibilist (and yet to keep a distance from what she considers as David Lewis's extravagance), this view of potentialities that are more real than mental or "abstract entities" but less real than actualities, is not the right way to take, let alone a clear and decisive enough one. Moreover, contrary to her view, subatomic particles are clearly actualities.[4] If

---

and in their substantiality and universal metaphysical significance as well. In my view, actualities, being time-variant, are also quite different from Lombardi's actualities, being time-invariant or even "timeless" (Lombardi and Castagnino 2008, p. 431).

[3] Indeed, relying upon Heisenberg view on the status of "potentia" as "something standing in the middle between the idea of an event and the actual event, a strange kind of physical reality just in the middle between possibility and reality," Kastner considers her proposal concerning "possibilist realism" (or possibilist transactional interpretation) as "merely chooses to take seriously Heisenberg's concept of potentia and follow it where it leads" (Kastner 2010, p. 91). Cf.: "something standing in the middle between the idea of an event and the actual event" (Kastner 2013, p. 36). In contrast, individual pure possibility is a real, mind-independent possibility, which is equally real as actuality is, even though, unlike actuality, individual pure possibility is exempt from any spatiotemporal and causal conditions or restrictions. Note that, unfortunately, John Cramer himself, the father of the transactional interpretation, does not follow Kastner's possibilist approach to his interpretation. Cramer states: "We note that Ruth Kastner's 'Possibilist Transactional Interpretation' … treats quantum wave functions as being real objects only in an abstract multidimensional Hilbert space, from which transactions emerge in real space. The possibilist approach is not incorrect, but we consider it to be unnecessarily abstract" (Cramer 2015). In contrast, I believe that Kastner is in the right (possibilist) direction, which Cramer, alas, considers as simply "unnecessarily abstract." Yet, unfortunately, her approach is not clear and decisive enough. As I see it, metaphysically speaking, one has to clearly decide between fully-fledged possibilism and actualism.

For elaborating on the idea that an empirical complex reality is an actualization of a virtual-potential quantum state, see Schäfer 2006a, b, 2008, and Schäfer et al. 2009. This approach, too, follows, to some extent, Heisenberg and Aristotle.

[4] Kastner's ontology of subatomic particles may gain support from some physicists and philosophers of physics. For instance, Casey Blood states: "in spite of all expectations, *particles are not necessary* to explain any observation. That is, there is no evidence that photons, electrons, protons and so on exist as particles, separate from the wave function—with 'particle' being defined here as a carrier of mass, energy, momentum, spin and charge localized at or near a single point" (Blood 2011, p.1). Zeh is even more radical over this matter (see, for instance, Zeh 2010, p. 1481). Décio Krause and Otávio Bueno comment on Blood's view in a more moderate way: "We believe that, as

the Higgs boson, for example, had not been discovered as an actuality in CERN (by means of indirect observation), it would have remained simply a theoretical possibility (or, a Heisenberg-Kastner's possibility). In CERN, the Higgs boson was discovered as an actuality, whereas before this discovery, this subatomic particle was predicted and its pure possibility was discovered by means of a theory. Subatomic particles are actualities, whereas quantum entities are quantum individual pure possibilities. These pure possibilities are the necessary fundamental conditions for the existence and identity of the quantum actualities, which are subatomic particles and their states. As macro-physical entities are compounds of subatomic, atomic, and molecular particles, quantum individual pure possibilities are eventually the fundamental possibilities of all physical entities, microphysical and macro-physical alike.

In fact, my view on quantum possibilities is different from all the acknowledged possibilist views known to me, let alone of actualist ones.[5]

---

far as current quantum physics (QFT) goes, he [Blood] is right. But, of course, his criticisms are directed against the standard concept of particle, described by classical physics. However, Redhead's point still stands: some concept of particle remains (whether such particles are thought of as 'epiphenomena of fields,' 'field excitations,' or whatever)" (Krause and Bueno 2010, p. 271). Thus, Krause and Bueno "acknowledge that contemporary physicists deal with a particle concept (although not with the classical one). This is the concept we are concerned with. We will speak of protons, neutrinos, electrons, and quarks as they appear in physical theories (albeit as a secondary species of objects), and we will consider which properties they have, which logic they obey" (op. cit., p. 273). The unavoidable conclusion from Zeh's interpretation of quantum mechanics and from his idea of its universal validation or expansion all over the updated physics makes classical or non-quantum physics simply redundant, as if a relics of what he considers as a scientific religion. I cannot believe that such a conclusion can be really defended. Classical physics and quantum physics are both necessary for our knowledge and understanding of physical reality. Moreover, all macro-physical phenomena are not quantum phenomena and quantum physics cannot describe or explain them. Only by means of classical physics including its theory of relativity, can we adequately grasp and understand the macro-physical reality.

[5] For other, quite different, possibilist approaches to quantum reality consult Lombardi and Dieks 2017, Lombardi and Castagnino 2008, Lombardi et al. 2011, and Suárez 2004. For an actualist approach, in contrast, consult Dieks 2010 and Bueno 2014. Referring to an actualist approach that is also an empiricist one, Bueno comments: "To recognize real modality in nature is to recognize the existence of non-actual, merely possible phenomena, and also to recognize necessary connections in reality. These are, of course, anathema to empiricism ever since Hume" (2014, p. 3). Dieks refers to the possibilists who think that "quantum mechanics makes possibilities as ontologically serious as actualities" (Dieks 2010, p. 119), whom he strongly opposes. This kind of quantum possibilism, which is realist about possible *worlds* (cf. Bueno 2014, p. 6, 8, and 12), is sharply different from the metaphysical platform on which I rest the following discussion. Of great interest is Diederik Aerts's approach, introducing "a well-defined proposal for the nature of a quantum particle, namely that it is not an object but a concept" (Aerts 2010, p. 16). With Aerts's view, as abstract, non-local concepts, existing outside of space-time, pulled into space-time through measurements (op. cit., p. 28), quantum particles are mind-dependent, individual pure possibilities. This view is certainly not compatible with mine. According to Aerts's approach, localization or collapse "means here rendering concrete or becoming more concrete" (ibid.). In contrast, individual quantum pure possibilities, as I see them, are particular and by no means abstract. As Jeffrey Bub claims, "the idea behind a 'modal' interpretation of quantum mechanics is that quantum states, unlike classical states, constrain possibilities rather than actualities" (Bub 1997, p. 173; cf.: "the change in the quantum state $|\psi\rangle$ manifests itself directly at a modal level—the level of pos-

On a metaphysical possibilist platform, I will try to illuminate the quantum reality as consisting of individual pure possibilities and their relationality, a reality which the reality, subject to classical physics, actualizes. This platform—panenmentalism—is realist about individual pure possibilities. Do not let the name "panenmentalism" misleads you; the metaphysics under consideration is both realist and nominalist and by no means idealistic of whatsoever sort. A good synonym for "panenmentalism" is "panenpossibilism". Panenmentalism is a novel possibilist metaphysics that is neither rationalistic, nor empiricist. It is also principally different from Kantianism. In panenmentalist terms, individual quantum pure possibilities and their relationality are not mere thought-constructions; rather they are ontologically robust, at least as actualities are. Panenmentalism is possibilism *de re*, not *de dicto*. Individual pure possibilities are *possibilia de re* (Gilead 2004b).

On the grounds of the above, individual pure possibilities are not ideal-like entities. Nicholas Saunders suggests that "to mathematically single out one state at the act of measurement (or decide the fate of … [Schrödinger's] cat) is to introduce idealike entities into the theory. These have the role of filling up all of the possibilities that exist in the superimposed wave-function" (Saunders 2000, p. 527). As this approach asserts, "whenever a measurement is made, the universe branches out into as many varied versions as there are possible results to that measurement. At the end of the experiment there exists one universe in which Schrödinger's cat is alive and one in which it is dead" (ibid.). Undoubtedly, it is an interesting interpretation but I do not accept its tacit metaphysical basis that possible worlds or universes exist.[6] Instead of possible worlds, the metaphysical view—panenmentalism—to which I refer here stipulates an infinitude of individual pure possibilities, each of which relates to all the others. No need exists to assume a possible world in which Schrödinger's cat is dead and another in which it is alive. The pure possibilities of being dead or of being alive are open to that cat, whether it actually exists in our one, actual world or not. Finally, instead of questionable or obscure "ideal-like entities," panenmentalism refers to individual pure possibilities and their relationality, for each pure possibility necessarily relates to all the others from which it differs. Such relationality is necessary because no two individual pure possibilities could be identical as they are *different from* each other and, hence, they necessarily *relate to* each other. As the principle of the identity of indiscernibles is valid for individual pure possibilities, two allegedly identical individual pure possibilities are, in fact, one and the same individual possibility. Against the background of the above, panenmentalism can suggest a possibilist interpretation of quantum physics without relying upon the idea of possible worlds at all.

---

sibility rather than actuality"). Cf. Tomasz Bigaj's critique concerning "the being of a disposition involves something that is merely possible, not actual" (Bigaj 2012, p. 216).

[6] In the current chapter, I use the notion "possible world" only in the ontological sense, not in the semantic one. Likewise, in his famous paper, "Six Possible Worlds of Quantum Mechanics," John Bell relates the many worlds interpretation to the notion of possible worlds (Bell 1987, pp. 192–193). This paper by Bell and some other papers about worlds or universes in quantum mechanics are included in an anthology devoted to the issue of possible worlds (Sture 1989).

John von Neumann's formulation of quantum mechanics (known as the orthodox quantum theory) also refers to "the multitude of possibilities" (Nicholas Saunders 2000, p. 528). With von Neumann's interpretation, "at the act of measurement, the wave-function, or mathematical description, of the quantum system collapses into only one of the possible outcomes that formed part of the initial superposition" (ibid.). Thus, the complete mathematical description of Schrödinger's cat "maintains the dual dead-and-alive nature up until the point of measurement, at which it is collapsed from these two possibilities into only one" (ibid.). An event is "something that distinguishes between the different possibilities inherent in the superimposed wave-function and results in one particular possibility being selected" (ibid.). This is more lenient to panenmentalism.[7] Note that the possibilities under discussion are, first of all, metaphysical or ontological. Namely, their status is not merely epistemic but only in the secondary sense, relying upon their primary, metaphysical sense. Whether we know these possibilities or not, they exist independently of our mind, thus our mind has to discover them as metaphysical or ontological.

To the extent that our present knowledge is concerned, it is short of a completely adequate knowledge of quantum reality, possibilities, and probabilities. We have two principle ways to explain this short of knowledge philosophically: the Kantian way and the Spinozistic one.

The Kantian way is to argue that we do not know anything about reality as it is in itself but only by means of our forms of knowledge. Quanta, wave-functions, non-locality, collapse, measurement, and the like belong to our epistemic concepts and quantum phenomena. Thus, according to this way, they have nothing to do with the physical reality as it is in itself. They have only to do with the *relationship between* us, as knowing persons, and reality. I do not endorse this view. I think that quantum physics is about reality as it is in itself, otherwise what is the point about the whole debate: If we do not know reality as it is in itself but only as a phenomenon, there is no problem at all in assuming that the observer changes the observed reality. According to Kantian view, any observation changes reality; actually, any observation constructs reality as a phenomenon. Notwithstanding, we have a problem with this view, and we would like to have a possible solution to it.

The Spinozistic view is more plausible at this point. Uncertainty, undecidablity, and measurement are part and parcel of our first kind of knowledge—*imaginatio*—which is simply *inadequate partial* knowledge of reality as it is in itself (*ut in se est*). For Spinoza, mistakes or errors have no absolute ontological status. We mistake

---

[7] Nevertheless, with no collapse. As Dieks puts it: "In accordance with what was just said about the implausibility of collapses, most modern treatments of quantum mechanics do without them. In these no-collapse interpretations (e.g., decoherence approaches, many worlds interpretations, modal interpretations) states that are superpositions of different possibilities are endemic. But as we have just

observed, even if one does accept the occurrence of collapses one must acknowledge that quantum states generally contain the actual and possible alike. This is therefore a typical feature of quantum mechanics" (Dieks 2010, p. 121). Note that Dieks opposes the possibilist interpretations of quantum mechanics and endorses rather an actualist, Humean one (op. cit., p. 130: "according to the Humean view on laws, only actual things and events exist").

or err only because we have partial knowledge, which is a part of reality as it is in itself! Because we know only a part of it, which we mistake for a whole, and because we draw a borderline between this part and the rest of reality, we imagine and misconceive reality, namely, conceiving it only in the first kind of knowledge, *imaginatio*. To emend this, we need to embed that part in the reality as a whole, which is a comprehensive, total system, the whole of nature (namely, Spinoza's substance).

Ontologically and traditionally speaking (say in Newtonian mechanics), physical reality consists of elementary particles, some "atoms" (in Latin "individua," namely "those that are indivisible," or "the indivisibles"), whereas the physical phenomena of this reality consist of sensible data, such as colors, sounds, tangible qualities, smells, and the like. This is the classical-Atomist distinction between *physis*—reality as it is in itself—and *nomos*—our convention in perceiving it: According to the *physis*, only atoms and void do exist, whereas, according to the *nomos*, there are sensible qualities and the like. According to my view, as it is presented in this Chapter, the relationship between quantum possibilities and classical physical reality is not between reality as such, as it is in itself, and its phenomena; the relationship is not between *physis* and *nomos,* but between fundamental or primary reality, consisting of quantum pure possibilities, on the one hand, and its *actualization* in what the classical physics has discovered so far, on the other. What my view and the Atomists' one both share is the assumption that the "foundations" of physical reality consist of elementary entities and their relationships. Yet, the Atomists consider elementary particles as actualities, whereas I consider elementary entities as individual pure possibilities. Subatomic particles are actualities and, as such, they are distinct from the quantum entities, which are their individual pure possibilities. Sub-atomic particles are micro-parts, elementary parts, of the macro-actualities, which are the direct objects of our observations and experiments.

To return to quantum physics, there is no way to prove of course that this scientific achievement has reached a completion. Probably, we have only a partial picture of reality. As individual pure possibilities, quantum entities must be, at least ontologically, distinct and different one from the other, no matter what the epistemological standing of quantum physics is. Since each pure possibility is different from any other pure possibility, in the light of panenmentalism, quantum metaphysics is committed to the principle of the identity of indiscernibles, even if quantum physics, as a theory under this or that interpretation, does not discern their individuality or, else, cannot apply the principle of that identity to the quantized excitations of the field.[8]

---

[8] For the debate about the validity of Leibniz's principle for quantum physics, consult Paul Teller 1983. Teller surveys and analyzes the attempts to subject quantum reality to Leibniz's principle concerning the identity of indiscernibles. His aim is "to further whet the appetite for work on what I think are some of the most fascinating interpretive problems we face today" (op. cit., p. 319). Teller's example of the merger of two waves on the rope into one wave (op. cit., p. 309) does not refute Leibniz's principle, rather the contrary: it justifies it. The *product* of the merger is a *homogeneous* whole whose "parts" are indistinguishable (in fact, it does not consist of parts). Hence, it is a single, one wave and not two combined identical or indiscernible waves. Furthermore, superposition is not a merger of waves (contrary to op. cit., p. 317), and unlike Teller's merger, it comprises the individual pure possibilities (in terms of states or values) that are open to the entity in

Nevertheless, we should not ignore the following comment: "Schrödinger ... wrote of indistinguishable particles as 'losing their identity', as 'nonindividuals', in the way of units of money in the bank (they are 'fungible'). That fitted with Planck's original idea of indistinguishable quanta as elements of energy, rather than material things—so, again, quite unlike classical particles" (Simon Saunders 2013, p. 350). This appears to be the received view, even though Simon Saunders himself mentions an opposing view, considering particles as distinct individuals. As elements of energy, are quanta really indistinguishable, nonindividuals, and fungible? Not at all. As long as quanta are measurable and countable, they are individuals.[9] Only individual entities can be counted and measured. It only appears to be the case that we do not discern any distinction between quanta, when we count or measure them. Notwithstanding this, in fact, whenever we add one quantum to another, whenever we count it, or measure it, we are obviously distinguishing between it and any other quantum. For instance, suppose that it is the 32nd quantum in our count. This makes it *different* from the 31st quantum or the 30th quantum, or from any other quantum in our counting. Each of these quanta certainly makes a difference, as there is a clear difference between counting 888 quanta, 887 quanta, or fewer. Even one quantum makes a difference. For this matter, it is not important if we count a quantum, as permutable, in another order and if we cannot identify it from other quanta after the counting. Even if we know nothing about its identity over time, in counting it, we distinguish between it and any other quantum in our counting. Any click of the apparatus sounds "exactly identical" to another, yet such is not the case, as each click makes a difference, signifying a difference. Any count requires the homogeneity of the counted individual units but, on the contrary, the homogeneity under discussion does not eliminate the individuality of each counted unit. No quantum is fungible, as long as *it*, and not another quantum, is counted and measured. It is *this* quantum that is measured and counted, not another one. The individuality and "thisness" under discussion rest upon the individual pure possibility of the actual quantum, to which we can *refer* and which is the *object*, the individual object, of our counting or measuring. Note that Leibniz's principle converts numerical differences into qualitative ones. If we do not discern any qualitative difference or any difference of properties between two apparently "identical" drops of water, the numerical difference between them, namely, that they are two and not one, is sufficient to discern between them and to conclude that they are not identical. Even though their place is permutable, there is a *real* difference between them and they are not identical. The first is not the second and vice versa. The same is the case with subatomic or atomic particles. The principle of the identity of the indiscernibles is thus valid for quanta and any subatomic or atomic particle. Finally, note that Quine, on whose

---

discussion, while each of these possibilities is a discernible individual. If actualization occurs, only one of these individual possibilities is determined or chosen.

[9] On counting photons or quanta, see, for instance, Fossum 2016. Nevertheless, as for counting photons, we should not ignore the following fact: "the clicking sound produced when a photon is caught by a detector says two things: yes, a particle was detected; but sorry, the way you detected it killed it, and its energy was converted into an electric pulse" (Orozco 2007, p. 872).

theory of identity Simon Saunders relies, was a declared actualist. My possibilist arguments are not affected by actualist ones.

Hence, contrary to Saunders's claim that "indistinguishability is essential to the interpretation of quantum fields in terms of particles" (op. cit., p. 340), I insist that because the individual pure possibility of each particle cannot be the individual pure possibility of another particle, Leibniz's principle of the identity of the indiscernibles is valid for actual quanta and particles, as much as it holds true for their individual pure possibilities (namely, pure quanta). Given that in our observation there is a specific number of quantum actual entities, what determines this number if not the sum of their individual pure possibilities?! To count them—their countability—necessarily requires that each of them has a distinct individual possibility of its own!

Before proceeding further, let me say something about the metaphysical status of wave-function and entanglement.

Some quantum mechanical scientists as well as philosophers of physics assume that entanglement and wave-function reflect adequately *actual quantum reality*. Though these quantum pure possibilities really exist, they do not exist as actualities, namely they do not spatiotemporally exist. My metaphysical interpretation distinguishes between individual pure possibilities, which are quantum metaphysical possibilities, and actualities, which are the objects of classical physics (in the branch of quantum gravity). Thus, as I see them, entanglement and wave-function are not actual quantum states[10] but pertain to quantum individual pure possibilities that may be accepted or rejected on theoretical grounds. On the basis of metaphysical considerations, any actual entity or event—an actuality—is inescapably subject to spatiotemporal and causal conditions, otherwise no actuality can exist (what otherwise would makes it actual, then?). In other words, spatiotemporal and causal conditions are necessary for the existence, recognition, discerning, and identification of any actuality. In contrast, did entanglement and wave-function reflect actual states and entities, could this be compatible with those necessary conditions for any actual existent? By no means, for, on possibilist interpretation, it is impossible to locate spatiotemporally any quantum entity or state and to link it with a causal chain. Only if one assumes, on actualist grounds, that quantum possibilities are, in fact, actualities existing in different worlds or universes, is one allowed to consider quantum states and entities as actualities and, thus, as subject to spatiotemporal and causal

---

[10] Cf.: "If we take $\Psi$ as the 'real' entity which fully represents the actual state of affairs of the world, we encounter a number of difficulties... . it appears to be impossible to understand how specific observed values $q$, $q'$, $q''$, ... can emerge from the same $\Psi$" (Laudisa and Rovelli 2013). Moreover, to cite Lombardi and Dieks, "quantum mechanics does not correspond in a one-to-one way to actual reality, but rather provides us with a list of possibilities and their probabilities... . In modal interpretations the event space on which the (preferred) probability measure is defined is a space of *possible* events, among which only one becomes actual .... By contrast to actualism—the conception that reduces possibility to actuality ... some modal interpretations, in particular the MHI [the modal-Hamiltonian interpretation], adopt a possibilist conception, according to which possible events—*possibilia*—constitute a basic ontological category ... The probability measure is in this case seen as a representation of an ontological propensity of a possible quantum event to become actual" (Lombardi and Dieks 2017).

conditions or restrictions. In my realist view about individual pure possibilities, quantum entanglement and wave-function are not actualities, neither do they reflect actualities. They are epistemic pure possibilities, which reflect ontological pure possibilities and their relationality.

According to this view, classical physics actualizes quantum pure possibilities and their mutual relationality. This means that the quantum realm is wider or more comprehensive than the actual realm of classical physics. This holds true for the realm of pure possibilities as a whole: The realm of pure possibilities comprises the whole realm of actualities, while the realm of actualities is a limited and confined part of the realm of pure possibilities. This part exists within the realm of pure possibilities.

This special relationship of comprising gives my metaphysics the name "panenmentalism" or "panenpossibilism," as panenmentalism presumes that the purely possible is the mental (or the nonphysical, to be distinguished from both physical and psychical), whereas the physical-actual consists of its conditioning and limiting under spatiotemporal and causal conditions or determinations.

To demonstrate that quantum reality consists of pure possibilities and not of actualities, we must first show that, at least under some interpretations, unlike actual quanta or subatomic particles, quantum entities are not spatiotemporally and causally determined or conditioned. Only pure possibilities are not determined or conditioned so, whereas any actuality is spatiotemporally and causally conditioned or determined. We are unable to determine such conditioning insofar as quantum entities are concerned. It is impossible to locate spatiotemporally a quantum entity, or you can never determine how to locate it spatiotemporally.

In other words, my answer to the notable question—Where is the border between quantum physical reality and classical physical one?—is that the border lies where we should distinguish between individual pure possibilities, pertaining to the quantum reality, and their actualities, eventually belonging, in contrast, to classical physical reality. This answer clearly differs from the approach concerning the so-called collapse of the wave-function. It rests, instead, upon the metaphysical distinction between quantum individual pure possibilities (including their relationality) and their actualities in the classical physical reality.

Notwithstanding, there is a strong connection between classical physics and quantum mechanics, otherwise we should divide nature into two separate and independent parts (as if "two Substances"), whereas the whole of our physical theory consists of universal laws, valid for nature as a whole. In its own way, the panenmentalist approach to quantum mechanics implies a unity between classical, non-quantum physics and quantum mechanics. Given that any individual pure possibility and its actuality constitute a *unity* (as they are inseparable), considering quantum reality as the purely possible substratum for classically physical reality, as the actualization of quantum reality, establishes a unity with it. This conclusion should be regarded as physically fruitful. Nevertheless, there is no received physical theory yet to show what the nature of such unity is, even though there is a novel promising suggestion how an adequate solution to this problem may be achieved.[11] In any case, the present state of physics does not exclude the *metaphysical* necessary linkage or conditioning of classical physics and quantum physics.

Panenmentalism has made an attempt to explicit this metaphysical linkage, and like other metaphysical explications, it is universal. I postulate that the quantum reality is *fundamentally all there is physically speaking,* namely, that the ontological "foundations" or "building blocks" of which physical reality *ultimately* consists are *quantum pure possibilities.* These possibilities are individuals, namely the elementary entities (indivisible entities) of which physical reality as a whole consists.

## 11.3  The Relationship Between Quantum Pure Possibilities and Classical Actualities

The relationship or linkage between quantum pure *possibilities* and classical mechanical *actualities* is *actualization.* In other words, all classical physical states are fundamentally actualities of quantum pure possibilities, of micro pure possibilities. In other words, all classical mechanical states are fundamentally quantum possibilities spatiotemporally and causally conditioned or determined. As such, these possibilities are not pure; they are actual instead. In panenmentalist terms, any actuality is subject to spatiotemporal conditions or restrictions. Thus, locality indispensably characterizes any actuality. In contrast, locality necessarily does not characterize pure possibilities, which are necessarily non-local, for pure possibilities are necessarily exempt from any spatiotemporal conditioning. Therefore, in the event that non-locality is indispensible for quantum mechanics, it concerns quantum pure possibilities instead of quantum actualities.

As a result, instead of dealing with "collapse" and some other such terms, we may from now on deal with measurement, determination, and observation of *actualities,* which are observable in the reality of classical physics. No measurement takes place in the micro, quantum reality, as it is in itself; only its actual manifestations are measurable. Though non-locality is inferred from observations of empirical phenomena, what scientists eventually refer to, in such cases, are the quantum

---

[11] Consult a recent fascinating idea put forward by Leonard Susskind (2016) and Maldacena and Susskind (2013). This novel idea suggests considering the two well-known papers by Einstein and his collaborators (Nathan Rosen and Boris Podolsky—ER and EPR respectively) as two *complementary descriptions* of physical reality. The first is the description that general relativity provides us with, and the second is the one with which quantum mechanics endows us. Susskind and Maldacena suggest treating the two papers as ER = EPR, which does not claim that the two irreducible descriptions are one and the same but that they are united or unified into a complete, coherent, and intelligible picture of physical reality. This holds true also for panenmentalism, for the description of reality as consisting of individual pure possibilities and their universal relationality is irreducible to describing it as a system of actualities and vice versa. These two descriptions are united but by no means identical. Such an approach leaves reality as rich as possible and does not commit it to the poverty that reductionism inevitably implies.

pure possibilities and not anything actual, which must be, as any actuality, local (otherwise, what is actual about it?). What we observe are quantum possibilities spatiotemporally and causally determined; nothing purely possible is observable. Anything in the classical physical reality is spatiotemporally and causally determined and conditioned, as classical physics shows. Hence, I suggest that "collapse" should be replaceable by "actualization."

Given that Einstein was perfectly right: God(-Nature) does not play dice, yet this Spinozistic truism holds true not only for classical physics, which, to the extent that physics is concerned, is fully subject to complete causal determinism; quantum reality too is deterministic. As my view clearly demonstrates (Gilead 2009), necessity holds for the purely possible reality as a whole. Thus, it certainly holds for the realm of quantum pure possibilities as a whole. It is *a priori* cognizable, namely, theoretically fixed and discovered, and each quantum pure possibility necessarily relates to all the rest. This is precisely, for instance, the famous involvement of every electron with the state of every other electron in the universe as an "ontological commitment of quantum mechanics" (French and Redhead 1988, p. 245)! This involvement has nothing to do with causal and spatiotemporal conditions or determinations.

Hence, in the classical physical reality there is no influence "from a distance" as the influence is subject to the constant $c$, whereas in quantum reality the involvement of each entity with all the rest is no other than the universal relationality of any pure possibility to all the rest. Nonlocality "in" the spacetime is a result of the nonlocal nature of quantum wave-functions (especially when there is entanglement), and this nature consists of the relationality of individual quantum pure possibilities.

Nonlocality does not appear to be compatible with causality and with actualities as such. It has to do, instead, with the relationality of individual pure possibilities, which, by definition, are not subject to any spatiotemporal and causal conditions or restrictions. As relying upon relationality, nonlocality reveals the ways in which pure quantum possibilities relate to one another. In empirical reality, in contrast, actualities are interconnected by means of causal relations. Nevertheless, this does not preclude nonlocality, whenever quantum individual pure possibilities are concerned. These possibilities relate to one another regardless of any locality, causality, or spatiotemporally. From a panenmentalist viewpoint, each thing is connected with all the others, as each individual pure possibility necessarily relates to all the others. Hence, in empirical, actual reality—the reality of classical physics which actualizes the reality of quantum mechanics—*nonlocality reflects only the relationality of all the relevant quantum individual pure possibilities*.

The non-causal involvement of each electron with the state of any other electron in the universe is beautifully compatible with the panenmentalist universal relationality of each pure possibility, which, owing to its *difference*, necessarily relates to all the others. Panenmentalist quantum metaphysics is thus committed to the principle of the identity of indiscernibles. Furthermore, according to this metaphysics, quantum possibilities, like any other pure possibilities, *must* be discernible *individuals*. This necessity is strictly metaphysical, whereas the physical reality as actual is contingent. In other words, physical-actual reality as a whole could not exist at all, whereas no pure possibility could not exist. Hence, the existence of each pure possibility is necessary, whereas that of anything actual is contingent.

Ontologically speaking, the realm of quantum pure possibilities precedes or is prior to the realm of physical actualities as a whole. In contrast, historically speaking, the discovery of quantum physics occurred well after the discovery of classical physics. Nevertheless, this does not make any difference for the priority of the quantum reality, which is ontologically, metaphysically prior to the physical-actual reality, which is well described by classical physics. If these conclusions are right, as a result, quantum mechanics does not actually describe actual physical reality but, instead, it rather describes the purely possible reality that makes actual physical reality possible. Quantum pure possibilities are the ontological grounds for the actual physical reality.

Are there "quantum events"? Is a quantum "event" an actuality? If it is spatio-temporally and causally determined and if it is measurable or observable (indirectly, by means of the observed consequences or indications), the quantum event is an actuality; if not, it remains purely possible. In such a case, it should not be considered an event at all but merely a determination of a pure possibility. If it is an actual event, it is an actualization or an actuality of a pure possibility. What determines which of the open possibilities under one and the same pure possibility-identity actually takes place must be an actualization. The act of measurement thus gives rise to the quantum event, which means that the *actualization* of either of the possibilities that are open for the cat—being alive or being dead—is eventually determined. Instead of being under either mathematical description, the event is an actualization of either pure possibility comprised in one and the same pure possibility-identity of that cat. Any actual living cat could be a dead one, as the actual existence of the cat is contingent only. The two pure possibilities, that of life and that of death, are open to any cat. The quantum case is different from the classical one, as the quantum case is actualized as a classical case, never as a quantum case. Thus, the quantum case always remains only purely possible. All the observations, directly or indirectly, that physicists make are of the classical, macro phenomena which are, eventually, the actualities of quantum states, i.e. of quantum pure possibilities.

Whenever physicists indirectly observe, measure, and weigh subatomic particles, they are analyzing the particles, isolating, and locating them only because they take part, however tiny in size, in the macro-physical reality, the only reality in which such observations and measurements can occur. We are not subatomic particles, interacting with other subatomic particles; we are macro-physical (and mental) entities, existing in macro-physical reality, in which we perform all of our observations and experiments and which is subject to classical, non-quantum physics. It is only because of the interaction with this reality that quantum observations can be performed and analyzed. Classical physics subjects all macro-phenomena, in which quantum possibilities take fundamental parts, to general relativity. Otherwise, subatomic particles would have no mass and no gravity and they could not have been measured.

An event deserving its name is an actuality, and, with panenmentalism, anything actual is contingent, as necessity belongs only to the realm of the purely possible, *a priori* cognizable. As Saunders writes, "orthodox quantum mechanics consists of two fundamentally different processes: Deterministic evolution under the Schrödinger equation and indeterministic collapse at the point of measurement" to which alone can the concept of an "event" be applicable (Saunders 2000, p. 529). In panenmentalist terms, a quantum event too, if really an event and not a mere determination of a pure possibility, is contingent or indeterminist. For only its determination as a pure possibility is subject to deterministic necessity. Note that according to panenmentalism, necessity pertains only to pure possibilities and their relations (think of pure mathematics consisting of mathematical individual pure possibilities and their relationality). In contrast, the actual is inescapably contingent—anything actual could not have existed rather than existed; anything actual depends on endless actual circumstances. In contrast, as exempt from any spatiotemporality, individual pure possibilities are unchangeable—they could have not been different. Panenmentalism does deals with "determinism of pure possibilities" instead of determinism of actualities, which is, according to panenmentalism, impossible (there are various kinds of determinism, only one of which is causal). The same holds for quantum reality. As consisting of quantum individual pure possibilities and their necessary relations, it is deterministic contrary to actual reality that is inescapably contingent.

## 11.4   The Two-Slit Experiment, Interference, and Entanglement in the Light of Panenmentalism

Each individual pure possibility contains all the possibilities that are open to it as maintaining its identity. For instance, this Chapter could have been created or produced under quite different circumstances—all of which are possibilities that are open to it—and still keeps its identity. The scope of the pure possibilities that are open to such a particular entity is, in any event, more comprehensive, "larger" than its actuality, which contains only the actual possibilities that this Chapter actualized, actualizes, or will actualize. All the rest of its possibilities are non-actual and remain simply pure.

The very same holds true for a quantum individual possibility. All the quantum mechanical possibilities that are open to it as an entity are included in its wavefunction. Under proper circumstances, namely in the two-slit experiment, some of the possibilities that are open to a quantum entity are actualized, and this entity behaves as a wave and not as a discrete particle. But when the circumstances are different, and fewer opportunities of actualization are in fact open to this entity, for instance, when it is measured or when in the experiment there is only one slit, the entity is actualized ("behaves") as a discrete particle. The reason for this is that measurement is, in fact, a quantization, which means that the entity is forced to be actualized as a discrete particle and not as a wave. The measurement quantizes the actualized entity as a discrete particle and exhibits it as if it were isolated from the rest of the possibilities that are open to it. When other possibilities are open to the same entity, for instance, in the opportunity to enter two slits and not one, a part of the wave of possibilities that are open to it is actualized. In such a case, interference occurs, as some of the possibilities that are open to the same entity exclude some others. Binary possible states, though excluding one another, are open to the same quantum individual pure possibility under different actual circumstances. Another kind of excluding possibilities that are open to the same quantum entity is that of incompatible observables.

Incompatible observables pertain to the quantum realm but not to the realm yielding to classical physics. These observables indicate excluding measurements; whenever the measurement of a quantum position is determined, the measurement of the quantum momentum is undetermined and vice versa. Moreover, change in the order of the measurements implies a change in the measured values of the same observable. Considering a measurement as an actualization of the relevant value, it must depend on the actual circumstances under which the measurement is taken. Thus, the different and excluding possibilities of the incompatible observables that are open to a quantum entity as an individual pure possibility can be actualized, in the same actuality, under different circumstances. In any case, only one of such observables can be determined, namely, actualized under the circumstances of a particular measurement.

Entanglement rests upon two different panenmentalist principles—relationality and relatedness. Relationality maintains the difference between the entangled entities and subjects each of them to the principle of the identity of the indiscernibles, whereas relatedness maintains the similarity and the complementariness of the entangled entities. This difference and this similarity have to do with these entities as quantum individual pure possibilities and, hence, they do not rely upon any spatiotemporal and causal conditions or restriction. The information treasured in the entangled entities and which opens for them the possibilities to behave complementarily, despite any possible difference between them, does not depend upon the speed of light or on any means of communication. It depends only on the relationship between the *entangled* entities as quantum individual pure possibilities that are *related*. No spooky action at a distance is involved, for the related (entangled) pure possibilities are not restricted to any spatiotemporal and causal conditions, whereas distance rests upon spatiotemporality and action rests upon causality. These entangled individual pure possibilities are related in such a way that any change in the actuality of one of them must entail a complementary change in the actuality of the other without any involvement of action or causal connection between them. The change in the actuality is due to measurement, which actualizes one of the possibilities that are open to the measured particle. When measured, the other entangled (related) particle actualizes the complementary possibility *open* to it under the actual circumstances. The determination of the one determines also that of the other, not because of any influence or causation, but only because of the entangled-related possibilities of the two entities. The determination of the one fixes that of the other. All the rest is a matter of the actualization of the relevant possibilities. Note that the results of the measurement are unpredictable, for, like any actuality, they are contingent. What is necessary is the existence of the relevant pure possibilities and the (non-causal and non-local) connection (relationality and relatedness) between them, whereas the measurement is contingent. Hence, until the measurement occurs, the relevant determination of the actuality does not occur; once the measurement occurs, the other entity can actualize only the complementary possibility that is open to it under the actual circumstances. The connection involved is only between the relevant quantum individual pure possibilities.

In this way, panenmentalist realism about quantum individual pure possibilities and the understanding of the relationality and relatedness of these possibilities explain away quantum "absurdities" at the metaphysical level.

## 11.5   Yakir Ahronov: Free Will, Time, Teleology, and Kant

### 11.5.1   Aharonov

In a recent paper, elaborating on Yakir Aharonov's innovative ideas of quantum mechanics, Aharonov, Cohen, and Shushi claim:

> We show that within the two-state vector formalism,[12] although both future and past states of the system are known, genuine freedom is not necessarily excluded. We then define and quantify *weak information* that is the kind of information coming from the future that can be encrypted in the past without violating causality. (2016, p. 53)

Furthermore,

> … even in the two-state vector formalism where present is determined by both past and future events, the quantum indeterminism enables free will. (op. cit., p. 56)

And, finally,

> When information about a future event is buried under quantum indeterminism it cannot violate free will. Similarly, encrypted information, such as the one available through weak measurements, does not violate causality. The existence of free will in these time symmetric models was conjectured to resonate with a dynamical notion of time. (op. cit., p. 59)

Let us start with the general conclusions that we can draw from these innovative ideas:

1. There are two types of temporal direction and temporality: (a) from the past to the future; (b) from the future to the present and the past.
2. Temporality of type (a) and (b) both evolve deterministically and they both form the present.

---

[12] "Two-State-Vector Formalism (TSVF) … is a time-symmetric formulation of quantum mechanics, using two state vectors to describe a quantum system instead of the single one in mainstream Quantum Mechanics. When using two boundary states for a quantum system, its evolution seems symmetric in both time directions" (Aharonov et al. 2013, p. 1).

3. Like Aharonov's weak measurements or information, the determination of kind (b) does not violate the causality of the past states and events that are also subject to temporality of kind (a).
4. Notwithstanding the deterministic causality subject to temporality of kind (a), free will and free choices are compatible with temporality of type (a), provided that physics may acknowledge the possibility of temporality of type (b) and does not exclude it.
5. Determination by temporality of kind (b) opens up *possible* alternative choices for determination of kind (a).
6. Hence temporality of type (b) enables free choice and it is compatible with free will. Temporality of type (b) makes it possible for the agent to be partly free from the determinism of the past and it leaves some room for one's free will, as the following is the physical reality in which our freedom of choice is allowed:

In the context of time-symmetric formulations of quantum mechanics, it is argued for many years now [by Aharonov and his co-authors], that God plays dice in order to save free will … It is common philosophical practice to point out the tension between the concepts of free will and determinism.[13] One of the virtues of the Two-State-Vector Formalism is that it gives rise to a new refined version of determinism, which sheds fresh light on the relation between these seemingly conflicting ideas. Within this framework, while both backward and forward states evolve deterministically, they have limited physical significance on their own—the physical reality is the product of the causal chains extending in both temporal directions. The past does not determine the future, yet the future is set, and only together do they form the present. But the existence of a future boundary condition, and its deterministic effect, do not deny our freedom of choice. It is allowed due to the inaccessibility of the data (which was a requirement of causality, as shown). Examining the concept of free will from a physical point of view, we find it must contain at least partial freedom from past causal constraints … Being macroscopical objects composed of many microscopic particles, we enjoy benefits from both worlds: freedom from the quantum domain and determinism from the classical domain, ensured by robustness. (Aharonov et al. 2014, p. 11).

7. The states and events under temporality of type (a) aim at a destination that lies in the future.
8. Temporality of type (a) is subject to destiny.
9. This destiny can be interpreted in teleological terms.[14]

---

[13] Nevertheless, there are philosophers who have endorsed compatibilism, according to which determinism and free will are compatible. Consult McKenna and Coates 2016 (my note. AG).

[14] Eyal Gruss (2000), suggesting a teleological interpretation of quantum mechanics, mentions Aharonov's consideration of weak measurement and weak values. So far, I have not found an explicit mention by Aharonov himself of the term "teleology" in his publications, even though it is relevant to his conception of retrocausality. Instead, he uses the term "destiny". See, for instance, Aharonov and Gruss (2005). Aharonov, Popescu, and Tollaksen (2010, p. 32) point out a fascinating possibility in which the term "*finality*" replaces the term "destiny." Such a term's replacement

10. The aforementioned destination is an end for the events and states subject to temporality of type (b).
11. In concluding, free will is by no means an illusion. Free will and free choices are compatible with physical reality despite its deterministic nature.[15]

Even though Aharonov's abovementioned innovative ideas concern quantum mechanical topics, they are applicable to the realm of classical physics on the one hand and to morality and human actions on the other.

## 11.5.2   Kant's Duality Problem: Natural Causality and Causality of Freedom

According to Kant, all human beings are citizens of two apparently incompatible kingdoms or states—the Kingdom of Nature and that of Freedom. The Kingdom of Nature is subject to strict physical causal determinism, whereas the Kingdom of Freedom is subject to quite a different causality—the causality of freedom in which our free will accepts and follows the moral laws of our *autonomous* Human Reason, and this will is motivated only by the feeling of respect for these laws. No other motives, however vital, motivate our free will, whenever it is motivated by that respect.[16] In such a state, we are entirely free from nature's causal determinism. Our actions that are caused by our free will do not interfere with Nature's causal determinism, which belongs to a different Kingdom. Kant's total moral imperative—to bring the world nearer the state of "the supreme good on earth" or to promote the ideal of the Kingdom of God on earth (as claimed in *Religion Within the Boundaries of Mere Reason*) holds for the moral history of humankind. According to Kant, each of our moral actions aims at that universal moral end. In such an ideal state, all human beings would acknowledge one another as ends-in-themselves and by no

---

is very close to a concept of teleology. As for the consistency of destiny and free will, see Aharonov and Tollaksen 2007. Aharonov and Tollaksen show that destiny and freedom-of-will can "peacefully co-exist" in a way consistent with the aphorism "All is foreseen, yet choice is given". This destiny forces nothing upon us, and, thus, we are free to choose between possibilities.

[15] See: "This experiment sheds a new light on the age-old question of free will. Apparently, a measurement's anticipation of a human choice made much later renders the choice fully deterministic, bound by earlier causes. One profound result, however, shows that this is not the case. The choice anticipated by the weak outcomes *can become known only after that choice is actually* made. This inaccessibility, which prevents causal paradoxes like "killing one's grandfather," secures human choice full freedom from both past and future constraints" (Aharonov et al. 2013, p. 10).

[16] Note that Kantian free will is not unmotivated or undetermined. An unmotivated or undetermined will makes no sense and cannot challenge or outweigh the reasons for determinism. Kant's point is that free will is motivated or determined only by our respect for the moral laws. This respect is a strong motive to follow the moral laws even against our vital interests, for not following these laws implies self-contempt and self-betrayal, as each human being is a legislator, taking part in the legislation of the universal moral laws. This legislation indicates the practical use of our Reason.

means as means alone. The Kingdom of God or the supreme good on earth is thus the Kingdom of Ends.

Note that according to Kant, each moral act is discrete. It is entirely independent of one's other past, present, or future acts. Obeying only the moral law in each of such acts, each of them is an outcome of a particular independent decision of our free will. This kind of moral causality—causality of freedom—is entirely different from physical causality, in which each causal link is determined by the preceding one and there is temporal succession and continuity of all of them. In contrast, each of our moral decisions, which are made only according to the moral law, is, in fact, a "cause of itself" ("*causa sui*"), which is precisely the term that Aharonov and Gruss use to describe our free choices (2005, p. 6). According to Kant, we should consider each of our genuinely moral acts as a new start, as an independent beginning, even though each of these acts aims at a common, universal end, performing a moral progress toward it.

All this poses a major problem with regard to Kant's philosophy: How can we render the two distinct realms ("Kingdoms") compatible? How can we overcome this major duality? How can physical causal determinism be compatible with free will and free choices that determine our moral actions in the *physical* world, in which we live and act? How can these two so different types of causality—that of freedom and that of nature—be compatible and even united? After all, human moral history should proceed in the time of physical states and events, which is inseparable from harsh determinism in which there is apparently no room for free will and moral freedom. If human history is subject to moral ends, how can this be possible notwithstanding the determinism that is necessarily involved with history and any temporal event?

In an attempt to answer these questions and to solve the problem that lies behind them, I suggested a solution based on the distinction between two types of Kantian temporality. The first, as a pure form of our inner sense, is presented in the Transcendental Aesthetics of Kant's *Critique of Pure Reason* and in other parts of this Critique, whereas the second is my suggestion to construct a different kind of temporality—the teleological one, as an original variation of a Kantian theme (Gilead 1985). Kant's text suggests some themes that can (even should) undergo original variations suggested by the readers, attempting at deeply understanding and clarifying the difficult text. Teleological time is compatible with the official Kantian time and it is implicitly implied from his moral philosophy, as it is elaborated in the *Critique of Practical Reason* and in the *Critique of the Power of Judgment*. What motivated my suggestion for such a variation was the well-known fact that Kant considered two uses of our faculty or power of judgment: (1) The constitutive-determinant one, in which time as a form of intuition, the pure form of our inner sense, is constructed; and (2) the regulative-reflective one, in which teleology plays a vital role (the third Critique, the *Critique of the Power of Judgment*, is devoted to the concept of end, to Kant's teleology, which is regulatively-reflectively valid for the theory of beauty and for morality as well).

"Use" or "employment" is a term that serves Kant for various purposes. He mentions the different uses of Human Reason (the theoretical or pure use and the practical or moral one); the different uses of our intellect (in formal logic and in the transcendental one); and the two uses of the faculty or power of judgment—the constitutive-determinant one and the regulative-reflective one. Even though these uses or employments are entirely different from each other, we have one united Reason, one united intellect, and one united power of judgment. The question remains: How is each of these unities possible or even necessary? Now, we face again the hard problem of the unity of the abovementioned two Kingdoms despite their radical differences.

The major Kantian duality of the Thing-in-Itself (the realm of noumena) and the realm of phenomena is, in fact, the duality separating the two Kingdoms of which we are citizens: Causality of freedom (moral causality) belongs to the noumenal realm, whereas causality of Nature pertains to the phenomenal one. Each of these realms requires a temporality of its own, and these two opposing kinds of temporality appear to entirely exclude each other. This most problematic duality requires the ingenuity of Kant's interpreters. Kant himself appears to leave this problem for them.

Hence, about thirty-two years ago, I published my variation on this Kantian theme. And now, to my great surprise, some of the innovations in quantum mechanics of Aharonov and others are somewhat similar to the one I had suggested to solve a Kantian problem entirely independently and regardless of these innovations, let alone of quantum mechanics. Philosophy may precede natural science in various ideas and insights that eventually happen to be quite compatible with later natural scientific innovative ideas or discoveries. I believe that my variation on that Kantian theme sheds a new, surprising light on the fascinating idea that temporal retro-determination can be compatible with free will without interrupting the determinism involved in the classical time conception, which Newton, Laplace, and Kant acknowledged. In order to render Kant's conception of nature as deterministically subject to time compatible, nevertheless, with his conception of human moral freedom and human free will, I suggested my variation according to which Kant might have a conception of *teleological time*. This time serves the moral imperatives of Human Reason, which rendera possible our behavior as rationally moral agents, acting in reality that is subject to determinism and classical temporality.

Unlike classical time, in which each event in an earlier moment successively determines the event in the following moments, the moments of the teleological time are discrete and each is entirely independent of the others; each moment is determined only by the end and by its proximity to it, which is the target of our universal moral destiny. The remoteness from this end and the proximity to it, and not any succession, fix the order of the moments in the teleological time. Such moments are completely free from the determination by the succession of the classical physical time. In each discrete action, which occupies a discrete moment in the teleological time, we have to decide whether to submit this action to the moral law or not.[17] Each such action, taking place in a discrete temporal moment, is thus

---

[17] The *discrete* nature of our moral decisions, each of which is subject to the moral imperative that obligates each one of our actions quite independently of the rest of them, can be strikingly compared with the following: "We have a description where the choice is '*causa sui*' (cause of itself),

completely free, relies upon our free choice to respect and obey the moral law even in cases in which such an obedience clashes with our interests and needs (and even with our nature). Such moral actions of the humanity as a whole construct a moral history, each of whose moments contributes a special tribute to the progression of the human action in the teleological time. The moral end (*telos*) directs our moral behavior in a universal progression, aspiring to the common moral end, free from determinism and classical time. All our actions exist in such a time, yet they are not determined by it but by the common, universal moral end. To consider each moment in the series of our moral actions as a moment in a teleological time does not violate the status of this moment in the classical time that serves physical determinism. In this way, teleological time is compatible with the classical one. Physical determinism does not consider us as free agents, as self-causes of our moral actions, yet it does not prevent them from being free under the conception of teleological time. These are two different employments of our power of judgment concerning temporality—the constitutive one and the reflective-regulative one.

Let us now "translate" the abovementioned eleven points concerning Aharonov's novelty to my distinction between Kant's time as the pure form of the inner sense and the suggested teleological time:

1. There are two types of time—the constitutive one (time as the pure form of our inner sense) and the regulative, teleological time. The direction of the constitutive-deterministic time is determined by its past, whereas that of teleological time is determined by its future (Gilead 1985, p. 545).
2. Constitutive temporality pertains to a physical causal determinism, whereas teleological temporality serves quite a different determination, free of determinism, which, according to Kant, is the causality of freedom.
3. The determination by the teleological time does not violate the causality of the past states and events that are also subject to the constitutive temporality.
4. Notwithstanding the deterministic causality subject to the constitutive temporality, free will and free choice are compatible with the constitutive temporality, provided that we acknowledge the possibility of teleological temporality and do not exclude it.
5. Determination by the teleological temporality opens up possible alternatives of choice for the causality of freedom (or free agency).
6. Hence, the teleological temporality enables free choice and it is compatible with free will.
7. The states and events under the teleological-regulative temporality aim at a destination that lies in the future, which regulates the present and the past.[18]

---

while still being the choice of the agent-system. This would constitute a unique realization of the concept of genuine free will" (Aharonov and Gruss 2005, p. 6). Moreover, such a "*causa sui*" is perfectly compatible with the Kantian moral (generally Reason's) spontaneity and autonomy.

[18] Indeed, "the past in teleological time and its significance are 'determined' and organized according to the future, and the future (unlike that of deterministic time which is but a form of intuition) 'determines' and organizes the past. In other words, despite the linearity of the progress in teleological time, its past as a 'determinate condition' does not have any privileged status" (Gilead 1985, p. 545).

8. The teleological-regulative temporality is subject to destiny.
9. This destiny must be considered in teleological terms.[19]
10. The aforementioned destination is the asymptotic moral end for our actions and states subject to the teleological temporality.
11. In concluding, free will and free choice are compatible with physical reality, despite its deterministic nature, with no absurd or paradox involved. Free will is by no means an illusion.

As we can realize, my variation on that Kantian theme is quite compatible with the general conclusion that we can draw from Aharonov's aforementioned innovations. Pure metaphysics can precede ideas, insights, distinctions, and terms that later, most of the time independently, serve physics. The Kantian ideas shed quite a new light on Aharonov's innovative ideas and contribute greatly to their force. All these ideas, in physics and in metaphysics, reveal possibilities in which free will and free moral choices can live peacefully and even harmoniously with the reality in which we live and act and which is certainly subject also to physics and other natural sciences.

Of course, Aharonov's idea of the final state of the universe is an idea in physics, not in ethics or moral philosophy. Nevertheless, our free will enables us, as well as according to Aharonov, to decide which final state, which end, the universe may have also because of our decisions and actions. Hence, this end, this final state, can be also a moral one, following the idea of Kant, for instance. The final state or end of the physical universe does not impede our moral wish to turn reality into a realm in which each human being treats any other human being as an end-in-himself/herself and not merely as a means. Our freedom of will allows us to set such an end to the universe in which human beings make moral decisions. After all, at least, the final state of the universe, along the lines of Aharonov's view of it, is compatible with such a Kantian ideal of morality, as much as physical reality, despite its determinism, allows us to act morally, following our Reason's moral imperatives and only out of the motive of respecting these imperatives, especially the total-historical one.

---

[19] Cf.: "When we spoke of the future affecting the past, we meant a very recent past affecting a slightly earlier past. But quantum mechanics lets one impose a true future boundary condition—a putative final state of the universe. Philosophically or ideologically, one may or may not like the idea of a cosmic final state. The point is, however, that quantum mechanics offers a place to specify both an initial state *and* an independent final state. What the final state would be, if there is one, we don't know. But if quantum mechanics says it can be done, it should be taken seriously" (Aharonov et al. 2010, p. 32). To compare this "final state of the universe" with the Kantian Final End as the universal moral end would be even more fascinating. In any event, both Kant (explicitly) and Aharonov (implicitly) rely upon teleological considerations.

### 11.5.3   Another Metaphysical Contribution Concerning Alternative Possibilities for Free Choice

Aharonov's aforementioned innovations can have a further metaphysical support, which is quite different from a Kantian one.

I devoted a paper of mine to the problem of free will (Gilead 2005a) on panenmentalist grounds and on the panenmentalist view of the identitiy of the indiscernibles.

Aharonov has been one of the first scientists, if not the first one, to think about the possibility that if two particles appear to be identical, there must be a difference between them that must be revealed in their future. This is perfectly compatible with his idea of retrocausality and the determination by the future. In this way, in his treatment too, the problem of free will and that of the identity of indiscernibles are strongly associated.

As for the problem of free will, the free choice of a possibility among alternatives is a necessary condition for such freedom. The difficulty is that many philosophers and natural scientists dogmatically assume that as soon as we have made our choice, it must be clear that all the other "apparent" alternatives were in fact not alternatives at all. They looked to us like real alternatives only because of our illusion concerning free will. It is only in our mind that they are considered as real alternatives, whereas in a mind-independent reality there is no place for such alternatives. In such a view, the allegedly wrong belief that other alternative possibilities were open to our apparently free choice is simply a result of our ignorance about the actual state of affairs, according to which we have no choice but to perform the act that we were determined to do. Had we really known this actual state of affairs, we would not have believed that other, alternative possibilities were open to our predetermined choice. Such is the view of many philosophers.

In my view, nevertheless, having made a choice, the alternative possibilities that have not become actual, have not been excluded. They remain possibilities that were open to us (and some of them may remain open for us in the future). There must be a prejudice upon which we rest our false assumption that in fact they were never real possibilities but only in our mind or imagination. This assumption is that of actualism, a metaphysical view according to which all possibilities are in fact, in the final account, simply actualities or necessarily depend on them.[20] Contrary to possibilism, actualism wrongly and dogmatically assumes that pure, non-actual possibilities do not really exist (unless in our mind or imagination; they can exist only *de dicto*—in our concepts—never *de re*—in reality, outside of our mind). In reality,

---

[20] There are various definitions of actualism and possibilism. Christopher Menzel defines these terms as follows: "Actualists ... deny that there are any non-actual individuals. Actualism is the philosophical position that everything there is—everything that can in any sense be said to be—*exists*, or is *actual*. Put another way, actualism denies that there is any kind of being beyond actual existence; to be is to exist, and to exist is to be actual. Actualism therefore stands in stark contrast to possibilism, which ... takes the things there are to include possible but non-actual objects" (Menzel 2014).

according to actualism, only actualities, actual possibilities, spatiotemporally existing and causally functioning, are existents. According to the actualist, pure, non-
actual possibilities, which are independent of causal and spatiotemporal conditions
and of physical circumstances, could not exist (unless in our illusion or imagination,
and such possibilities are at most *de dicto*—mind-dependent—and never *de re*,
namely, never mind-independent). According to actualism, counterfactual possibilities are of actualities alone and they have no independent existence. They entirely
depend upon actualities, as their possible states.

Determination by the future, according to Aharonov, keeps the alternative mind-
independent possibilities open for the free choice of the agent, even when she in fact
chooses none of them. They are open to the future and by the future. Is Aharonov a
possibilist similar to me? I am not sure of that at all but one thing is certain—
Aharonov, too, appears to be quite sure that our free will is not an illusion but a real
fact about us, a fact that must be compatible with a new physics, according to which
the future is already with us, opens up alternative possibilities for our free choice.

Upon what grounds can we rest the conviction that our free will is not an illusion? Do we have well-established knowledge, or at least some knowledge, about it?

Whenever we are in pain, we do *not know* that we are in pain. Equally, whenever
we choose or act freely, we do *not know* that our will is free. In each of these cases,
the issue is not any kind of knowledge. Yet, there is no possible doubt about our
experience. Our free will, like our pain, is a matter of self-awareness, of experience
and feeling, not of any kind of knowledge, let alone a knowledge that is fallible or
susceptible to doubt and refutation. It is an irrefutable experience, free of any illusion and doubt. When I am in pain, nobody, even an omniscient or omnipotent
person, can deny that I am in pain or refute this fact, whereas knowledge is refut-
able. Whenever I am in pain, the proposition or the claim "I am in pain" is absolutely true and nobody can refute or deny it. One can suspect that I am pretending to
be in pain and thus my claim about my state at the time is false, but there is no possibility of refuting my experience—feeling or sensation—of being in pain. The very
same holds true for my experience when choosing and acting according to my free
will. This experience of mine is irrefutable and undeniable. Equally, if somebody
argues that my experience of pain is an illusion, such a claim is simply absurd.
There cannot be an illusion of pain; such a so-called "illusion" is undoubtedly an
experience of pain. No skeptical consideration of it, nor by myself neither by others,
can be valid whenever I have such an experience of pain. The very same holds true
for my experience (or awareness) of acting or choosing with no coercion, entirely
free, out of my free will. There is no valid skeptical consideration of such an
experience.

Nevertheless, what about the case in which I act under the influence of hypnosis?
Even in such a case, my free will certainly plays a decisive role, for no hypnosis is
possible without the free consent of the hypnotized person. Hypnosis is impossible
without autosuggestion, and thus it requires the free and independent cooperation of
the hypnotized person. Indeed, autosuggestion relies also upon the use of one's
unconsciousness, but this use involves one's free will. We may have unconscious or
repressed wishes, which under hypnotic suggestion would be fulfilled, but this is

done not against our free will but with its cooperation in circumstances in which we feel ourselves *free* to do something that under different circumstances (that is without the help of hypnotic suggestion) we would not do it at all. This can be chosen and done without relying upon knowledge but upon feeling or experience of being free, of freely wanting to do something. Hence, even if unconscious motives play a significant role in our choices and acts, this does not change the fact that we are free to choose and to act.

This freedom of will and of choice can be compatible with physics, as long as philosophers and physicists are aware of the possibilities that are still open for us, even in deterministic reality, because this reality is not subject only to actualist terms but equally to possibilist ones.

Panenmentalism subscribes to "determinism of pure possibilities" (Gilead 1999, pp. 24, 54–65 and 90–99; cf. Gilead 2009, pp. 68, 102, 175, 266, 267, 281, and 292), according to which determinism is a necessary entailment of pure possibilities and which is compatible with free will and free choice of pure possibilities. The combination "determinism of pure possibilities" may appear less astonishing to the reader if one is aware of the fact that causal determinism is not the only kind of determinism. There are other kinds. For instance, logical determinism is not a causal one. This kind of determinism rests, inter alia, on the distinction and relations between necessary and sufficient conditions. According to logical determinism, for instance, an admiral's official order is a *sufficient* condition for starting a sea battle. The order was given in the past, and the battle follows in the near future. The fact that the battle has already started is a *necessary* condition for the abovementioned order. Thus, a future event, as a necessary condition, determines the past event, as a sufficient condition. Here we encounter symmetrical determination as well as symmetrical temporal directions, partly analogously to Aharonov's temporal symmetry and determination.

Causal determinism is about actualities, whereas logical determinism, when conditions are discussed, relates to logical pure possibilities. Hence, because causal determinism concerns actualities, whereas logical determinism concerns pure possibilities, determinism can be compatible with individual pure possibilities as well as with actualities.[21] This means that both actualism and possibilism are compatible, but for different reasons, with determinism.

In my possibilist view, individual pure possibilities are as equally real as actualities. Quantum mechanics, according to the interpretation of Aharonov and

---

[21] Bas van Frassen (1989) suggests a modal notion of determinism. Referring to Russell's causal determinism, van Frassen considers the following modal approach to determinism: "Russell was wrong to concentrate on the actual history. To say that the past determines the future means not that the actual changes fall into a certain pattern, but that only certain possibilities are open to the system" (p. 253). As Van Frassen concludes, "to define determinism, we really need to pay the price of taking possibility seriously" (op. cit., p. 255). Nevertheless, this modal approach rests upon the notion of possible worlds, which panenmentalism, considering individual pure possibilities as the prior or fundamental existents, rejects.

some others, appears to me to yield to some possibilist views and to reject the tyranny of actualism.

Even though my panenmentalist view concerning free will does not rely upon any quantum mechanical considerations, it can support the view of Aharonov and others concerning free will and its compatibility with physical views.

### 11.5.4  Some Philosophical and Scientific Lessons

The abovementioned surprising similarities between ideas of apparently entirely independent disciplines, such as philosophy and quantum mechanics, require an explanation.

Philosophy is an inexhaustible source of possibilities of thought for natural sciences and other disciplines. Most of the time, scientists are not aware of the fact that some of their novel vital ideas are purely philosophical, being quite independent of empirical facts and empirical reality as a whole and yet are indispensable for scientific progress and breakthroughs. The reason why scientists are not aware of the purely philosophical nature of such possibilities of thought is that they consider them through natural scientific spectacles, which creates the impression that these pure possibilities are simply scientifically theoretical only and not also purely philosophical. A similar independence is the merit of pure mathematics that endows scientific thought with vital mathematical possibilities whose origin is by no means empirical. Yet, there is a difference: In the case of relying upon pure mathematics, scientists are aware of the fact that the possibilities under consideration are purely mathematical and not simply natural scientific. Philosophy and pure mathematics are both the *a priori*, non-empirical, grounds of natural science. These purely mathematical grounds and the purely philosophical ones are the origins of pure possibilities of new ideas, which are vital for science.

Pure possibilities of philosophical thinking are accessible not only to philosophers. In different ways, many of them are accessible to writers, artists, scientists, and mathematicians. Our reason, imagination, and creativity are not confined to actual reality. Our partial independence of actual reality and actual entities makes our access to philosophical pure possibilities possible.

The place of pure mathematics in quantum mechanics is crystal-clear and very well known. The relevance of quantum mechanics to philosophical issues is also quite known. Less known and less considered is the vital role that philosophy plays *within* quantum mechanics. In fact, the interpretations of quantum mechanics of Aharonov and others are replete with philosophical considerations or assumptions. In other words, they employ purely philosophical possibilities to explicate, interpret, and understand quantum mechanical findings. In other cases, such relying upon philosophy is implicit and not explicit. From time to time, the reader is impressed that Aharonov attempts to answer philosophical major questions, such as the possibility of free will and the nature of time, by means of theoretical physics and not by explicitly philosophical means. Yet, philosophical questions are too general, too universal, to be answered by means of one of the other human intellectual disciplines. Philosophy seeks universal answers that can throw light upon various intellectual disciplines.

There are no final philosophical answers to genuinely philosophical questions; no such answers can be accepted by all philosophers at all times. No specific science can provide philosophers with final answers to their questions. Philosophy is a fundamental intellectual discipline in which the individuality of any philosophical point of view is indispensable. Philosophy is unlike the natural sciences in which the individual point of view is not so indispensable, if at all. Nevertheless, a genuine, profound scientific drive always has some philosophical deep, most of the time hidden, roots. As basic theoretical sciences, natural sciences maintain the philosophical spirit—it is not a practical, useful aim that drives the basic scientist but her or his curiosity, out of pure love of knowledge and wisdom, to know and to understand the relevant facts that are subject to her or his scientific research. This pure curiosity is philosophical in nature; it is the pure wish to know and understand for the sake of pure knowledge and understanding. This is certainly the drive behind Aharonov's studies and research. This drive creates remarkable meeting points of his thought and that of Kant, or that of his and of a particular kind of possibilism, such as panenmentalism.

Quantum mechanics in the light of Aharonov's innovative ideas demonstrates how quantum mechanical reality, and even physics in general, actualizes philosophical pure possibilities, which had been discovered (there are theoretical discoveries as much as there are actual discoveries) quite independently of empirical knowledge. Such is the case of Aharonov's ideas about symmetrical time, the determination of the present by the past and the future, the possibility of free will, and the final state of the universe. These great ideas meet philosophical ideas and insights—philosophical pure possibilities—that had been discovered by philosophers long before such scientific theoretical discoveries and ideas or, at least, quite independently of them.

As much as philosophical pure possibilities shed light upon scientific theoretical innovations, such innovations shed light upon such possibilities. To examine, for example, teleological time, as a variation on a Kantian theme, the possibility of free will (challenging determinism or compatible with it), possibilism, and Kantian teleological morality through the spectacle of Aharonov's new physics, is fascinating and illuminating. Both philosophy and physics can gain great insights and advantages in maintaining profound dialogues between them.

Scientific philosophy, of which Reichenbach hoped to succeed in combining philosophical results with those of science and thus to unite science and philosophy (1962 [1951], p. viii), gains some support in this Chapter, which, contrary to Reichenbach, is not "free from metaphysics" (1944, p. vii).[22] Analyzing some of

---

[22] Notwithstanding, Reichenbach's empiricism did not impede his devotion to the clarification of great metaphysical questions. As Putnam put it: "Although Reichenbach was just as much of an empiricist as Ayer, empiricism, for Reichenbach, was a challenge and not a terminus. The challenge was to show that the great questions—the nature of space and time, the nature of causality,

Aharonov's innovative ideas in quantum theory and some independent philosophical ideas, I have demonstrated above how it is possible and fruitful to combine metaphysical ideas with those of scientific theory.

## 11.6   Conclusions

All my discussions in this Chapter end in the conclusion that quantum reality fundamentally consists of quantum individual pure possibilities and their universal relationality, which are discovered by means of quantum physics as a pure theory regardless of any particular interpretation. The physical reality that classical physics (non-quantum physics) investigates is in fact the actualization of the quantum reality. The metaphysical approach of this Chapter is not confined or limited to any particular interpretation of quantum physics. However, this approach is especially valid for the controversy between possibilist and actualist interpretations of it, supporting the possibilist ones but on an independent novel metaphysical basis, realist about individual pure possibilities and their universal relationality.[23]

---

the justification of induction (and also … the question of free will and determinism, and the nature of ethical utterances) could be adequately clarified within an empiricist framework, and not merely dismissed. In clarifying them, Reichenbach was guided by what many philosophers today would be comfortable calling a metaphysical picture (even if "metaphysics" was a pejorative term for Reichenbach himself)" (Putnam 1991, p. 62). Of course, my metaphysical view is entirely different from Reichenbach's empiricist metaphysics. As Putnam put it, "for Reichenbach *probability* is the foundation of both metaphysics and epistemology" (op. cit., p. 72), which is certainly incompatible with my metaphysical view. Notwithstanding, the illuminating possibilities of dialogue and meeting-points between metaphysics and physics, as Aharonov's idea of new physics demonstrates, is beautifully compatible with some of the ideas in Reichenbach 1962 (1951).

[23] I am indebted to Meir Hemmo for various helpful comments on an early version of this chapter. Although Meir and myself disagree over some essential points concerning quantum metaphysics, the dialogue between us about it has proven to be quite interesting and illuminating.

# Chapter 12
# Brain Imaging and the Human Mind

**Abstract** Brain-imaging technologies have posed the problem of breaching our brain privacy. Until the invention of those technologies, many of us entertained the idea that nothing can threaten our mental privacy, as long as we kept it, for each of us has private access to his or her own mind but no access to any other. Yet, philosophically, the issue of private, mental accessibility appears to be quite unsettled, as there are still many philosophers who reject the idea of private, mental accessibility. On panenmentalist grounds, I have attempted to refute such rejections and to establish this idea on firmer grounds. My arguments in this Chapter show that brain imaging allows no access to our mind and that psychical privacy is quite different from brain privacy, as the latter can be breached by brain imaging, whereas the former cannot. From a panenmentalist psychophysical view, a reduction of the mind to the body will inevitably fail, as there is a categorical difference between mind and body or brain, which is compatible with their inseparability. Brain imaging cannot enable one to "read" the mind or to breach our mental privacy, based upon the panenmentalist principle that a person's mind is a singular individual pure possibility, which is wholly different from any other individual possibility, pure or actual. Hence, there is no external access to one's mind. Each of us has exclusive access to his or her own mind, which is a singular individual pure possibility.

## 12.1 A Current Ambition Concerning Brain Imaging Technologies

More and more people nowadays appear to believe that brain-imaging technologies have posed the problem of breaching our brain privacy. Until the invention of those technologies, many of us entertained the idea that nothing can threaten our psychical privacy,[1] as long as we kept it, for each of us has private access to his or

---

The major part of this chapter was published as "Can Brain Imaging Breach Our Mental Privacy?" *The Review of Philosophy and Psychology* 6:2 (2015), pp. 275–291.

---

[1] The received term is "mental privacy," but since "mental" as a panenmentalist term does not pertain to singularity, privacy, and subjectivity, in this chapter, I replace the common "mental privacy" with the panenmentalist "*psychical* privacy." In this chapter, the attribute "psychical," instead of "mental," is ascribed to states, subjects, life, activity, and reality of psychical subjects.

© Springer Nature Switzerland AG 2020                                                      217
A. Gilead, *The Panenmentalist Philosophy of Science*, Synthese Library 424,
https://doi.org/10.1007/978-3-030-41124-4_12

her own mind but no access, epistemic or otherwise, to any other. Yet, philosophically, the issue of private, psychical accessibility appears to be quite unsettled, as there are still philosophers, following Alfred Ayer,[2] Donald Davidson,[3] and others, who reject the idea of private, psychical accessibility.[4] If brain privacy entails psychical privacy, brain-imaging technologies may threaten or endanger our psychical privacy, too.

Considering these technologies, Martha Farah claims:

> For the first time it may be possible to breach the privacy of the human mind, and judge people not only by their actions, but also by their thoughts and predilections. ... Neuroscience is providing us with increasingly comprehensive explanations of human behavior in purely material terms.[5]

In stating that, she relies upon a "reduction of mental [in panenmentalist term, psychical] to physical process," and on these grounds, she argues:

> The brain imaging work ... indicates that important aspects of our individuality, including some of the psychological traits that matter most to us as people, have physical correlates in brain function. ... is there anything about people that is not a feature of their bodies? ... The idea that there is somehow more to a person than their physical instantiation runs deep in the human psyche and is a central element in virtually all the world's religions. Neuroscience has begun to challenge this view, by showing that not only perception and motor control, but also character, consciousness and sense of spirituality may all be features of the machine. If they are, then why think there's a ghost in there at all? (op. cit., pp. 38–39)

Whether these paragraphs are representative or not, I consider them as reflecting a current ambition concerning brain-imaging technologies. Such an ambition relies on the assumption that brain privacy entails psychical privacy and, thus, that brain imaging can breach psychical privacy.

---

[2] Ayer 1971, pp. 199–205. Following Ayer and Arnold Zuboff (1981, pp. 202–212), Peter Unger attempted to refute the idea of the privacy of experience by means of the "zipper argument" (Unger 1990, pp. 177–184).

[3] Davidson 1989, 1991, 1994, 1996, and 2003.

[4] I have attempted to refute such rejections and to establish this idea on firmer grounds. See Gilead 2003, pp. 43–75; 2009; and especially 2011.

[5] Farah 2005, p. 34. Cf.: "mental privacy could face enormous new challenges, in both legal settings and beyond, as there has been no precedent for being able to look into the mind of another human being" (Tong and Pratte 2012, p. 502). Nevertheless, Farah et al. (2009, p. 119; cf. p. 126) somewhat limit her abovementioned claim. On the other hand, Valtteri Arstila and Franklin Scott argue that brain imaging does not threat psychical privacy *yet*, as it depends on the information that the subjects provide voluntarily about their psychical states (2011, p. 207). For a criticism of the relevancy of neuroimaging to the study of the mind, consult Coltheart 2006a and 2006b; and Tressoldi et al. 2012. For a methodological response to Colheart, see Roskies 2009. Roskies concludes: "There are limits to what imaging can tell us about psychology, and we have yet to determine what they are. One can acknowledge this while also accepting that neuroimaging can bear on questions of mind" (op.cit. p. 939).

## 12.2  Some Preliminary Doubts

To begin, I have two comments about this view. First, there is a major difference between psychical privacy and brain privacy. "Psychical privacy" does not mean "private property" or "private possession." As I will argue below, one's psychical states are subjective, namely, private, not in the sense of the privacy of a property or possession that belongs to one person only, but in the sense that they are accessible only to that person. The privacy of my brain means that it belongs to me only, whereas my mind does not belong to me; instead, as a person, I *consist* of my mind.

The second point is about the familiar metaphor "the ghost in the machine," which Gilbert Ryle and Arthur Koestler used in criticizing psychophysical dualism, especially the Cartesian one. This metaphor wrongly suggests that the idea that there is more to persons than their "physical instantiation" implies a psychophysical dualism that is aporetic, blocking from the very beginning every possible way that might lead us to any solution or reasonable treatment of the old psychophysical problem. Nevertheless, it is possible to avoid both Cartesian psychophysical dualism and any reduction of the psychical to the physical, of the mind to the body. If the psychical is irreducible to the physical, brain privacy does not entail psychical privacy. Moreover, if the psychical is irreducible to the physical, there is certainly more to persons than their bodies.

As a psychical being, I am a subject whose states are subjective, namely, the subject consists of subjective states. *If* all my psychical states are subjective,[6] I am the only one who consists of them; it is impossible to share them with other person(s). Still, are we allowed to conclude that as subjective each of my psychical states must be private or accessible only to me (as a psychical subject or person)?[7]

Though the reality of one's psychical states is subjective, this *reality* is beyond any possible doubt, and it is impossible to consider it as a fiction or an illusion. When I am in pain, there is no illusion in such an experience, and what someone might call "an illusion of pain" is pain no less, whose *reality* is beyond any possible doubt. Yet, based on some "objective" or intersubjective data, other people may suspect that I am really *not* in pain whereas I certainly am (even when there are no objective or external indications or data—behavioral, physical, or medical—that I am in pain). Though its reality is beyond any doubt, nobody except me can feel or

---

[6]Thomas Nagel, Geoffrey Madell, and John Searle rightly assume that subjectivity characterizes any psychical or conscious trait. See: Nagel 1979 and 1986; Madell 1988, p. 124; and Searle 1994.

[7]There are philosophers who dissociate subjectivity from private accessibility. An exception, for instance, is Madell, who considers subjectivity as a matter of privacy, namely, of what is epistemically, phenomenally, or experientially accessible only to a single subject (Madell 1988, p. 88; and 2003). Note that Nagel's view on subjectivity or the psychical does not imply an endorsement of the idea of private psychical accessibility: "I am not adverting here to the alleged privacy of experience to its possessor. The point of view in question is not one accessible only to a single individual. Rather it is a *type*" (1979, p. 171). According to Nagel, we practically do have access, though not a direct one but at least a partial one, to other people's minds; it is only other species, such as bats, to whose minds access is denied to us (ibid., p. 172).

experience this pain; moreover, nobody else has access, epistemic or otherwise, to it or to any of my experiences, *if* all they have is *merely* subjective, namely, *if* it is impossible to convert or reduce them into intersubjective or objective states. Below I will explain what it means that others are aware or know, on intersubjective or objective grounds, that I am in pain, though they have no epistemic access to any of my experiences and psychical states. In any event, the reality of this pain is entirely psychical, for with no consciousness or awareness, which is undoubtedly psychical, there cannot be any pain. Not only ecstatic or euphoric states of mind (which may be followed by some biochemical changes in my brain) but even various forms of abstraction or distraction of my mind may result in "killing" the pain. Suppose that my leg is broken, which is extremely painful; nevertheless, my mind can be entirely distracted from this state, if I am completely absorbed in something that is most important, attractive, or valuable for me at the moment, with no mind-independent change in the objective state of my body. In such a state, I would not feel any pain; I would *not be in* pain at all. Alternatively, I may be in pain for psychical reasons only, i.e. without any objective, physical grounds. Thus, my psychical state is different from my physical state at the same moment.[8] Being in pain is a subjective, psychical state, and it is irreducible to my physical state (in this case, my broken leg and the way it affects my nervous system and brain).

The reality of the subjective is beyond any doubt (at least no less than the objective).[9] Yet, is there any convincing way to reduce subjective states to objective ones, which are publicly accessible (accessible from the outside, from without)?[10]

Real, irreducible subjectivity necessarily entails privacy. Can two or more persons share anything subjective? This would turn such a thing into something quite different, into something *inter*subjective. Were subjectivity shared with others, it would have been redundant, meaningless, or insignificant. If we take subjectivity seriously as real and irreducible, which we should, we are not allowed to consider it as something that can be shared with others. Since my pain, like any of my psychical states, is subjective, there is no possible way to transfer it to another person, however close and intimate. Thus, my pain, like

---

[8] In Saul Kripke's words "the relation between … [pain and C-fiber stimulation] is not that of identity" (Kripke 1980, p. 154).

[9] The indubitability of the reality of the psychical-subjective can be along some Cartesian lines. Prominent defenders of the indubitability and irreducibility of the psychical-subjective and the first-person ontology are John Searle and Galen Strawson. Thomas Nagel's criticism of psychical reductionism and his defense of the irreducible reality of the psychical and the subjective have much force. The same holds true for the views of Colin McGinn (1983) and John Foster (1991), concerning, in different ways, the irreducible reality of the subjective and the psychical.

[10] Fred Dretske claims that as a "result of thinking about the mind in naturalistic terms[,] subjectivity becomes part of the objective order" (Dretske 1997, p. 65). On these grounds, he excludes private accessibility. One of the possibilities to refute such a naturalistic view is Madell's. Madell argues that "there is … no way in which phenomenal, or perspectival, or first-person awareness can be accommodated in a materialist framework" (Madell 2003, p. 125), a framework which is subject to objective viewpoint. In contrast, Galen Strawson and some other materialists or naturalists argue that the irreducibility of the subjective can be quite compatible with their views.

any of my psychical states, is strictly private and it is impossible to share it with any other psychical subject. Transference of a subjective state to another person is simply an illusion or worse. Indeed, as a psychical subject, I *consist* of my psychical, subjective states, and it is impossible for any other psychical subject to consist of them; otherwise, I would have a duplicate, which is also impossible: If two psychical subjects were doubles, this would reduce subjectivity to something quite different; more precisely, it would eliminate it entirely. Thus, we can conceive the possibility of a *physical* or biological cloning of human beings without contradicting ourselves, but it is impossible to think about psychical subjects as doubles without contradicting ourselves or without understanding what a psychical subject is and what subjectivity is. Thus, irreducible subjectivity entails strict privacy. In contrast, intersubjectivity or objectivity implies no privacy, unless in the sense of possession, property, or ownership, whereas subjectivity is a matter of consisting, of what the psychical subject or individual consists.

These are my preliminary doubts concerning our issue. Nevertheless, physicalists may still argue that subjective and intersubjective perspectives are either reducible to the objective one or, at least, allow us some access, a sort of an indirect one, to what is going on in the mind of other persons. In what follows I will attempt to show that my arguments cut the ground from under such physicalist or naturalist views, without begging the question, and, hence, their ambition to provide us with access to the mind of other persons must be frustrated.

## 12.3   What Does It Really Mean that We Know What Other People Think?

Even if the psychical is irreducible to the physical, still—the reader may argue—judging from my physical state, behavior, and on the grounds of intersubjective relationships, other people may be aware of the fact or even truly know that I am in pain and what my feelings, beliefs, expectations, hopes, or thoughts really are. Moreover, why not accepting the possibility that relying only upon brain imaging techniques, while entirely ignoring my expressions (verbal or otherwise), behavior, physical state, and the like, a neuroscientist may know for sure that I am in pain? Why should we not accept the possibility that in the future, relying only on brain imaging, neuroscientists will be able to tell what is going on in our mind? Or, in what state of mind are we? What are our thoughts, beliefs, volitions, emotions, feelings, inclinations, fantasies, and so on? Especially, what are our "propositional attitudes"? Are we telling the truth or lying? And, moreover, what are our unconscious thoughts, desires, and emotions? After all, our bodies and behavior may *reflect* quite sufficiently what is going on in our mind. The same holds true for our dialogues with other persons. Such dialogues may, furthermore, provide us with insights about our psychical states. Some of such insights may pertain to other persons whereas we may remain completely blind to them, while deceiving ourselves about our psychical states, ignoring or repressing them, and the like.

What does it mean that other people know or understand what is in my mind? They know nothing whatsoever *of* it intrinsically, but they may know enough *about* the intersubjective meaning and objective significance of what is on in my mind. My subjectivity certainly has some intersubjective and objective implications or imprints. And when my wife relates to what is in my mind, she does not refer to what is there but only to the imprints or reflections that what is there leave on our shared interpersonal and objective reality. Of course, she knows quite well what are my wishes, volitions, ideas, beliefs, and the like, *not intrinsically*, as they are solely in my mind, but only *relationally*, namely, in the ways these psychical states reflect on our shared interpersonal and objective reality.

When my wife tells me, "I know better than you what is on your mind this morning," she does not consider or "see" from a different perspective or viewpoint, indirectly or by inference, what is in my mind; instead, she considers herself as knowing and understanding better what is the *interpersonal* meaning and significance of what is in my mind (I use "interpersonal" to designate an intimate intersubjective relationship). She refers to these meaning and significance (which we both share in our interpersonal reality), not to what exists or occurs in my mind, which is entirely inaccessible to her. She knows me quite well as a subject, sharing an intimate, interpersonal, reality with her, but this does not allow her any access to my mind, to my private reality. She does not consider or "see" what is there from a different perspective or viewpoint. She, like other people, may know that I am in pain, not because they have any access to my brain, let alone to my mind, but because my psychical state reflects on the intersubjective and objective reality, which we share, that I am in pain. They may know about my pain, more precisely, of the intersubjective significance of my pain, judging from my behavior, expressions, verbal or otherwise, and other means of intersubjective communication. But none of them allows any access, epistemic or otherwise, direct or indirect, to what exists or is going on in my mind as it is in itself.

The crucial point is that once someone relates to my psychical states, such as experiences, pains, emotions, feelings, volitions, propositional attitudes of any kind, thoughts, and the like, one does not refer to these psychical states of mine intrinsically. One *refers* then to intersubjective or objective signals or imprints *relating* to what is there in my mind. In intersubjective relationships, language plays an indispensable role. I assume that, semantically and syntactically, nothing in any language is private. Even the most intimate word, "I," is not private, and no "private" name is really private—other persons may be called by the same name, as every person uses the same word "I." When we claim to capture what is in one's mind, which is strictly private, by means of language, we must fail, because no language is private. What we really capture by means of language is only the intersubjective meaning of what in one's mind. The referents to which intersubjective means of communication refer are in fact not psychical or private. Such means refer to the relationality of these referents, not to them as they are in themselves, intrinsically. They thus refer to the relations that they have with the psychical referents of other persons. One cannot break out of the intersubjective relationship to get into the other person's mind. One cannot have access to other minds by means of inter-subjective imprints or implications.

The relationship between the mind and the intersubjective or public-objective reality is analogous to the relationship between the mind as a "thing-in-itself" and its phenomena. This is a Kantian analogy but I use it with a principal reservation, as each of us has an epistemic access to his or her mind as it is itself, intrinsically, and not as a phenomenon. This is certainly not a Kantian view. Nevertheless, the anal-ogy holds true for the idea that the reflection or the imprints of the mind on the intersubjective and objective reality allows no epistemic access from without to the mind as it is in itself, namely, intrinsically. Thus, the intersubjective and public-objective reflections or imprints of the mind are strongly, even inseparably, con-nected with and related to the mind but, as in the case of the relationship between the thing-in-itself and its phenomena, the dependence and connection in discussion does not allow any epistemic "trespassing" beyond the phenomena to the thing-in-itself. As a result, we know nothing intrinsically about what is going on in the mind of other persons, but we have adequate access to the intersubjective or public-objective reflections or phenomena *of* what is going on there. This is all there is in what we really mean in claiming that we may know, in some circumstances, what other people think. This knowledge, however well-established, allows us no access whatsoever, epistemic or otherwise, to another person's mind. Even an imaginary omniscient observer, completely knowing what there is to know about a person, her behavior, bodily states, and intersubjective relationships, has no epistemic access to what is going on in her mind.

The father of psychoanalysis and modern psychotherapy also teaches us a similar lesson. Preliminary instructing an analysand, Freud demands:

> ... say whatsoever goes through your mind. Act as though, for instance, you were a traveler sitting next to the window of a railway carriage and describing to someone inside the car-riage the changing views which you see outside. (Freud 1913, p. 135)

The analyst is the traveler who is sitting inside the carriage and who can see absolutely nothing of what can be seen outside—unlike the traveler sitting next to the window, i.e. the analysand. Nothing of the changing views which only the analy-sand can observe is accessible to the analyst, namely, nothing of what is going on in the analysand's mind is accessible to the analyst. For this reason, the analyst needs the analysand's free, spontaneous report of what is going on in his or her mind. No other way exists for the analyst to know anything about this without such a report or description. Only the analysand has access to his or her inner, psychical reality. He or she is incapable of transmitting anything of that reality to the analyst. What is possible is the report, the description, which reveals the intersubjective meanings and significance of what occurs in that private reality. An intersubjective reality can be shared both by the analyst and the analysand. In contrast, they can share nothing of their psychical, inner realities. Note that this fine example is not about different point of views or perspectives. The analyst cannot observe what the analysand observes from a different point of view. There is no way for the analyst to observe the changing view, as he or she is sitting inside the carriage, with no access to the window. The analyst thus can analyze only the analysand's report, which is an inter-subjective linguistic or communicative object. No outside intrinsic analysis of what is going on in the analysand's mind is possible.

## 12.4  The Irreducibility of Individual or Personal Differences

Still, the reader may doubt as follows: perhaps all I have argued above refers to the indisputable fact that nobody else can undergo my experiences, can think, be aware of, feel, and so on in the same way that I, singularly, subjectively, or privately, think, am aware of, feel, and so on—but this fact does not prevent other people from having epistemic access to what is going on in my mind. For instance, my friend may know for sure that I am not in pain right now, that I believe that tomorrow will be a nice day, that I want to read my mail as soon as possible, plan to visit my friends the day after tomorrow, and so on, without considering all these from my "point of view" but in a different way from that in which I relate to or experience any of these psychical states. The point that Farah and others make is that they have, at least some, epistemic access to what is going on in a subject's mind, while analyzing his or her brain imaging. On the grounds of such an analysis or "reading," they know or will know, in the foreseeable future, much more and even with certainty what is going on in the subject's mind.

Unlike other authors but along the lines of Freud's example mentioned above, I do not consider the subjective, intersubjective, and objective merely as points of view or perspectives. Instead, to begin with, these are three kinds of *reality*, each of which is irreducible to the other and which are yet necessarily connected. It is quite wrong to consider *what intrinsically I myself* think, feel, or want as something that can be considered or seen from other viewpoints or perspectives. What I think, feel, or want, being entirely private or subjective, cannot be shared with others and, thus, cannot be considered from other viewpoints or perspectives but is accessible to me alone. In contrast, intersubjective and objective kinds of reality are, by their nature, shareable by various persons and are accessible to them.

Nevertheless, there have been philosophers, call them "physicalists" or "natural-ists" of some kinds, who would argue in opposition to me that in fact there is only one kind of reality, there is one nature—objective, physical reality. All the rest, what I call the subjective and the intersubjective, are simply fictions, epiphenom-ena, or even illusions. Suppose for a moment that they are right; nevertheless, phe-nomena of all kinds and illusions have their own kind of reality—they affect us, quite strongly sometimes, and their effects on us are very real. The question is whether as epiphenomena or illusions (as it were), the subjective and the intersub-jective should be considered as perspectival only and, if so, they are simply points of view on physicalist or naturalist grounds (as if the subject were merely a sort of "a cerebral eye" instead of "a mind's eye"). In this manner, points of view are con-sidered spatiotemporally only, that is, as modes or states of one and the same physical reality. Were such the case, the subjective and the intersubjective could have been in fact reduced to states or modes of the objective-physical. Nevertheless, to ignore personal, individual differences or to reduce them to spatiotemporal perspectives is simply to ignore the very idea of psychical life in any of its forms, materialist or otherwise. No physicalist or naturalist reduction of the psychical is allowed to ignore these differences, as they are real also from any physicalist or naturalist point of view.

Suppose, with the physicalist that the subjective is merely a state or activity of my brain, and suppose that the relevant brain cells and the connections between them can be copied or transferred to another brain.[11] In such a case, neuroscientists would have to overcome the immunity system of the other brain to avoid a rejection of the transplantation. In other words, they would have to overcome the individual differences between the brains (which their immune systems indicate) and to neutralize, rather, nullify the identity of each of them, as if they had become one brain instead of two. Nevertheless, physicalism or naturalism has to acknowledge the identity of each brain, for these theories have attempted to rest on physical or natural grounds all the individual differences, which are undeniably real, of persons. To deny such differences, to deny the identity of each brain, which is different from that of any other brain, leaves personal differences, which are undoubtedly real even if they could have been reduced to the physical, outside these theories. Hence, there is no way, even for the physicalist or the naturalist, to entertain the idea of transplantation of brain cells, such as those governing the experience of illusion or of subjectivity, from one brain to another unless by neutralizing or eliminating the identity of each brain. Thus, even according to physicalists or naturalists, the physical-biological differences between us are not perspectival, let alone spatiotemporally perspectival. This holds true even for clones: cloning rests on genetic identity but this leaves room for many differences in implementing the same genetic code in each of the clones, each of which still maintains its physical-biological uniqueness despite their common genetic identity. In concluding, even ardent physicalists or naturalists cannot deny, on physicalist grounds, the irreducible individuality pertaining to each person *as a physical entity*.

To decipher, let alone correctly, my brain imaging tests, a neuroscientist must rely not only upon many other brain imaging tests but also, and first of all, on the reports and reactions of the relevant subjects, and these are clearly intersubjective grounds, which are indispensable for neuroscience. Neuroscientists cannot dispense with them by relying only upon the findings of brain imaging. Secondly, the subjective factor (whether of the examiner or of the subject) also plays an indispensible and irreducible role in deciphering the signals that the tests exhibit. The subject must be aware of what is going on in his or her mind and, without such awareness, he or she cannot answer the examiner's question: for instance, "What are you thinking right now?" The subjective and the intersubjective factors in such studies are

---

[11] It is beyond the scope of this chapter to discuss Peter Unger's zipper arguments or Alfred Ayer's arguments against the privacy of experience. I have discussed them elsewhere. See Gilead 2011, pp. 71–91.

indispensable and irreducible, and the brain imaging technique cannot do without them and has no substitutes for them to decipher the brain signals. Suppose that while my brain is being scanned by means of fMRI, the examiner says nothing to me and I keep silent, yet he thinks: "The subject right now is thinking about a white rose." The only way to confirm this as much as possible is not to rely only upon other tests of other subjects or even upon many other scans of myself but to ask me again in each of the tests what I am thinking about. Thus, the intersubjective communicative factors as well as the subjective factors are indispensable and irreplaceable for obtaining such information. Again, we should not ignore the fact that human brains, not to mention human minds, are different one from the other. No one can be sure that the signals in one's brain imaging have the same meaning and significance as those of another brain. Furthermore, owing to the variability or plasticity of the brain, even the same cerebral signals of the same brain may have quite different meanings or significances at different times or under different circumstances.

When it comes to the psychical, individual differences are overwhelming. Think of a fantastic possibility, far from being actual, that a signal in the brain imaging of great number of subjects previously to my tests was deciphered as "the subject is now thinking about a white rose," and suppose that this signal appeared previously hundred of times in my brain imaging tests while I was, in fact, thinking about a white rose—even this does not exclude the real possibility that this time, when the signal appears, I do *not* think about a white rose or any other flower. One of the reasons for this is that there is not any known specific law bridging the mind with the brain: because of the inescapable singularity (namely, subjectivity) of each psychical subject, there is no law for his or her physical actualizations. Thus the singularity (subjectivity) of my thinking of a white rose cannot be recognized in the activity of my brain, for my brain shares some common traits with other brains, whereas my singularity shares nothing in common with that of another person. Even additional further data of such tests including recurrent patterns in the findings will not allow scientists to discover specific laws governing the brain-mind bridge. Suppose that neuroscientists can allow "bridge principles or auxiliary assumptions that enable one to infer function from location" (Roskies 2009, p. 932); nevertheless, to locate brain activity correlating to psychical functions allows no epistemic access to one's mind. Judging from the behavior of a person, we can know that right now he is thinking, peacefully or anxiously, about something, and yet we can have not even the slightest idea about what he is thinking. Could brain imaging give us more information about that? As long as two subjects cannot share one and the same subjectivity, and in this sense they are entirely different one from the other, no such putative principles or laws can govern the psychophysical unity. The physical, biochemical, and biological similarities between our brains do not reflect the singularity of each of our minds, which even a physicalist theory of the mind should not overlook. Moreover, owing to cerebral flexibility ("plasticity"), changes of the same person, as well as individual differences, the signal may appear in the subject's brain without relating at all to the same psychical state despite the well-established fact of the psychophysical unity or inseparability. Even if neuroscience is able to find statistically high correlations between brain imaging findings and some psychical states, as have been reported by the subjects to the examiners, the information about these correlations cannot be considered as an established knowledge of what there was in the subject's mind. Against this background, the correlation in question must be intersubjectively interpreted or deciphered anew in any single case and it entirely supervenes on the inescapable individual psychical differences of the subjects involved. In each case, neuroscientists have to consider the individual, though necessary, psychophysical connection, for it must be different from one person to another. The brain is subject to physical, biochemical, and physical laws yet, owing to the singularity of each person, the subjective, the psychical, is anomalous, whereas the intersubjective is subject to rules (such as those of syntax and semantics). These rules are different from the laws of nature to which the objective is subject. The necessary psychophysical unity is anomalous specifically, namely, though it is subject to some common or general traits, there are no specific laws that govern this unity as long as the psychical, as subjective, is independent of any rule and law. Indeed, the subjective, also in physicalist terms, is not subject to rules and laws, which are general and can be shared.

The indispensability of an interpretation of the brain-imaging signals discloses the conventions within which neuroscientists interpret these signals, and this interpretation or deciphering strongly depends on subjective and intersubjective factors, which are entirely and inescapably independent of the tests themselves. To the extent that metal functioning or states are concerned, the machinery of fMRI or any brain imaging is a useless tool without these factors. The machine which enables neuroscientists to watch the images of the subject's brain does not exempt them from the indispensable intersubjective relationship—they have to maintain a dialogue with the examined person to enable them to decipher or interpret the images or signals of his or her brain. No dialogue is maintained between subjects and machines; dialogues are maintained between subjects—dialogues are intersubjective, whereas machines are objects and cannot be considered as subjects.

Intersubjective, let alone interpersonal (which is intimately intersubjective), relationships allow us much more and better information *about* the psychical life of a person, namely, about its reflection or bearing on the intersubjective or interpersonal reality, than any brain imaging can do. On the abovementioned grounds, intersubjective relationships reflect the subjective much better and clearly than any brain imaging can do. Yet this reflection allows no access from outside, epistemic or otherwise, to one's mind.

My conclusion is that, to the extent that our psychical life is concerned, neuroscience cannot ignore the subjective and the intersubjective on which this science depends from the outset and which, as such, are beyond the reach of any brain imaging. Nor can it ignore personal differences and identities, which are not subject to laws. Brain imaging cannot depend on itself alone; it is a human-created phenomenon, which can be informed, deciphered, or interpreted only by persons who are psychical subjects.

## 12.5   Psychical Possibility and Its Physical Actualization; Psychophysical Inseparability or Unity

Farah mentions "physical instantiation," as if the body were the physical instantiation of the mind.[12] Does the body instantiate the mind? Do bodily or, rather, brain manifestations concretely represent the mind? Is the mind something abstract that bodily manifestations concretizing? All these expressions are misleading, first, for psychical states are no less concrete than physical states, as we feel or experience them both quite concretely and should not conceive any of them as abstract entities. Secondly, mind-body relation, connection, or unity is not like the relation between type and tokens, between abstract, general entity and concrete ones.

Instead of "instantiation," we might prefer to use "actualization." As I have suggested, the body actualizes the mind and an adequate conception of the psychophysical unity can be quite satisfactorily termed as that between *a pure possibility and its actualization.*[13] Any psychical possibility is concrete; it should not be considered as abstract at all. The actualization under discussion is quite different from the "realization" that functionalists commonly use and that allows multiple realization of the same function. In contrast, I treat the pure possibility in question as an individual possibility, which rules out any multiple actualization of the same possibility in different individual actualities (bodies). The brain actualizes psychical possibilities, or psychical possibilities are actualized in the brain. Hence, there is a necessary connection between mind and body, which ensures their unity or inseparability—this is the unity between the psychical possibility and its actualization. This unity should be sharply distinguished from identity—mind and body are not identical and yet they are inseparable. We cannot refer to psychical activity without physical actualization. Thus, any psychical activity is a brain activity (though not every brain activity is a psychical activity), but this does not mean that the brain and the mind are the same; it means that they are inseparably united. There is a difference

---

[12] Unfortunately, "instantiation" plays a similar role in the current discussion regarding to what extent brain imaging has to do with the mind. See, for instance, Roskies 2010, p. 659. Roskies's paper fluctuates between various, even conflicting, terms concerning the psychophysical relation: the brain is the "material *basis* of the mind" (ibid., p. 635); the mind is "*realized* in the brain" (ibid., p. 640); and "the mind *is* the brain" (ibid., 653; all italics are mine).

[13] See Gilead 1999; 2003; and 2009. Studies of the brain thus illuminate greatly how the brain *actualizes* psychical states. Yet, none of these studies can serve as a substitute for a psychological, intersubjective study or for any study that relates to the mind or to the psychical as such.

between the purelypossible and the actual, though no actuality is separable from its pure possibility. The purely possible is not subject to any spatiotemporal and causal conditions, whereas the actual is inevitably subject to them. Equally, there is a difference between the psychical and the physical (or the bodily), though no psychical subject is separable from its actualization (whenever such actualization exists), which is inescapably physical. I treat the actual and the physical as identical to the extent that both are subject to the same spatiotemporal and causal conditions. It is clear enough that psychophysical unity or inseparability does not entail psychophysical identity or reductiveness: it is impossible to reduce a pure possibility to its actuality, as each of them is subject to quite different conditions and as quite different properties are ascribed to each of them. Nevertheless, the pure possibility and its actuality are inseparable: psychical possibility, the mind, and its actuality, the body, are inseparable.

On these grounds, we can see the images of the brain's functioning on a computer screen, but we cannot see the images of the psychical states that this functioning actualizes. Anything physical is actual and can be publicly accessible. By means of imaging techniques, we may have access enough to anything that takes place or occurs in one's brain, but this does not entail that by means of such techniques we may have access to other person's mind. No physical accessibility can replace psychical accessibility. If indeed the psychical is privately accessible, as I think the case necessarily is, no physical accessibility, such as by means of brain imaging, can elude or overcome private, psychical accessibility.

Thus, we are facing two kinds of reality concerning our existence, each is irreducible to the other—the psychical and the physical. The former is private-subjective, the latter is public-objective. Brain imaging is valid only for a part of the latter. Since no reduction of the psychical subjective to the physical-objective is possible, no brain imaging, which is an objective means of access to my brain, has any access to my psychical states as such. Yet, these two kinds of reality are absolutely inseparable one from the other.

Given the aforementioned considerations, character, consciousness, and "sense of spirituality" are not features or traits of our brain at all. They are "features" of our mind. Otherwise, we would reduce mind to brain, and such a reduction begs the question, suggesting with no proof at all that brain imaging entails mind reading. By "mind reading," I understand an epistemic accessibility to other mind(s). The question is not whether this accessibility is direct or indirect but whether it is possible at all. It is obvious that the first-person accessibility pertains exclusively to each person, but this is not the point that I make in this Chapter. Were access from the outside to other minds possible, another person would not be aware of my psychical state, for instance, in the same way that I am aware of it, but he or she would have been aware only of *what* it is (which, my argument concludes, is impossible, too).

Any reduction of the mind to the body or to the brain requires a solid justification. I do not think that such a justification has been suggested so far. If the arguments—that brain imaging allows or enables access to the mind or that brain imaging entails mind reading in the epistemic sense—rely upon the presumption that the mind is reducible to the body or the brain, these arguments are groundless.

## 12.6   Why Is There No Psychical Inspection Either from Within or from Without?

Indeed, any of our psychical states cannot be inspected by means of brain imaging. Only the actualization of our psychical states as brain states or activity can be inspected in this way. Nobody can observe the images on the screen as though they were images of our psychical states. Even we do not observe our own psychical states. To consider our "introspection," reflection, or "inner perception" as an obser-vation or inspection creates another misleading metaphor. There is no homunculus in one's mind to inspect or observe any "inner theatre." We, as persons or psychical subjects, consist of our emotions, volitions, thoughts, and other psychical states; we can never watch them. Our psychical referents are not objects subject to any inspec-tion or observation. We experience them, we are thus made of them, and they are our psychical beings, the modes or modifications, namely, states, of our psychical real-ity. We are incapable of watching the psychical states of other people, as much as none of us can watch his or her psychical states. No "little spectators" exist in our minds (or heads) watching a kind of an "inner theatre." "Inner theatres" and "inner spectators" are but fabrications or fantasies, which have nothing to do with our psy-chical reality or the reality in which we are living. Private, psychical accessibility in no way entails the absurd, misleading analogy of an "inner theatre," privately and exclusively watched, as it were, by a single spectator. There is no separation and, in some sense, no distinction between our psychical states and us as their "observers," better, experiencers. We, as persons or psychical subjects, *are* our "psychical states;" we consist of them, of our experiences, of our private reality, which nobody else can share with us. Hence, I cannot have another person's experience (Gilead 2011, pp. 71–91), neither can I transfer to him or her anything of my mind, as much as he or she is not made of or consists of my experiences, psychical states, and anything psychical "of" mine.

People can observe what happens to my body. By means of imaging techniques, they can observe and inspect what exists or occurs in my brain, but there is no means in the world, actual or possible, by which one can inspect or observe what is "in" my mind. Unlike brain imaging, mind reading is simply impossible, unless in some fic-tions. In science fiction, unlike in real science, one can imagine that there is a means of inspecting, observing, or imaging from the outside what is in one's mind. In real-ity, to have any access to what there is, one has *to be this* mind.

In reality, you can watch the visual images of your brain states and activity on the screen in front of you. Unlike your thoughts, feelings, and experiences at the time of the imagining, there is nothing private, subjective, or personal about any of those images. They are transferrable to any other computer, can be duplicated, as well as publically watched, whereas none of your psychical states is like that. Not one of them is transferrable nor duplicable, nor can it be watched, privately or publicly. In reality, nothing psychical can be subject to visual imaging. No one, especially you, say, as a neuroscientist, can identify these images as images of your *psychical* states.

On the abovementioned ontological or metaphysical grounds, it is quite invalid to argue that maybe only because of the current technological limitations can we not *yet* inspect other minds. When the brain imaging technologies are entirely developed, it is not possible that in the future we will be able to read other people's minds. All we can do and possibly will do is to inspect, directly or indirectly (by means of imaging), what is going on in their *brains*, not in their minds. Minds and brains are not the same. On the grounds of materialist or physicalist reductions of the mind to the brain could one assume that only because of the current state of our technology can we not yet inspect other people's minds or inspect the images of what is going on there. No technology can or would turn our mind into an intersubjective or public reality. No technology can annul subjectivity. We can have visual images of facts and processes taking place in our *brains*, but, except in our imagination or fantasies, we can have no such image of what exists or is going on in one's *mind*. Hence, no future technology could change this ontological or metaphysical fact or make possible inspecting of other minds.

## 12.7 Why Our Body and Behavior Do Not Allow Any Epistemic Access to the Mind

Even if we reject any reduction of the psychical to the physical, of the mind to the brain, and assuming that they are inseparable one from the other, why does this inseparability not enable brain imaging to read our mind, namely, to have some epistemic access to my mind?

If, on the one hand, the views that assume the mind to supervene on the body or on the brain are not reductionist (which I doubt), such views, too, cannot imply mind reading from brain imaging. For even if our psychical traits or states were completely dependent (namely, supervene) on our brain (which I do not accept), such dependence (namely, such supervenience) would not enable psychophysical identity. Thus, inspecting brain imaging does not imply inspecting anything psychical even according to such views. If brain imaging allows access to the brain, this does not mean that it allows access to the psychical states "supervening" on the brain. If, on the other hand, supervenient views of the mind are reductionist, the question remains whether there is a solid justification for any reduction of the mind to the body or the brain, whereas I have shown above that there are solid reasons to *reject* any reduction of the mind to the body or to the brain. If there is no such a justification, these views fail to prove that brain imaging implies mind reading or that this technique allows access to the mind.

Nevertheless, you may argue that what we inspect by means of brain-imaging techniques is not our mind but the *behavior* of our mind, which, like our behavior in general, intersubjectively or objectively reflects or indicates what going on in our mind. As much as we can observe, diagnose, or infer from one's behavior what one's state of mind actually is, this argument goes, we can expand our observing or watching of one's behavior by means of those techniques, and, thus, enable access to one's mind.

First, watching one's behavior, cerebral or otherwise, does not allow the observer any access, epistemic or otherwise, to one's mind. Our behavior is external and can be publically, intersubjectively, or interpersonally (that is, intimately) watched or observed, whereas nothing psychical or privately inner is reducible to anything external, such as behavior. Anything behavioral is observable, whereas nothing psychical can be watched. Nobody can infer from one's behavior what there is in one's mind intrinsically.

Second, is our brain's activity a behavior at all? Can we inspect or observe cerebral behavior or actions? Brain imaging is valid for cerebral facts, events, activity, functioning, occurrences, processes, or states; it is not valid for actions. Though some of the events and states of one's brain are relevant to one's action, they do not take part in such an action. Thus, if one decides to stretch out her hand to take a piece of bread, but is incapable of doing so (because of paralysis following spinal damage, for instance), she does not perform an action at all, even though her decision to stretch out her hand is (physically) actualized in her cerebral events or states. Thus, inspecting or observing our brains by means of imaging techniques is not of our actions or behavior. Actions and behaviors must be external to the brain, notwithstanding its relevance to them.

In concluding, observing our cerebral events and states is unlike our inspecting or observing one's behavior. For this reason, *even* a behaviorist approach cannot really help one in implying psychical privacy from brain privacy, psychical accessibility from brain accessibility, or mind reading from brain imaging.

The situation becomes more crucial when it comes to morality and ethics in judging one's action or behavior. Observing and inspecting cerebral facts, events, activity, functioning, occurrences, and states, we do not observe values, moral attitudes, rules, or laws, all of which are entirely different from the former or from anything cerebral. As nothing cerebral is behavioral, we cannot even *infer* from the former what the latter possibly are. Moral attitudes are special psychical states, which no brain imaging can detect or capture.[14] Judging from our behavior, one can tell whether we are following a moral rule or not. If moral rules are transformable

---

[14] If such indeed is the case, brain-imaging techniques cannot display or detect any of our "alien attitudes," namely, attitudes that our moral judgments, values, declared or conscious moral views, oppose. For an opposite view, see Räikkä and Smilansky (2012). Räikkä claims that "at least in the future, brain imaging and other applications of neurosciences may violate people's right to privacy, in particular, their right to conceal parts of their inner life" (2010, p. 5). Räikkä and Smilansky assume that "it is likely that fMRI and similar technologies will provide all kinds of information about people's inner lives" (2012, p. 527).

to laws—to use a Kantian distinction, if maxims are transformable to moral laws—according to one's behavior, we may judge whether one is following a moral law. In contrast, if there is no such a thing as a cerebral behavior (there are only cerebral events or activities), no brain imaging can serve us in knowing whether one is following a moral rule or law. Even more, no such imaging can reveal to us what are one's moral attitudes and, especially, values.

Nevertheless, assuming that mind and brain are not identical one to the other and rejecting any reduction of the mind to the body, why does psychophysical inseparability not enable brain imaging to read the mind? If the connection between mind and body is the most intimate, direct, deepest, strongest one we can imagine, why should this not enable brain imaging to read the mind on the basis of its inseparability from the brain? Such cannot be the case, for the fact that the brain and the mind are inseparable does not allow external, non-private access to the mind. As long as the mind is not reducible to the brain or identical to it, even their inseparability does not allow external access to the mind. The mind and the mind only is accessible to itself, whereas it is only the physical component or part of the psychophysical unity that is accessible from the outside. Hence, the psychophysical inseparability does not imply that brain imaging entails mind reading. In other words, it does not imply that brain imaging allows access to our mind.

## 12.8 Psychical Reflection and Clinical Picture as Opposed to Imaging

Psychical reflection cannot be subject to brain imaging or to any kind of imaging. When looking at our reflection in the mirror, we can see, watch, observe, or inspect our mirror image. In contrast, when reflecting about our thoughts, for instance, we do not watch or see them. Instead, we relate to them directly, not to any reflective image of them. This reflection is a way in which we relate to our psychical states, and this way is entirely different from seeing something or seeing an image of something. Except for having their name in common, psychical reflection and reflection pertaining to the mirror image of one's face or body are entirely different. While reflecting on our thoughts, no images are required to mediate between us and our thoughts. To reflect on our thoughts and psychical states, we need no device or mediation of any kind. Reflection is a trait or a relationality of our thoughts and psychical states. Hence, there is no imaging whatsoever, let alone brain imaging, concerning our psychical reflection.

Furthermore, one does not read one's own mind. We can read a message, note, letter, or book that someone wrote. In such cases, we know the meanings of some written or printed signs or symbols, but the way we relate to our mind or to other's mind is certainly not like that. There is nothing in our mind to be read, not by ourselves and not by other people. In contrast, the fMRI images can be read, but the way of reading them is entirely different from the way in which we read one's message, book, and the like. By means of such writings, we communicate with other persons, but by brain imaging we communicate with no person at all; at most we "communicate" with or relate to a person's brain, whose electronic images we inspect. Our relation with objects—and the brain is an object—is entirely different from our communication with other minds, which is an intersubjective or interpersonal communication.

Imagine that in the future instead of writing a message to your best friend, who is far away from you for the time being, you will mail her a copy of your brain-imaging test to inform her what your thoughts and feelings concerning your friendship actually are. Instead of writing her, for instance, that you miss her very much, you would suggest that she examine your brain imaging, as a *reliable* test of your real feelings about her. Is this not a mere fantasy that no sober person can take seriously, however far brain-imaging technologies may advance in the future?

Thus, no person can inform us about the intersubjective meaning of his thoughts and psychical states by means of his brain imaging. Only by means of his behavior, action, or expression can he inform us so, whereas his brain activity is not a behavior, action, or expression. The brain expresses nothing; only persons can express something. Psychical states or traits should not be pertained to the brain; instead, they pertain to persons, who are psychical subjects. Thus the brain does not intend, think, reflect, feel, or will; the brain actualizes these states of mind, and all we can subject to brain imaging are physical actualities, none of which is identifiable as psychical.

Still, as actualized in one's brain or as a cerebral activity, reflection is subject to brain imaging, like any cerebral states, events, functioning, and the like. Nevertheless, this imaging reveals nothing, and cannot reveal anything, about the nature and the content of one's reflection as a psychical state. It may indicate what the brain activity is, while one is reflecting on one's thoughts, feelings, desires, psychical states, and the like. Our cerebral activity tells nothing about our psychical states as such; it "tells," instead, about the actualization of these states in our brain. From the behavior of a friend, you may guess or infer what her intention is, but her behavior is not an image of what is in her mind, and you cannot observe, "see," or experience what is there as you have no access there. Brain imaging tells the examiners much less than the behavior of the subject. It tells nothing about one's intention, reflection, and any of her psychical states as such.

A polygraph detects physical symptoms that may indicate that a subject is lying. But, by means of any polygraph, one cannot read the subject's mind. Brain imaging, whether at present or in the future, can serve (or disserve) us in detecting brain symptoms, not in reading one's mind. It may serve us much better and more accurately than a polygraph but the difference is not in principle.[15] One's blood pressure, pulse-rate, skin's electrical conductivity, rate of breathing, and the like are bodily symptoms whose psychical meanings and significance are a matter of interpreta-

---

[15] Littlefield 2009; cf. Cf. Wolpe et al. 2010. For a somewhat different view see Bles and Haynes 2008.

tion, not of experiencing, observing, inspecting, and the like by others. None of these is a means of watching or inspecting one's lies, for instances. The same holds for any change in the human brain that is detectable by means of imaging technologies. No such technology can "read" one's lies or what one has in mind.

The doctor must ask the patient about his state, and there is no way of replacing the patient's answers by means of any brain imaging. Suppose that instead of asking the patient whether he is in pain, or whether he is better today, or how does he feel just now, or what he suffers from, the doctor, if not relying upon the patient or upon his subjective attitude, would ask for a further test—brain imaging. Is the doctor entitled, or will the doctor be entitled in the future, to ask the patient nothing and to get the reliable answers from reading the patient's brain imaging? I do not think so. The doctor cannot know about the patient's "inner" states except for relying on the patient's answers (whether she believes the patient or not). In many cases, the doctor may get the clinical picture not by relying first upon the physical examination and additional tests but, indispensably, on the patient's reports. For instance, in diagnosing stable angina pectoris, the detailed report of the patient is sufficient to draw the clinical picture for the doctor who will then prescribe the needed medications, especially in case of emergency. The physical tests may objectively confirm this picture (if there is enough time to perform them), but there is no substitute for the patient's report about his state, sensations, and experience. This kind of information is indispensible for any serious medical diagnosis, and there is no behavioral, physical, or brain-imaging substitute for it. No brain imaging can or will able to tell how the patient feels. There are cerebral, physical, or behavioral indications for some of his psychical states, yet all these indications allow no access to the psychical states as such but only to the ways in which they are physically actualized.

Nothing can be a substitute for subjectivity; nothing intersubjective or objective can replace it. To have access to the subjective state of a person, one has to be that person, and, as no two persons can be identical, no person has access to the mind of other person. Even if human cloning is possible in the future, this would be a biological cloning, by no means a psychical one, for no person, by definition, is duplicable; only objects, no psychical subjects, are duplicable. Perhaps in the future, scientists may duplicate a human body, but not any human mind.

Now, although we know the difference between a brain's electronic or visual images and the brain itself, some consider them, for the purpose of diagnosing, as if they were not images, as if we could actually see and inspect the brain itself. Some wrongly believe that the images on the computer screen are the pictures of our brain or its activity, as much as our photos are considered as the pictures of ourselves.[16] In any event, some pictures of some persons are quite similar, as much as some brain images of several subjects look quite similar. No such similarity might reflect the

---

[16] See Roskies 2007. As she writes: "an important similarity exists between the informational relation photographs bear to their subjects and the informational relation holding between brain activity and neuroimages. Images generated by both techniques are causally and counterfactually dependent upon the phenomena they are meant to reflect. However, there is not parity in the two cases" (p. 867).

subjectivity or singularity of each one of us as a psychical subject, as a person. Nothing which is physically unique about any of us (such as one's fingerprints) reflects or indicates his or her singularity. It is only a physical indication of the *actualization* of one's individuality. What is objective about any of us, such as one's brain states and activity, can be seen and inspected, subject to this or that kind of watching and observation. In contrast, what is subjective about any of us has inter-subjective implications and it may reflect on our relationships with other people, but it does not subject to any brain imaging, observation, or inspection.

One may object: But this is the imaging test *of your* brain, not of any other brain, so it reflects your subjectivity in an objective way! This objection is blind to a principal difference between mind and brain and, thus, it ignores the idea that subjectivity pertains to the mind only. While my brain belongs to me, it is incorrect to state that my mind belongs to me. I do not possess my mind, whereas my brain belongs to me and it is inalienable (unless one believes that in the future we will be able to transplant one's brain into other person's body; such a belief is incompatible with what I argued above about personal differences even at a biological level of actualization). As a person, I do not possess my mind; as a person, I *am* my mind. I consist of my mind. Nevertheless, the objector may further argue: you consist *also* of your brain. Arguing so, he or she makes another psychophysical mistake. It is the actualization of my mind that my brain consists of! Thus, I consist of my mind whose actualization as *my* brain and body *belongs* to me.

## 12.9    The False Dualistic Picture of the Psychophysical Unity

This does not mean that our mind is a ghost in our brain-machine. This false dualistic picture of the psychophysical unity has nothing to do with my conception of the brain as an actualization of the mind. The brain is not a machine that the mind drives, leads, operates, or the like. Though brain and mind are not identical, they are inseparably united. No such unity exists between a machine and something or someone who operates it. The brain is not a machine or a device "in the hands" or "under the control" of the mind. The mind cannot operate on the brain, as it cannot operate at all without the brain, as there is no actualization or activity of the mind without it. Any activity of the mind is bodily and cerebrally actualized; there is no psychical activity that is not physical. Causes are actualities only. According to my view, actualities are physical or bodily and there are no other forms of actualization because the physical and the actual are subject to the same spatiotemporal and causal conditions. This means that there are no psychical causes; only physical causes exist. Thus, the relationship between mind and body is not causal, and, thus, it is not between an operator and his machine. Only the "machine" (namely, the brain) is the operator in this case. The psychophysical unity is the unity between a psychical possibility and its actualization, which is bodily or physical. It is quite senseless to say that the mind is in the body (or in the brain). Such a spatial metaphor misses the psychophysical unity entirely. The mind is inseparably united with the body.

This unity leaves the psychophysical *distinction* intact. Though all our movements cannot be performed without the body, including its neuromechanism in particular, there is all the difference in the world between our voluntary, intentional movements and involuntary movements whose causes have only neurological *significance* (as in Tourette's Syndrome), as we know quite well about the difference between actions, which are actualities of our psychical states (such as intentions, reasons, desires, and decisions), and movements that have only a bodily or physical significance. Even more apparent is the distinction between our mind and body, and in either case, psychophysical difference does not entail psychophysical separation, and the categorial differences I ascribe to mind and body are valid without weakening, even slightly, the inseparable connection between them. Psychophysical unity does not entail psychophysical identity, rather the contrary. It also excludes the psychophysical dualism to which the metaphor "the ghost in the machine" refers. The ghost in the machine is a myth, which is not valid even for a Cartesian psychophysical view. The analogy between sailor/ship and mind/body is inadequate even in Descartes' view, for no such analogy can reflect the special intimate or direct (immediate) connection between mind and body.[17] Furthermore, such an analogy is incompatible with mind-body inseparability.

## 12.10 Real Physicalism and Brain Imaging

Still, there is quite another possibility to treat the psychophysical question. This view, entitled "real physicalism," follows neither psychophysical reduction nor psychophysical dualism and treats the physical and the psychical not as opposing each other but as pertaining to the same fundamental psychical stuff (Galen Strawson et al. 2006). According to this view, psychical-experiential phenomena, as concrete, are physical phenomena but the former are irreducible to non experiential physical phenomena. Psychical-experiential phenomena, according to "real physicalism," "have essentially" internal, private, and subjective characterizations. In other words, "real physicalism" actually ascribes private accessibility to psychical-experiential states or phenomena, which means that this view does not allow external, intersubjective or objective, access to the psychical-experiential. My conclusion is, therefore, that "real physicalism," too, actually does not allow brain imaging any access to any of our psychical-experiential states, such as being in pain. According to this view, too, breaching our brain privacy does not entail breaching our psychical privacy.

Suppose that, following a similar view and unlike my view, our brain is made of a kind of stuff that can have subjective-private states and, thus, it can be in pain. This

---

[17] "Nature also teaches me, by these sensations of pain, hunger, thirst and so on, that I am not merely present in my body as a sailor [or a pilot] is present in a ship, but that I am very closely joined and, as it were, intermingled with it, so that I and the body form a unit" (Descartes 1984, p. 56; AT 81).

pain is accessible solely to this particular piece of stuff, and it is not transferable to any other piece of stuff. Why? Because, in any case, pain is psychical-subjective and not physical-objective; moreover, pain vitally requires consciousness, which, like any psychical state, is irreducible to the physical-objective. Along these lines, if the brain thinks, feels, wishes, intends, and the like, and if brain imaging allows external access to the brain, it does not allow access to the feelings, wishes, intentions, and any other psychical states, none of which can be inspected, observed, or be watched; each of which is subjective. If a materialist or physicalist (such as Galen Strawson) still endorses the view that the psychical-subjective is irreducibly real and, hence, private, he or she, too, has to reach the conclusion that brain imaging cannot enable access to our mind.

## 12.11  Conclusions

My arguments above clearly show that brain imaging allows no access to our mind and that mind privacy is quite different from brain privacy, as the latter can be breached by brain imaging, whereas the former cannot. We should not worry whether brain imaging can or will be able to read our mind. We have nothing to worry about regarding our psychical privacy, for there is no external access to one's mind. Each of us has exclusive access to his or her own mind. I also show above that a reduction of the mind to the body inescapably fails, as there is a categorial difference between mind and body or brain, which is compatible with their inseparability.

# Chapter 13
# Neoteny and the Playground of Pure Possibilities

**Abstract** Neoteny—the retention of juvenile traits in human adults, traits that characterize to some extent our juvenile or fetal evolutional ancestors—has been acknowledged, especially recently, as a decisive factor in human evolution. Such juvenile traits were insightfully understood by an eminent psychoanalyst, Donald Winnicott, in revealing how playing, for instance, plays a decisive role in the mental growth of children and, no less, in human coping with reality and in developing our culture, sciences, philosophy, and arts. From a panenmentalist viewpoint, I explicate what are the profound philosophical grounds for the great contributions of neoteny for us.

## 13.1 The Role of Neoteny in Human Evolution and Life

As the *Oxford English Dictionary* defines it, neoteny (from the German *Neotenie*) is "the retention of juvenile characteristics in a mature organism," especially "the appearance of ancestral juvenile characteristics in the adult stage of a descendant, as an evolutionary process in which somatic development is retarded."

Jules Bemporad ascribes the continual curiosity, playfulness, and plasticity of human adults to our neoteny, that is, to the "expansion of juvenile characteristics into adult life" (Bemporad 1991, p. 46). Following Stephen Jay Gould (1977),[1] Bemporad writes that we may be considered as slowly developing apes whose prolonged infancy allows us "to internalize and develop a much more complex behavioral and cognitive repertoire and who persist in displaying juvenile features well into adult life" (ibid.). Playfulness is striking among these juvenile features, and though playing is ascribable to all mammalian brains, as neuroscientist Jaac

---

A first version of this chapter was published in the *International Journal of Humanities and Social Sciences* 5:2 (February 2015), pp. 30–39.

---

[1] Gould's insight that "neoteny has been a (probably *the*) major determinant of human evolution" (Gould 1977, p. 9) has been strongly criticized by Shea (1989), McKinney and McNamara (1991), Godfrey and Sutherland (1996), and others. Nevertheless, the "case for neoteny" still gains strong support. See, for instance, Somel et al. 2009, p. 5743; Bufill, Agusti, and Blesa 2011, p. 729; McNulty 2012, p. 489; Zollikofer and De León (2013), p. 28; Hawkes 2014, pp. 33–34; and Cohen 2014.

Panksepp mentions, humankind is still an especially playful species possibly because "we are neotenous creatures who benefit from a much longer childhood than other species" (Panksepp 1998, p. 287).[2]

Indeed, neoteny—the retention of juvenile traits in human adults, traits that characterize to some extent our juvenile or fetal evolutional ancestors—which could be shallowly and wrongly considered as a mere retardation,[3] has been acknowledged, especially recently, as a decisive factor in human evolution.[4] Hence, immunologist Irun R. Cohen has acclaimed neoteny as a major factor in making us human (Cohen 2014, §147: "Neoteny has made us human"). The saltatory evolution from the ancestral *Pan* (a chimpanzee-like primate) to the human relies upon neoteny, as "neoteny generated a leap in the developmental landscape from *Pan* to humans" (op. cit. §151). Cohen thus highlights the neotenal leap, which inevitably characterizes our unique evolution as cultural beings. Human culture owes much to neoteny because

> Human culture has developed a new environment to solve the human's natural deficiencies. The ever-learning human brain in its collective communication has learned to pamper the fetus-like human body in the protective isolation of a life-long womb—the womb of human culture. The human survives, despite being an ever-immature chimp, by leaving the natural environment, for which it is so ill prepared, to construct a virtual womb … that it can inhabit till it dies. Human culture, like the womb that protects the developing fetus, provides housing, heating, air conditioning, ample processed food, and protection from the hazards of the natural environment, the environment to which the *Pan* progenitor was so well suited. The

---

[2] Furthermore, "play is an index of youthful health. … The period of childhood has been greatly extended in humans and other great apes compared with other mammals, perhaps via genetic regulatory influences that have promoted playful 'neoteny'" (Panksepp 1998, p. 298). Having discussed neoteny, psychologist David Bjorklund writes: "There is no other species that demonstrates curiosity and play into adulthood to the extent that *Homo sapiens* do. … Novelty and the unknown are typically avoided in adult animals, with the notable exception of humans. In fact, what academics do for a living is often termed as *playing* with ideas. Intellectual curiosity, or play, is a hallmark of the human species and likely a necessary component to invention" (Bjorklund 1997, p. 158). On the significance of human neoteny (defined as "the mechanism behind the epistemological plasticity of human cognition") for comparative epistemology, the cultural evolution of mind, and ethics, consider Shaner 1989, pp. 70–90. Shaner fondly refers to the Buddhist maxim "Retain the childlike mind" (op.cit., p. 80).

[3] Or, better, neoteny instructs us to look at this apparent "retardation" quite differently: "In many cases, important evolutionary changes are brought about by retardation of development, not by acceleration. This is reflected by the concept of *neoteny,* which means literally 'holding youth' or the retention of embryonic or juvenile characteristics by a retardation of development" (Bjorklund 1997, p. 155). Gould highlights "the undeniable role of retardation in human evolution" (Gould 1977, p. 9) and he considers human neoteny as an evolution by retardation (op. cit., p. 355).

[4] For innovating discussions of neoteny and human biological evolution and especially of neoteny and the human mind (also with regard to imagination, playing, and creativity, in reference to Newton, Darwin, and Einstein as examples), we are in debt to Ashley Montagu (1955, 1956, and 1989). As he concluded, "As persons we are designed to grow and develop our childlike behavioral traits through all the days of our lives, and not to grow up into fossilized adults" (Montagu 1989, p. 300). Furthermore, with the facts of neoteny, Montague may be on solid ground "in thinking of women as biologically more advanced than men," namely "of the female's natural biological superiority" (ibid.).

human has evolved as an unfit chimpanzee to build a new reality suited to its sorry physical specifications; the half-baked chimpanzee has cooked up a new world. (op. cit. §148)[5]

To create such a virtual womb or home, such an artificial environment without which we could not survive, we could not rely only upon our instincts or any ready-made, actual data[6]; we certainly needed quite different capacities—intellect, ingenuity or creativity, and, first of all, imagination. These capacities have enabled us to transcend the limitations and boundaries of our natural state, of our being animals, however highly developed. These uniquely human capacities rendered it possible for us to free ourselves from the bondage of instincts and actual data and to create a suitable environment, a womb-home, in which we could not only survive but also create science, art, literature, and philosophy, all of which are solely human creations. Hence, our debt to neoteny is even greater for we owe it not only our survival but also our human advantage and superiority. In a sense, our neoteny has rendered it possible for us to transcend our biological and genetic limitations. Though some of our genes are certainly involved in our neotenal being and capabilities, our neoteny has made it possible for us to transcend our genetic limitations. It has allowed our brain extraordinary and long-termed flexibility. As Irun Cohen puts it,

> Human brains manifest neotenic gene expression; the human brain does not complete the development of its networks till relatively late in life. It is interesting that this neotenic gene expression is not uniform, but characterizes specific parts of the brain, such as the prefrontal cortex—the areas we use to think conceptually and learn. The human brain manifests its plasticity—its ability to form new networks—far beyond the time limits of the brains of chimpanzees and other creatures. Brain maturation is marked by closure of the window of opportunity for making new networks; the mature brain stops developing; it just knows what it already knows. The delay in maturation affords the human brain with continuing flexibility and plasticity, and allows us to keep on learning throughout our lives. It is no accident that unusually creative brains often are associated with playful, ever-curious immature and playful personalities. Indeed, curious brains manufacture experience; we call it play. Playful curiosity is a mechanism by which brains organize their connections. Brains learn as they play. *Pans* and other mammals stop playing when they mature and close their brains; humans just never stop monkeying around. Play is the basis of creativity. Curiosity is not merely a strategy for solving problems; curiosity is fundamental to building brains. (op. cit., §147)

---

[5] Cohen's humorous and playful language here may remind one of the ideas of anthropologist Claude Levi-Strauss, who found that all human beings at any known time have done some cooking, which means that they have had to process their food according to recipes that have not been ready-made or naturally given but have had to be contrived by human imagination and ingenuity. This approach reminds very much of the Kantian idea that any given, empirical data has to adjust itself to the *a priori* forms of the Human Reason, otherwise we could not know empirical phenomena. All the more, according to Kant, our instincts, drives, and (passive) emotions should be subject to the moral imperative of our Reason. Thus, our morality is possible because we are not enslaved to our drives, (passive) emotions, and instincts but can liberate ourselves from them. That novel idea of Levi-Strauss can be easily associated with neoteny.

[6] Cf.: "Education is the proof of our relative freedom from instinctual determinism, and its power—either negative or positive—is based on what both John Dewey and contemporary neuroscientists call the 'original plasticity' of the young, which is a primary aspect of *neoteny*, or the extraordinarily long period of relative immaturity in humans" (Kennedy 2014, p. 103).

Thus, neoteny is associated with all our intellectual and artistic capabilities, such as conceptual thinking and learning, imagination, and creativity, not only in forming new networks but in any area in which we create something new. The flexibility and plasticity, owing to neoteny, keep the window of opportunities (better, of new pure possibilities) open for producing new networks in our brain. Neoteny keeps us always immature to some degree, which means that we cannot exhaust our creative and imaginative capabilities, as if we remain children all our lives. We should not relinquish our juvenile imagination if we want to make not only art but also philosophy and science. Hence, testifying about his way of thinking, Einstein said in a famous interview: "I am enough of an artist to rely upon my imagination. Imagination is more important than knowledge. For knowledge is limited, whereas imagination embraces the entire world, stimulating progress, giving birth to evolution" (Viereck 1929, p. 117). In this vein, when Einstein was asked by a librarian who was also a mother of young children, what kind of books he recommended them to read to cultivate their scientific talents, he advised her to let them read imaginative, fanciful works, as imagination is essential for doing science. Einstein was well known for the productive use of his imagination for revolutionary scientific discoveries by means of thought-experiments. Knowing, as natural scientists have known since Galileo had revealed that the Book of Nature must be written in mathematical language, that mathematics is the language in which nature reveals to us her secrets and wonders, Einstein could imagine or visualize how Maxwell's field equations would manifest themselves to a boy riding alongside a light beam (Isaacson 2007, p. 7). To watch the wonders of the world as if from the point of view of a child playing on the beach of a huge ocean was quite typical of an earlier genius of a scientist—Newton.[7] A fresh scientific look at natural phenomena, such as evolution, requires a lot of imagination and thus it also heavily relies upon our fortunate neoteny. What could be fresher than the viewpoint of a child?

Human saltatory evolution and human creativity and imagination are perfectly compatible. As Einstein put it, "initially there is a great forward leap of the imagination" (as quoted by Isaacson 2007, p. 549). One cannot exaggerate the role of such leaps of the imagination in Einstein's innovations.[8] Leaps of imagination are inevitable to break out from the confines of conventional knowledge and wisdom. Such is the nature of our childlike ingenuity and imagination.

Yet what really does make such leaps possible not simply for the sake of entertainment and playing but also for the sake of actual scientific discoveries? After all, breaking out from the confines of conventional knowledge and wisdom may lead to fantasies that are not valid for actual reality and allegedly cannot serve the test of reality which natural science investigates. Why may the leaps of our imagination and playing not lead us astray from the truths about actual reality but, instead, reveal

---

[7] As Newton saw himself: "I seem to have been only like a boy playing on the seashore, and diverting myself in now and then finding a smoother pebble or a prettier shell than ordinary, whilst the great ocean of truth lay all undiscovered before me" (Westfall 1980, p. 574).

[8] Cf.: "Einstein's visual imagination allowed him to make conceptual leaps that eluded more traditional thinkers" (Isaacson 2007, p. 93). Cf. op. cit., pp. 2, 4, 6, 7, 92, 360, 379, and 387.

us the profound or hidden truths of nature? The answer to this intriguing question should wait till the third section of this Chapter.

## 13.2  Winnicott: Playing and Reality

Turning now to the insights of the eminent psychoanalyst, Donald Woods Winnicott, we can shed more light on Cohen's statement that "play is the basis of creativity" (Cohen 2014, §147). Undoubtedly, we owe much of our understanding of the child's psychology and equally of the adult's one to the work of Winnicott. To understand his most significant contribution to the psychoanalytic insight concerning playing and actual reality, we should consider first his debt to Freud at that matter. According to Freud, the "transference playground" with its "almost complete freedom" (Freud 1914, p. 154) is a typically mental creation, in which no current actual, external reality but inner, psychical reality is revealed. The playground—on which the freedom to conjure up the "unthinkable" or repressed possibilities and meanings prevails—is a fiction or an artifact, the mental phenomenon of transference, and is also "a piece of real experience" (ibid.). Freudian psychoanalytic therapy consists in replacing the analysand's neurosis by an artificial illness which is the transference-neurosis, and from which the analysand can be cured (ibid.). Transference "creates an intermediate region between illness and real [actual] life through which the transition from the one to the other is made" (ibid.), and it is of a provisional nature (ibid.). At this point, we are acquainted with the inspiration for one of Winnicott's most fruitful ideas.

Weaning the analysand from the illusions and fantasies involved in transference characterizes Winnicottian psychoanalysis, too. Like Freud, Winnicott thought that transference is a transitional phenomenon, and Winnicott's "intermediate area" is in the heart of our discussion at this section. This area is the third between inner, personal, subjective, psychical reality and the actual world in which the individual actually lives, and which can be objectively perceived (Winnicott 1971, pp. 102–103; cf. 1975, p. 231). Inner and external reality are mingled or diffused in this intermediate area (or what he also calls "potential space"), so reality-testing must be suspended; it is the suspension of our testing of actual reality as it truly, objectively is. To this intermediate area belong transitional phenomena, such as transference and playing. As transitional, these phenomena are an inevitable phase in human growth and development. Nevertheless, as neoteny teaches us, such phenomena continue to remain with us as adults serving our creativity and imagination in art, religion, philosophy, imaginative living, projects, creative scientific work, that is, in scientific theories and discoveries (Winnicott 1975, pp. 241–242), and cultural experience in general (Winnicott 1971, p. 102). Regarding external, actual reality and reality-testing, the child, the analysand, the spectator, the artist, the scientist, or the philosopher must be eventually weaned from their fantasies or illusions at least to some extent. In other words, to be sane enough, we must eventually not forget the real distinction between inner and external, actual reality. For this reason, transitional, not objective, phenomena populate the intermediate area. Winnicott could not have denied that the ability to distinguish between inner-psychical and external-objective reality is a necessary condition for sanity: "Should an adult make claims on us for our acceptance of the objectivity of his subjective phenomena we discern or diagnose madness" (Winnicott 1975, p. 241). Being able to distinguish in this way is also a necessary condition for disillusionment (op. cit., pp. 238 and 240), for weaning (op. cit., p. 240), and for "objective perception based on reality-testing" (op. cit., p. 239). This is valid not only for psychotherapy and for the need to be weaned from the transference, but no less for scientific sanity as well as for our capability to enjoy works of art of any kind. Whenever we do not distinguish the reality of such work—such as literature, theatre, and cinema—from actual reality, we may lose our capability to enjoy them. Thus, no sane person is unable to distinguish between the intermediate area and external, actual reality, even when he or she, suspending reality-testing, does not actually make the distinction at that moment (in order, say, to enjoy a cinematic experience; but this enjoyment is involved with a suspension of the reality-testing not with abolishing it).

The analysands' ability to use the analyst depends on their "ability to place the analyst outside the area of subjective phenomena" (Winnicott 1971, p. 87). Similarly, Winnicott insists, "in examining usage there is no escape: the analyst must take into account the nature of the object, not as a projection, but as a thing in itself" (op. cit., p. 88). Indeed, "relating may be to a subjective object, but usage implies that the object is part of external reality" (op. cit., p. 94). Winnicott rightly emphasizes "the patient's attempt to place the analyst outside the area of omnipotent control, that is, out in the world" (op. cit., p. 91). This must be a difficult disillusionment which antecedently requires destruction of the object *in fantasy* (op. cit., p. 93), namely within the boundaries of the intermediate area (or "the potential space"). Winnicott concludes, "[i]n this way a world of shared reality is created which the subject can use and which can feed back other-than-me substance into the subject" (op. cit., p. 94). He defines the intermediate area as "a resting-place for the individual engaged in the perpetual human task of keeping inner and outer reality separate yet interrelated" (Winnicott 1975, p. 230). The ability to distinguish between inner and external reality must not contradict Winnicott's assumption that "the task of reality-acceptance is never completed, that no human being is free from the strain of relating inner and outer reality" (op. cit., p. 240); or his assumption that the "matter of *illusion* is one which belongs inherently to human beings and which no individual finally solves for himself or herself" (ibid.). In Winnicott's view, we are doomed to everlasting wandering between illusions and disillusionment, and any disillusionment confronts us with external, actual reality. In this wandering, Winnicott's intermediate area, or "the potential space" (1971, p. 100), contributes much to further saving of open pure possibilities in art, science, various projects, in experiencing love and friendship, and in many other kinds of human creativity, including philosophy.

Playing like a child, our imagination entertains with new, fantastic ideas, and we can free ourselves from the bondage of actual reality as the received knowledge and understanding perceive it. Without such a freedom, no human evolution, no culture, no scientific progress, and no artistic creation could exist. Note that I am speaking about the bondage of actual reality as the received knowledge grasps it, not of this reality as such. No scientific progress could be achieved without relying upon data with which only actual reality and actual experience can provide us. Nevertheless, any confinement to the actual experience, which we have had so far, may block scientific progress; moreover, it may block any human progress, which inescapably requires new projects, novel ideas, and new concepts, which are not confined to the actual, empirical reality as our actual experience so far has made us acquainted with. Without intermediate areas for playing and entertaining with novel pure possibilities and new ideas, no human progress can be made. Instead, slavery and confinement to our so-far actual experience would alas prevail.

## 13.3 Panenmentalism, Saving Pure Possibilities, and Neoteny

Discussing the objects of our thought, philosophers may habitually prefer to refer to "abstract objects." Instead, panenmentalism considers the objects of mere thought as individual pure possibilities, which are specific or concrete enough.

Hence, the objects of our mere thought, imagination, and creativity are individual pure possibilities. This implies that we are endowed with a non-empirical accessibility to these possibilities. The only way for us to have access to actualities is by means of empirical experience, observations, and experiments, whereas we have quite a different access to individual pure possibilities: we have such an access thanks to our imagination, ingenuity, and intellect, which are not confined to actual reality and empirical data. We can imagine and reason beyond the boundaries of actual reality as we empirically know it at the moment. As in genuine arts and in pure mathematics, we can most specifically and concretely imagine possibilities of whose actual existence we have no evidence. Hence, we may contrive projects, which—in comparison to actual reality—may be considered as mere fantasies or dreams and yet, having contrived them, we would turn to implement, realize, or actualize them in the actual circumstances under which we live. In this way, some of our dreams, fantasies, and imaginary or fabulous projects, all consisting of individual pure possibilities, would come actually true. But without contriving such possibilities, without discovering them beforehand, no such implementation could have become real. The same holds true for works of art. The ability of the artist to make personal discoveries of individual pure possibilities, of specific and concrete pure possibilities, is necessary for rendering these works actual.

We should enlarge instance such as this to include culture as a whole. Culture is not a gift of nature or ready-made; culture is human-made, an outcome of implementation of human projects consisting of individual pure possibilities, accessible to human imagination, ingenuity, and intellect, which are not enslaved to actual reality and empirical experience. This human ability to transcend these reality and experience is the sign of our neoteny. It is in the nature of children to imagine and fantasize many things that are merely individual pure possibilities and not actualities;

it is also in their nature not to accept anything actual as it is. They ask questions. They wonder why things, even the most usual and ordinary things, are the way they actually are. Actual facts rarely provide answers to their incessant questions. They do not consider actual reality and empirical experience as necessary but as contingent; for them, actual things could have been quite different. Human neoteny means that Winnicottian intermediate area (or "potential space") is most significant not only for children but equally, even more, for adults. Children, while immersed in playing in their potential space, frequently suspend reality-testing (actuality-testing, in panenmentalist terms). And so do we, while playing for our relief and pleasure and also while making mathematical, theoretical scientific, or artistic discoveries. Like children, adults need such a playground with individual pure possibilities. Neoteny is our actual, biological *capability* to have ever-open access to the realm of pure possibilities and to transcend the confinements of actual reality as we have actually known it so far. Panenmentalism replaces Winnicottian potential,[9] intermediate space with the realm of individual pure possibilities. These possibilities are the specific and concrete objects of our mere thought independent of actual reality.

Thought-experiments play a major role in scientific investigations and progress (again, Einstein's use of them is a good example). These experiments use only individual pure possibilities and the relations between them; otherwise, such experiments would not be called "thought-experiments" but rather experiments with actualities and observations concerning them. Our imagination, ingenuity, and intellect are the essential factors of constructing thought-experiments. Against the background of the above, thought-experiments, too, owe much to our human neoteny.

We know too little about the psychology and cognition of animals, first of all because we cannot communicate with them by means of any natural or artificial language. Animals, particularly mammals, undoubtedly have minds, are subject to some psychology, endowed with emotions, and the like. But we know nothing or almost nothing about their thoughts, whether or not they have human-like consciousness, imagination, or any access to individual pure possibilities. As I see it, we know for sure that we have such access but we are incapable, at least in the present state of science, to ascribe any such access to animals, however high they may be on the scale of evolution. So far, we are allowed to ascribe with certainty such access only to us, human beings. This access is our main superiority over any other existing creatures. This is the meaning and significance of our leap, our singular leap, in the history of evolution as a whole. To the best of my knowledge, we are the only creatures that transcend actual biological circumstances and have access to the province that is governed by our imagination, ingenuity, and intellect, to the realm of individual pure possibilities transcending actual reality and empirical experience.

---

[9] Panenmentalism replaces "potential" by "purely possible," as anything potential must depend on the actual, which is antecedent to the potential and makes it possible (to use an Aristotelian example, the mature oak produces acorns, which contain the potential of being mature oaks). In contrast, individual pure possibilities do not depend on anything actual. Panenmentalism thus treats Winnicottian potential space or intermediate area as a "room" enough for individual pure possibilities.

Our accessibility to pure possibilities opens up actual reality and empirical experience for our knowledge and understanding. By means of our scientific theories, some of which are truthful fictions—fictions that serve us in discovering some actual truths (see Gilead 2009)—we have access to the deepest secrets hidden in actual reality, in nature. Scientists construct models, contrive theories, and make predictions, all of which consist of individual pure possibilities and the way they relate to each other (i.e. their relationality), to discover the laws of nature, the structures of matter, and the natural order in general. All these models, theories, and predictions are products of our playing with and entertaining ourselves with individual pure possibilities and their relationality.

Thanks to neoteny, children and adults alike are able to consider and imagine actual reality as contingent: for us, many actual things, events, and even actual history could have been different. Panenmentalism ascribes contingency to any actuality, facts, and events. In contrast, it ascribes necessity only to individual pure possibilities and their relationality. Thus, logical, mathematical, and many theoretical pure possibilities are subject to necessity, and the reason for that is that any individual pure possibility could not be different. As free from the bounds of spatiotemporal and causal conditions, each pure possibility is unchangeable, atemporal (or, if you like, eternal), indestructible, and was never born or created. The fate of any actuality is quite different—it was produced, created, or born, it is inescapably destructible, temporal, and transient. Thanks to our neoteny, to our actual accessibility to the realm of individual pure possibilities, we can liberate ourselves, at least in our imagination but certainly in advancing art, philosophy, science, and technology, from the bounds and confinements of actual reality and empirical experience. In this way, we transcend history and actual nature.

Moreover, thanks to our neotenal imagination we can strip any actuality[10] of all its spatiotemporal and causal conditions and of the actual circumstances under which it exists. This leaves us with the identity, the individual pure possibility, of that actuality.

We can now have a concluding answer to the question that I raised at the end of the first section above: Why may the leaps of our imagination and playing with individual pure possibilities not lead us astray from the truths about actual reality but, instead, reveal us the profound or hidden truths of nature? Our access to the realm of individual pure possibilities inevitably serves us in identifying, knowing, and understanding actual reality as it truly is. The reason for this is that actualities, of which actual reality is made, can exist only as actualities of individual pure possibilities. For instance, all physical entities are actualities of mathematical-physical individual pure possibilities and their relationality, otherwise no modern natural science could be possible. In Galileo's terms, the Book of Nature is written in a mathematical language, which holds true for any progress that natural science has made since its very beginning as a modern science. Generally speaking, the

---

[10] Which is undeniably an actual possibility, for everything that actually exists is also actually possible, although not purely possible.

indispensability of individual pure possibilities is not only logical (there is no actuality which is not logically purely possible) and epistemological (excluding some individual pure possibilities may result in ignoring their actualities and misidentifying them, as was in fact, for instance, the case of quasicrystals), but it is also ontological. Were pure possibilities nonexistents, actualities would have been nonexistents, too. Thanks to our epistemic accessibility to the realm of individual pure possibilities, we can recognize, identify, know, and understand actual phenomena as they truly are. In this way, the realm that is accessible to our imagination, ingenuity, and intellect and the realm of actual reality, accessible to our perception and empirical experience, are strongly connected, and our imagination, ingenuity, and intellect help us inevitably in testing actual reality and not only in our liberty from it. As long as we do not ignore the distinction between the realm of individual pure possibilities and that of actual reality, our imagination and the independence of actualities are most useful for us and they should be considered as our great advantage.

Cohen rightly writes: "Curiosity is not merely a strategy for solving problems; curiosity is fundamental to building brains" (Cohen 2014, §147). In panenmentalist terms, the pure possibilities of curiosity are fundamental to actualizing our brains. Curiosity, playing, imagination, and creativity owe an inevitable debt to neoteny. A neotenal leap made possible the evolution of the human brain. Our brains actualize neotenal pure possibilities.

Neoteny is also our capability of imaging and knowing that actual things could have been different and that it is in our hands to create a future that will be different both from the past and the present. Moreover, it is in our capability to implement that, to render our fantasies and dreams actual. Thanks to our neoteny, we can change nature, for better or for worse, in the image of the individual pure possibilities that are open to us.[11]

---

[11] I am grateful to Professor Irun R. Cohen for letting me read the manuscript of his book and for allowing me to cite from it *in extenso*. This most enlightening text has introduced me to the wonders of neoteny.

# Chapter 14
# Stanley Milgram's Experiments and the Saving of the Possibility of Disobedience

**Abstract** Milgram's experiments have exposed the bitter truth that, against their moral standards, the great majority of subjects actually obeyed malevolent authorities and are ready to cause great suffering, even death, to innocent victims. The reason for such unexpected and shocking behavior can be clearly explained in the light of panenmentalist philosophy, according to which individual pure possibilities and their relations are as real as actualities and, normally, persons are free to choose between alternative pure possibilities in whatsoever circumstances. Nevertheless, whenever persons ignore the singular individuality of other people, such persons can cause most evil, entirely immoral deeds, to the others simply because impersonal authorities order them to do so. Hence, panenmentalism reveals the philosophical conditions because of which obedience or defiance to malevolent authority is possible.

Stanley Milgram's famous experiments have demonstrated that a great majority of subjects—about two thirds of them—may obey even the cruelest and immoral orders whose origin is impersonal authorities. As Milgram states:

> Subjects have learned from childhood that it is a fundamental breach of moral conduct to hurt another person against his will. Yet, 26 subjects [out of 40] abandon this tenet in following the instructions of an authority who has no special powers to enforce his commands. To disobey would bring no material loss to the subject; no punishment would ensue. It is clear from the remarks and outward behavior of many participants that in punishing the victim they are often acting against their own values. Subjects often expressed deep disapproval of [electro]shocking a man in the face of his objections, and others denounced it as stupid and senseless. Yet the majority complied with the experimental commands. (Milgram 1963, p. 376)

> *Subjects would obey authority to a greater extent that we had supposed.* (Milgram 1965, p. 61)

> A reader's initial reaction to the experiment may be to wonder why anyone in his right mind would administer even the first shocks. Would he don't simply refuse and walk out of the laboratory? But the fact is that no one ever does. (Milgram 1974a, pp. 4–5)

---

A first version of this chapter was published in *Journal of Social Sciences* 12: 2 (2016), pp. 88–98.

A. Gilead, *The Panenmentalist Philosophy of Science*, Synthese Library 424,
https://doi.org/10.1007/978-3-030-41124-4_14

To focus only on the Nazis, however despicable their deeds, and to view only highly publi-
cized atrocities as being relevant to these studies is to miss the point entirely. For the studies
are principally concerned with the ordinary and routine destruction carried out by everyday
people following orders. ... The dilemma posed by the conflict between conscience and
authority inheres in the very nature of society and would be with us even if Nazi Germany
had never existed. ... the demands of democratically installed authority may also come into
conflict with conscience. (Milgram 1974a, p. 178)

A substantial proportion of people do what they are told to do, irrespective of the content of
the act and without limitations of conscience, so long as they perceive that the command
comes from a legitimate authority. (op. cit., p. 189)

Because the experiments and the debate around them are so famous, this study will
not present them again. The author of this Chapter takes it for granted that the
learned readers are sufficiently familiar with Milgram's publications about these
experiments. It should be mentioned that Milgram's experiments are based upon a
deception of the subjects involved in it. They were deceived to believe that they
were really administering electroshocks to the victims involved, which was not the
fact at all. This, to refer at least to one example, raised some severe criticism (see
Mixon 1972, 1989).[1] Nevertheless, the electroshocks that the subjects—the
"teachers"[2]—experienced at the beginning of the experiments (to show them what
it feels like to receive the electroshocks that they would administer the victims)
were quite real. Why should they not assume that because these shocks were real,
then the shocks that they administer the victims—the learners—were equally real
and even more painful? Hence, the deception in question has not altered the *real*
psychological state and situation that the experiments meant to study.

Mixon carried out an alternative of a fully staged and equipped version of
Milgram's experiments: "This ... version of the Obedience study resembles in some
ways a stage play performance in which one of the actors improvises" (op. cit.,
p. 148; and see Milgram 1974a, pp. 198–199 as to his experiments). Although such
a staged version, too, cannot reflect or represent genuine a real situation as it is, still,
as the little performance—*The Murder of Gonzago*— in Shakespeare's *Hamlet* (Act

---

[1] Mixon challenges Milgram's experiments as follows: "Milgram attempted by deception to make
his subject believe that he was taking part in a learning experiment and that the consequences of
the shocks he was delivering were real" (op. cit., p. 149). Mixon raised "the possibility that
Milgram's subjects might also have assumed that safety precautions were in effect; if such be the
case then Milgram's judgment that obedient subjects behaved in a 'shockingly immoral' fashion
must be reappraised" (op. cit., p. 156). He also argued that "since an experimenter cannot legitimately order a subject to actually harm another, legitimate destructive obedience cannot be studied
in the context of normal experimenter commands in the conventional experiment" (Mixon 1972,
p. 175) and, finally, "not only was the research design built on an elaborate deception—a series of
lies if you like—but many of the subjects in it had shown alarming signs of emotional disturbance.
In other words many were led by lies to experience something extremely disturbing. Could the
findings possibly justify the lying and the suffering?" (Mixon 1989, p. x).

[2] From now on, when referring to Milgram's experiments, this chapter uses the words, *teachers* for
the persons who administered the electroshocks, and *learner* for the person who was supposed to
suffer the shocks, without inverted commas.

3, Scene 2, entitled "The Mousetrap") demonstrates, a staged version can serve as a good enough means to reveal the hidden truth in the mind of the subjects involved (cf. Milgram 1974a, p. 198). In contrast, Mixon did not hide or deny the humane and moral approach of Milgram as a distinguished social psychologist. Mixon is quite right in concluding that "it is clear when reading Milgram's study that he wanted his subjects to disobey, that he valued the act of saying no to someone issuing an inhumane command and that he considered those who obeyed immoral" (op. cit., p. 151).

Most interestingly, in Mixon's alternative acting experiments, with the exception of one participant, "the behavior of all naive actors deviated in no respect from the behavior reported for Milgram's deceived subjects" (ibid.). Indeed, the results are very much the same. Thus, Mixon's experiments, too, challenge us with the same philosophical question: what may cause quite ordinary human beings to behave in such an immoral way?

Among Milgram's critics, it was Mixon who was especially attentive to philosophical arguments challenging Milgram's conclusions (for instance, Mixon 1989, pp. xiv–xv, referring to the philosophical critique by Patten 1977). Furthermore, and more importantly, to the extent that possibilism is concerned—namely the metaphysical approach according to which not only actualities but pure possibilities, too, do exist—Mixon rightly ascribed an opposite approaches—actualism and determinism—to B. F. Skinner's behaviorism (Mixon 1989, p. 3) and by application to Milgram's social psychology. Actualism excludes pure possibilities, whereas panenmentalism fully accepts them. Against this background, I do not think that Mixon's criticism, both philosophically and empirically, did justice to Milgram's research, especially to the extent that Mixon wrongly ascribed determinism to Milgram's study (Mixon 1989, p. 3). As the reader will realize, Milgram's *in*determinism is consistent and, according to this view, we enjoy a freedom of choice. At this point, Mixon was certainly wrong about Milgram's approach. In what follows, this Chapter will relate Milgram's study to tacit panenmentalist assumptions.

This Chapter attempts to make Milgram's results and conclusions more understandable on a purely philosophical basis. Thus, it does not challenge his results and conclusions but take his words for them. This study finds his arguments, based on a veridical empirical basis, valid and sound.[3] I am not an experimentalist psychologist, sociologist, or anthropologist. My viewpoint is simply philosophical, and because almost any domain in our life and knowledge is based upon some philosophical assumptions, most of which are implicit or totally unconscious, it is vital to expose these assumptions and to examine them in detail and as explicitly as possible.

In praise of Milgram, he was well aware of the *philosophical* significance or reflection of his experiments: "The inquiry bears an important relation to philosophic analyses of obedience and authority (Arendt 1958; Friedrich 1958; Weber

---

[3] For a recent appraisal of "the Milgram Paradigm" after 35 years, see Blass 1999.

1947)" (Milgram 1963, p. 372).[4] In fact, Milgram's experiments imply some basic philosophical problems, such as: (1) The nature of our choices and responsibility of them; (2) Why do people habitually intend to obey what appears to them as an authority? (3) To what extent are we free to disobey an authority? (4) Even if our decisions are inescapably determined by some causes (on the basis of our education and upbringing), to what extent are we free to choose and to be responsible for our choices? All these are extremely vital philosophical questions and should be treated as such.

Note that in Milgram 1965, the philosophical tenor is more explicit. For instance, Milgram puts the problem, with which his obedience experiments deal as follows: "If $X$ tells $Y$ to hurt $Z$, under what conditions will $Y$ carry out the command of $X$ and under what conditions will he refuse?" (Milgram 1965, p. 57). In other words, this is an implicit or shorthand phrase for "for all $X$, $Y$, and $Z$, whenever $X$ tells $Y$ to hurt $Z$, under what conditions will $Y$ carry out the command of $X$ and under what conditions will he defy it." In this phrase, "all" signifies a *universal quantifier* ($\forall$). This universal quantified phrase is typically philosophical, because the conditions under discussion are not only actual, which yield to the empirical and experimental observation, but *also purely possible*, which yield to our imagination and philosophical consideration and not to empirical facts and inductive reasoning. The language and terms of Milgram's paper of 1965 clearly have a philosophical tenor (see especially footnote 4 on p. 58, which is a purely philosophical analysis of the paper's terminology).

I am not at all sure that what Hannah Arendt, the famous social philosopher and journalist, named "the banality of evil" can help us very much to philosophically clarify Milgram's alarming experimental results. Because we all are routinely subject to bureaucratic ruthlessness all over the world, the banality of evil is not sufficient to explain the results and conclusions of Milgram's experiments about human obedience to authority. Many of us are victims of ruthless bureaucratic authorities, a fact that may make us revolt against and even fight it. This worldwide prevalent human state cannot explain the obedience to authority in the way that Milgram's experiments demonstrated. Such ruthlessness provokes subjects to doubt, revolt,

---

[4] Cf. 1965, p. 57, the first paragraph; Milgram 1974a, pp. xi–xii: mentioning the philosophical problems of freedom, and p. 2. The following passage bears a clear philosophical tenor: Milgram's attempts at "carefully constructing a situation that captures the *essence* of obedience—that is, a situation in which a person gives himself over to authority and no longer views himself as the *efficient cause of his own actions*" (Milgram 1974a, p. xii; my italics). The italicized terms are primarily philosophical. See also: "Many of the subjects felt, at the philosophical level of values, that they ought not to go on, but they were unable to translate this conviction into action" (Milgram 1974b, p. 568). Issue 3 of *Metaphilosophy*1983 devoted some space to a philosophical exchange between Milgram and Morelli over Milgram's experiments. See Morelli 1983 and Milgram 1983. Milgram's answer is much more convincing, both scientifically and philosophically, than Morelli's criticism. Morelli made various philosophical distinctions most of which, it appears, made no difference concerning the issue under discussion. For another philosophical treatment of Milgram's experiments see Patten 1977, which, sadly, appears to be misleading and useless in challenging Milgram's approach and conclusions.

and defy the authority in question. They must be aware of their right and duty to transfer such defiance from such cases to their own life in which they should *defy* an authority in any case of ruthlessness or immoral orders. Such ruthlessness may teach its victims to disobey in such cases. We thus need for more and different explanations, philosophical in nature, of the phenomenon of the obedience to a malevolent authority.

The problem of choice, the question of determinism, the aim and function of closing up of pure possibilities contrary to opening them up are some of the focal points that panenmentalism has discussed widely and profoundly.

Especially relevant to our issue are the panenmentalist treatments of cruelty (Gilead 2015c) and the significance of life (Gilead 2016). These treatments are based on the panenmentalist conception of personality as a singular psychical pure possibility and not as an actuality only (namely, only as a body).

Panenmentalism assumes psychical determinism, namely, it assumes that everything in our mind is necessary and never contingent, for each entity or distinction in our mind has meaning and significance.[5] Since psychical reality consists of psychical pure possibilities and not of physical actualities,[6] then free choice of pure possibilities is certainly possible side by side to psychical determinism (see especially Gilead 2005a). In contrast, an actualist view—according to which nothing but the actual exists—as this view is also deterministic (e.g., Spinoza's philosophy), is not compatible with free choice and with free will. As panenmentalism defends the possibilities of free will and free choice, moral responsibility gains a strong support from panenmentalism, according to which the possibilities that we did not choose are no less real than the ones we did. Thus, we normally enjoy a real free choice between real possibilities.

This Chapter will try to show how Milgram's study of the obedience to authority, however malevolent or ruthless the authority may be, can be clearly based upon panenmentalist grounds. The same holds true for understanding and explaining defying of such an authority on panenmentalist grounds.

From a panenmentalist viewpoint, cruelty rests upon the denial of the singular individuality of its victim (Gilead 2015c). This singularity has an absolute value. Each person, each human being, is a singular psychical subject who is different from any other such subject and whose inner, psychical reality is accessible only to him or to her. Cruelty ignores and despises this individuality.

---

[5] In this way, panenmentalism adopts an assumption that Freud explicitly made. See Freud 1901, pp. 242 and 253, and Freud 1910, pp. 38 and 52. Freud was a compatibilist, namely, according to his psychoanalytic theory, free choice and psychical determinism are compatible. On different grounds, panenmentalism is compatibilist too. Moreover, it assumes libertarian, yet motivated, free will (see Gilead 2003, pp. 131–156, 2005a).

[6] For this reason, there are irreducible differences between mind and body, as each has the properties that the other could not have (for instance, the body has spatiotemporal properties that the mind could not have). Hence, panenmentalism rejects any reductionistic physicalist approach. As long as our choice is between pure possibilities, the unchosen possibilities are no less real than the chosen, actual possibility. This means that we could always choose otherwise and that, in principle, our choice is free.

As for Milgram's obedience study, it is not only the singular individuality of the victim (the learner) that is denied or ignored; it is also that of the authority (the experimenter) and that of the teacher as well.

Bearing this in mind, consider the following passages:

Another psychological force at work in this situation [in which the experiments took place] may be termed "counter-anthropomorphism". ... [which means] attributing an impersonal quality to forces that are essentially human in origin and maintenance. Some people treat systems of human origin as if they existed above and beyond any human agent, beyond the control of whim or human feeling. The human element behind agencies and institutions is denied. Thus, when the experimenter says, "The experiment *requires* that you continue," the subject feels this to be an imperative that goes beyond any merely human command. He does not ask the seemingly obvious question, "Whose experiment? Why should the designer be served while the victim suffers?" The wishes of a man—the malevolent designer of the experiment—have become part of a schema which exerts on the subject's mind a force that transcends the personal .... "The experiment" had acquired an impersonal momentum of its own. (Milgram 1974a, pp. 8–9)

There is a fragmentation[7] of the total human act; no one man decides to carry out the evil act and is confronted with its consequences. The person who assumes full responsibility for the act has evaporated. Perhaps this is the most common characteristic of socially organized evil in modern society. ... There was a time, perhaps, when men were able to give a fully human response to any situation because they were fully absorbed in it as human beings. (op. cit., p. 11)

The ... inhumane actions performed by ordinary Americans in the Vietnamese conflict ... do not appear as impersonal historical events but rather as actions carried out by men just like ourselves who have been transformed by authority and thus have relinquished all sense of individual responsibility for their actions. (op. cit., p. 180)

Finally, and most important:

Something far more dangerous is revealed: the capacity for man to abandon his humanity, indeed, the *individuality* that he does so, as he merges his *unique* [in panenmentalist terms, singular] *personality* into larger institutional structures.... Each *individual* possesses a conscience, which to a greater or lesser degree serves to restrain the unimpeded flow of impulses destructive to others. But when he *merges his person into an organizational structure, a new creature replaces autonomous man*, unhindered by the limitation of *individual morality*, freed of human inhibition, mindful only of the sanctions of authority. (op. cit., p. 188; my italics)

Indeed, what is so typical of Milgram's obedience experiments is that there is an authority, represented by an experimenter whose name and identity are irrelevant, as well as a teacher and a learner whose individuality does not really matter (though

---

[7] Fragmentation plays a crucial role in the behavior of the subjects who showed signs of remorse and conflicts but, nevertheless, continue to obey the experimenter until the end of the experiment. For instance: "Mrs. Rosenblum is a person whose psychic life lacks integration. She has not been able to find life's purposes consistent with her needs for esteem and success. Her goals, thinking, and emotions are fragmented. ... she failed to mobilize the psychic sources needed to translate her compassion for the learner into the disobedient act. Her feelings, goals, and thoughts were too diverse and unintegrated" (Milgram 1974a, p. 84).

Milgram 1974a analyzed some cases of individual teachers[8]). In his reports, most of these subjects and victims, teachers and learners, are *abstracted from their individual personality*, let alone from their singular individuality, but the circumstances, in which the experiments occurred, are highlighted.

This treatment by Milgram is not accidental—as an experimental psychologist, he had the conventional habit to abstract from the individual differences of the subjects and to make statistical generalizations. Furthermore, as Milgram assumes, "in certain circumstances it is not so much the kind of person a man is, as the kind of situation in which he is placed, that determines his actions" (Milgram 1965, p. 72) and "the individual, upon entering the laboratory, becomes integrated into a situation that carries its own momentum" (op. cit., p. 73). Milgram's experiments are generally not explorations in the field of personality that is devoted to the research of the "motives engaged when the subject obeys the experimenter's commands" (ibid.). Instead, his experiments "examine the *situational variables responsible for the elicitation of obedience*" (ibid.; my italics). And Milgram emphasizes:

> ... whatever the motives involved ... action may be studied as a direct function of the situation in which it occurs. This has been the approach of the present study, where we sought to plot behavioral regularities against manipulated properties of the social field. Ultimately, social, psychology would like to have a compelling *theory of situations* ... and then point to the manner in which definable properties of situations are transformed into psychological forces in the individual. (op. cit., p. 74)

The point is that, in general, this kind of experiment does not consider the psychological forces or motives of the individual as such, who becomes simply a statistical

---

[8]Despite Milgram's acknowledged aim to discover situational conditions of obedience and defiance, "the conviction that obedient subjects were behaving in an immoral fashion has focused attention on *individual morality*. ... Individual conscience should somehow be able to triumph in situations where in an authority tells a person 'to act harshly and inhumanely against another man'" (Milgram 1974a, p. 852, my italics). Cf. Milgram 1974a, pp. xi–xii and 44–55, analyzing the individual subjects and their experiences: "We need to focus on the individuals who took part in the study not only because this provides a personal dimension to the experiment but also because the quality of each person's experience gives us clues to the nature of the process of obedience" (op. cit., p. 44).

Considering an alternative kind of obedience experiment, Mixon claims: "however much I would welcome a world of men and women with conscience enough to resist and wisdom enough to recognize inhumane commands, I feel committed to discovering how to specify conditions in which even the weak and foolish cannot threaten the world with their destructive obedience" (Mixon 1972, p. 172). Thus, Mixon's alternative to Milgram's experiments ignores personal differences: "An All and None procedure can discover and specify the situationally dependent conditions in which people, *no matter what their personal characteristics*, will *as a rule* obey or defy destructive commands" (Mixon 1972, p. 175; my italics). Nevertheless, if individual conscience and resistance really matter, there are no specific conditions in which any subject may defy or obey an authority. Individual conscience means that it is in our hands to disobey even under the most difficult and harsh circumstances. In contrast, it is the assumption of torturers that any human being must be broken under a "competent" interrogation and pressure, psychological or physical. There are quite enough examples to demonstrate that these agents are quite wrong. Although many victims have been broken under torture, even not a few, quite ill and physically wrecked persons, have shown resistance to the harshest and most ruthless tortures.

item. Notwithstanding, society is an abstract entity, whereas individuals are real entities. Milgram states:

> ... the person entering an authority system no longer views himself as acting out of his own purposes but rather comes to see himself as an agent for executing the wishes of another person .... I shall term this *the agentic state* .... This term will be used in opposition to that of *autonomy*—that is, when a person sees himself as acting on his own.... In this condition the individual no longer views himself as responsible for his own actions but defines himself as an instrument for carrying out the wishes of theirs. (Milgram 1974a, pp. 132–133)

In contrast, "residues of selfhood, remaining to varying degrees outside the experimenter's authority, keep personal values alive in the subject and lead to strain, which, if sufficiently powerful, can result in disobedience" (op. cit., p. 155).

In fairness to Milgram, he personally was very interested in the individual personality of the disobedient persons. Consider the descriptions of the defiance of the subjects by the names "Professor of Old Testament," "Jan Rendaleer, Industrial Engineer," the post-factum remorse of "Morris Braverman, Social Worker" (Milgram 1974a, pp. 47–54) and, especially, "Gretchen Brandt, Medical Technician" (op. cit., pp. 84–85) who re-mentioned the equality of the learner and herself as free-willed human beings. Indeed, defiance to malevolent authority is based on considering the learner as an individual human being deserving of moral and humane consideration, treatment, and jugement. As for Mrs. Brandt, Milgram comments: Her "straightforward, courteous behavior in the experiment, lack of tension, and total control of her own action seems to make disobedience a simple and rational deed. Her behavior is the very embodiment of what I had initially envisioned would be true for almost all subjects" (op. cit., p. 85). In fact, Milgram was quite frustrated to the extent that almost two thirds of the subjects obeyed the malevolent authority!

Nevertheless, in fact, in Milgram's experiments there are almost only statistics, focusing around the age and occupation of the subjects involved. By its nature, statistics do not make individual or personal distinctions. For this reason, Freud refused to use statistics in his psychoanalytic studies (Freud 1917, p. 460). According to Freud, statistics does not consider the differences between individual analyzands, which makes it useless for the psychological inquiries that consider such differences very seriously. The distance from the identity of the person involved, either as teacher or as learner, was maintained also by employing a lottery to *pick at random* (which is contrary to choose) the teacher and the learner (Milgram 1974a, p. 373).

The whole frame of the experiments rests upon the irrelevancy of the singular individuality of the persons who obeyed the experimenter. Furthermore, the teachers ignored or disregarded the singular individuality of the learner. This disregard or indifference created an arena for an act of cruelty toward him. This is most vital to understand how cruelty is possible even in a laboratory setting and not only in life outside the psychological or social laboratory. To ignore the individuality of the persons involved, to disregard the singularity,[9] to deny the irreplaceability and

---

[9] Which is actualized in a unique body; there are not two human bodies that are identical, all the more so human minds, each of which is singular and not even similar to any other human mind.

absolute value of *each* person—any of these renders cruelty possible and even begs it. When functions (such as authority) and circumstances replace persons—each of whom is a singular individual—cruelty raises its head. This is one of the tacit major lessons of Milgram's experiments (though he did not focus on cruelty). These are the built-in conditions of the experiments, which, from the outset, simply ignore the identity of the persons involved.

In Milgram's reports of his experiments, notwithstanding, there were a few cases in which the individuality and exceptional reactions were distinctly mentioned; for instance: "On one occasion we observed a seizure so violently convulsive that it was necessary to call a halt to the experiment. The subject, a 46-year-old encyclopedia salesman, was seriously embarrassed by his untoward and uncontrollable behavior" (Milgram 1963, p. 375).[10] It is really typical of humane behavior to be attached to the singular individuality of the person who is no more simply one of the "subjects" of a psychological experiment but an individual, singular person. This still anonymous 46-year-old encyclopedia salesman is a singular human being, not simply one of the participants in this experiment. He was not a Führer, a Leader, or a Celeb, he was just a singular human being. Such a being, acknowledging the singular individuality of each of the learners and teachers, including himself, cannot behave cruelly. Once a person acknowledges the singular individuality, the dignity and the absolute value of any of his or her potential "victims" or the learner, such a person cannot behave or relate cruelly to any of them. This acknowledgement is both emotional and cognitive.

Note that torturers do not hide their crimes; it is the identity of the torturers that is concealed. They try as much as they can to spread the message that they, in the name of a state or an agency, use such brutal means in order to terrorize the targeted population.[11] In contrast, the identity of the torturers is kept as an absolute secret. The torturers not only know that their atrocity and crime against humanity justify a severe punishment, which they try to elude; they also know or feel that they are highly immoral agents, actually criminals against humanity, for they would not wish such a treatment for themselves. They cannot universalize an imperative such as

---

[10] Or, one of the observers attending the experience behind a one-way mirror, reports: "I observed a mature and initially poised businessman enter the laboratory smiling and confident. Within 20 minutes he was reduced to a twitching, stuttering wreck, who was rapidly approaching a point of nervous collapse. He constantly pulled on his earlobe, and twisted his hands. At one point he pushed his fist into his forehead and muttered: 'Oh God, let's stop it.' And yet he continued to respond to every word of the experimenter, and obeyed to the end" (Milgram 1963, p. 377). It is quite exceptional that Milgram mentioned some individual features of a reported obedient person.

[11] Contrary to the received view, the vital aim of torture is to terrorize the targeted population; it is not to extract life-saving information (Gilead 2003, pp. 97–111, 2005b). We have good reason to believe that such means do not extract veridical or reliable information, on the contrary—torture causes hallucinations and psychotic states in which the tortured informer cannot be reliable. It is only the torturers who claim the "success" of their methods to extract reliable information from the tortured persons; there is no independent, academic evidence for this claim.

"You have the right to torture a suspect under whatever circumstances." There is no moral justification for torture and it is morally absolutely forbidden.

Torturers fit very well Milgram's experiments. In democratic states, secret police sometimes has a "lawful" authority (or defenses by the law) to commit such crimes, especially when the population is under the threat of terrorism. Some such torturers try to represent themselves as "saints" that are forced to do dreadful deeds in order to save civilization as a whole, their country, or innocent civilians, defending them against barbaric, dreadful terror, and the like. No terror, and torture is a typical kind of terror, can fight back justly or efficiently any terror, however awful. Most of the torturers obey authority. Obedience to malevolent authority is the mother of torture.

Torturers believe that there is a prize for any human action, belief, ideology, fight and the like. They are convinced that torture can break any person, however determined, devoted to one's object or conscientious he or she may be. They compare their job as torturers to breaking into a safe to get access to its hidden vital contents. However, there is no access from without to any human mind, for each human mind is singular and thus it is not accessible from without, by other person. Thus, torturers can destroy, mentally and physically, their victims but they cannot destroy their singularity, which they try to deny.

The very nature of obedience is at least to restrict, sometimes to an extreme extent or even silence, the singular individuality of the obedient person. Any kind of fascism rests upon this. There is no fascism without obedience and sacrificing the singular individuality of the persons involved (except, perhaps, for the Leader, the Duce, or the Führer whose singular individuality is absurd or simply a caricature).

Most shockingly,

> Upon command of the experimenter, each of the 40 subjects went beyond the expected break off point. No subject stopped prior to administering Shock Level 20 (At this level—300 volts—the victim kicks on the wall and no longer provides answers to the teacher's multiple-choice questions). Of the 40 subjects, 5 refused to obey the experimental commands beyond the 300-volt level. Four more subjects administered one further shock, and then refused to go on. Two broke off at the 330-volt level, and 1 each at 345, 360, and 375 volts. Thus a total of 14 subjects defied the experimenter. (Milgram 1963, p. 375)

Indeed, the significant majority the teachers in Milgram's experiments were not aware of the *mere, pure possibility* of defying the instructions. Unconsciously, they excluded such a possibility and strongly believed that they had to obey the experimenter. They had received a decent education, instructing them to obey the authority of parents, teachers, judges, officers, and the like and to be good citizens. Without obedience there must be anarchy, and we are educated to believe that anarchy is the worst enemy of the civilized order (one may really wonder whether anarchy is really the real enemy of humanity and civilization). We are raised to believe that to disobey, to defy, is to behave criminally, selfishly, immaturely, without responsibility, consideration, cooperation, and the like.

In Milgram's experiments, if the teacher turned to the experimenter for advice or instruction whether he or she should continue to administer electroshocks despite the yelling and the protests of the learner, the experimenter responded, inter alia, with the following prod: "It is absolutely essential that you continue" (Milgram

1963, p. 374). Whenever the subject said that the learner refused to go on, the experimenter replied: "Whether the learner likes it or not, you must go on until he has learned all the word pairs correctly. So please go on" (ibid.). Whenever the teacher still refused or hesitated, the experimenter urged him or her in a firm but not impolite tone in the following manner: "You have no other choice, you must go on" (ibid.). Yet *if*, after this prod, the subject still refused to obey the experimenter, "the experiment was terminated" (ibid.).

Hence, whenever the teacher did not exclude the *pure, mere possibility* that he or she had the free choice not to continue but, instead, to stop the experiment (despite the prods of the experimenter) and the teacher chose to actualize this possibility, any of the experiments, with no exception, was terminated. However, in any case in which the teacher excluded this possibility, even as a pure or mere one, he or she obeyed the authority of the experimenter, or even that of the impersonal experiment (prod 2 reads: "the experiment requires that you continue") despite scruples and inconvenience or even real stress if there were such. Indeed, the main point is that most of the teachers in the experiments did not really consider the mere possibility not to obey in this case. They, in fact, excluded it, even as a pure, mere one. Only those who did not obey, considered such a possibility seriously and eventually decided to actualize it and rendered it actual.

Excluding pure possibilities may be one of the main obstacles, if not the prime one, to block scientific progress. In various publications of mind (Gilead 2013, 2014a, b, 2015a, b), I have referred to some of these amazing, sometimes even shocking or tragic, phenomena. The dogmatically harsh reaction of Pauling to Shechtman's discovery of quasi-crystals ("There are no quasi-crystals, there are quasi-scientists!") (Gilead 2013) is a most notorious example. Whenever, owing to dogmatic attitudes, for instance, vital pure possibilities have been excluded, scientists could not recognize, study, or understand the actual phenomena that they encountered. Think about the reaction of Ignaz Semmelweis's colleagues to his conjecture (in 1847) that microorganisms should be considered as the cause of puerperal (childbed) fever. Or, think about the more recent opponents to the idea that a micro-organism causes stomach ulcers. All such opponents simply excluded these possibilities, even as pure ones (that is, simply on theoretical grounds and sometimes even on subjective or prestigious ones), and thus delayed scientific progress. The most negative reaction, skepticism, and ridicule by Semmelweis's colleagues led to his mental illness that ended in his confinement to an insane asylum until his death in 1865. Had he lived in Rome, at the time of Bruno, he would probably have been burnt to death in the Campo de Fiori for his heretical ideas and praxis.

Another intriguing aspect of excluding or closing possibilities has to do with the fact that most of the subjects (of the teachers) did not doubt that the experimental situation was a real life one, a real process of learning (op.cit., p. 375). Such credulity is based upon closing or excluding possibilities, whereas skepticism, doubts, critical mind, and humor have to do with saving or opening up pure possibilities. Another interesting point is the gap between the shocking actual results of the experiments and the predictions of some experts, including Milgram himself and his colleagues. Most of them believed that only very few subjects would obey the experimenter until the end of the experiments, whereas the actual results were proved the contrary. This shows that there is, and that there has indeed always been a gap between reality as purely possible or expected, and as actual.[12]

There are many cases in which we exclude pure possibilities for our choices, attitudes, approaches, actions, or behavior. In the case of Milgram's experiments, the subjects who continued with the experiment until its planned end, first of all excluded the possibility—the mere, pure possibility—that they should not obey the experimenter and listen to the voice of their conscience or feelings. In each of these cases, the pure possibilities in consideration are individual, as the choice has to do with an individual decision concerning an individual learner.

The Nazis used prods, such as those employed by Milgram's experimenters to persuade soldiers, torturers, informers, and other collaborators to obey their instructions or orders. Obviously, they used other, much harsher, methods. What is common to all these cases and those of Milgram's obeyed subjects is the *excluding of the individual pure possibility* that such an instruction or command should not be obeyed, that malevolent authority should be defied particularly in this case.

Panenmentalism has made a great effort to draw the attention of the readers to the heavy price that we have to pay in closing or excluding vital pure possibilities for our knowledge and morality. Indeed, whenever we try to reach a decision, we have to exclude some possibilities, especially pure ones. Equally, whenever we wish to get to the truth of the matter, we have to exclude some possibilities, for instance: Is it a micro-organism that causes stomach ulcers or is it stress or a particular diet? If all or both relevant possibilities are kept open, we cannot make any progress in knowing and understanding more and better the relevant phenomena or their causes and in making decisions what we have to do in practice. On the one hand, we need to be familiar with as many different possibilities as possible in order to choose the right one among them, yet, on the other hand, we have to close or exclude some of them in order to choose one of them. This is a vital balance that should be maintained. Nevertheless, without saving vital pure possibilities we cannot make any progress, epistemic, scientific, or practical.

The subjects in Milgram's experiments who refused to obey, acted, unknowingly, according to a panenmentalist imperative—never close or exclude an individual pure possibility that can be vital for your free and decent choice. Beware of closing pure possibilities on dogmatic basis, on grounds of prejudice, and the like. In contrast, those who obeyed the experimenter against their conscience or feelings, in fact

---

[12] Furthermore, "many of the subjects, at the level of stated opinion, feel quite as strongly as any of us about the moral requirement of refraining from action against a helpless victim. They, too, in general terms know what *ought* to be done and can state their values when the occasion arises. This has little, if anything, to do with their *actual behavior* under the pressure of circumstances.... values are not the only forces at work in an *actual*, ongoing situation" (Milgram 1974a, p. 6; my italics). Values, moral imperatives, and "the ought" are not actualities; they are not facts. Instead, they are individual pure possibilities pertaining to the mental (while taking part in an intersubjective reality) or to the psychical (while taking part in the subjective, inner reality of a psychical subject or a person).

closed a most vital pure possibility for them as well as for us or for our society as a whole.

Dictatorship or fascism rests upon excluding vital possibilities and open possibilities in general: as if there were one leader, one nation, all strongly united until there is no differentiation to be made in the nationalistic totality, one common state of mind, one ideology, one way of conduct, and the like. In this way, the obedience to tyrannical, dictatorial, or fascist authorities is similar even equal to the obedience of the subjects to the experimenter in Milgrom's most shockingly illuminating experiments.

The motives of Milgram were clear enough and well expressed, loud and clear: As a son of a couple of Jewish refugees from the Nazi occupation in Eastern Europe, he knew quite well what was the horrible price that humanity in general and the Jewish people in particular have had to pay because of blind obedience to authorities without any question, doubt, or criticism. Or, in panenmentalist terms, without considering other individual pure possibilities that are open to one's choice and praxis.

Individual pure possibilities play a more vital and decisive role in psychological experiments than that which what meets the eye. Dixon and others have drawn our attention to the significant difference between experimental reality, in which a deception of the subjects plays some vital role, and actual reality. For instance, he explains: "By using role playing the situation can be faced squarely with the open acknowledgement that the actual consequences are not 'real'" (Dixon 1972, p. 169).[13] Thus, a psychological experiment, as it should be, is more a playing with or entertaining individual pure possibilities and their relationality than facing actualities. In Milgram's experiments, the subjects did not administer real electroshocks to the "victims"—the learners—but they believed that they did so. This was a deception ("a false fire"), as was "The Mousetrap," which Hamlet instructed the players to play in order to reveal or display the crime of his uncle, the King. Thus, the psychological experiments achieved what they precisely should. Such is the case because, according to panenmentalism, when it comes to psychical reality, *psychical pure possibilities rather than actualities are what our mind consists of.* We have to distinguish between three kinds of reality: subjective, intersubjective, and objective. Subjective reality consists of psychical pure possibilities of which our mind consists; intersubjective reality consists of mental pure possibilities shared by people who live in the same society, speak the same language, and the like; finally, objective reality consists of actual, physical possibilities. When it comes to psychical reality, pure possibilities rather than actualities are vital. In social life, it is our intersubjective pure possibilities and their relationality that are crucial; when it comes to one's psychical reality, it is psychical, subjective, personal pure possibilities that are concerned.

---

[13] For Milgram's answering back the deception argument, see Milgram 1974a, pp. 173–174. In summing up, "the majority of subjects accepted the experimental situation as genuine; a few did not. Within each experimental condition it was my estimate that two to four subjects did not think they were administering painful shocks to the victim" (op. cit., p. 173).

According to panenmentalism, dreams, expectations, fears, anxiety, hopes, projects, images, meanings, thoughts, and so on are not actualities; rather, they are psychical or mental pure possibilities. The same actuality has different psychical meanings—psychical possibilities—for different persons. Moral imperatives, conscience, fear of punishment, obedience, defiance, and the like are not actualities but pure possibilities first. Such possibilities and not only actual behavior play a decisive role in Milgram's experiments. The same holds true for the gap between pure possibilities and their actualities.

In summing up, panenmentalist principles and terms shed such an enlightening light on Milgram's experiments, which makes it possible to understand better why ordinary human beings may obey malicious commands against their better education, feelings, and conscience. Panenmentalism reveals the prime philosophical conditions in which such persons may behave in this inhumane and uncivilized way. This raises some hope that philosophical awareness of the meaning and significance of such obedience can change the reality, which Milgram's experiments reveal, for the better, both morally and practically. Such awareness can make it possible for us to choose otherwise, following our morality, conscience, and humane obligations.

# Chapter 15
# Singularity and Uniqueness: Why Is Our Immune System Subject to Psychological and Cognitive Traits?

**Abstract** Immunologists use psychological and cognitive terms to describe and explain the behavior of our immune system. Do they use them metaphorically or literally? In this paper, I show that on the grounds of panenmentalist psychophysical assumptions, the uniqueness of each person (or self) as an individual organism necessarily corresponds to the singularity of each person as a psychological subject, which is a singular individual pure possibility. On the basis of these assumptions, immunologists, irrespective of their various conceptual frames, are entitled to ascribe psychological and cognitive traits to our immune system and its behavior. Immunologists are allowed to do so because each immune system of any higher, unique individual organism corresponds to psychological traits, which are ascribable only to persons, each of whom is a singular being. This correspondence is necessarily compatible with the psychophysical unity or inseparability. Furthermore, the psychological or cognitive traits pertain to the immune system require no consciousness. In the case of artificial immune systems, in contrast, the application of psychological or cognitive terms is only metaphorical, for each such system is not unique but it is replicable. Only the immune system of unique individual organisms that, as psychological subjects, are singular beings—i.e., persons—can be subject, literally or non-metaphorically, to psychological and cognitive terms.

The aim of this Chapter is to demonstrate that immunologists are allowed to use, literally or non-metaphorically, cognitive and psychological terms in describing, understanding, and explaining our immune system.

## 15.1 My Psychophysical Assumptions

Let us begin with some psychophysical assumptions (hereafter "my psychophysical assumptions," or in other words, panenmentalist psychophysical assumptions).[1]

---

[1] These assumptions are established, explained, and elaborated in my possibilist metaphysical theory, called panenmentalism. They serve here as insights to throw some light on the justification of immunologists in using psychological and cognitive terms while describing and understanding

© Springer Nature Switzerland AG 2020

A. Gilead, *The Panenmentalist Philosophy of Science*, Synthese Library 424,
https://doi.org/10.1007/978-3-030-41124-4_15

Each one of us is a person (self). As a psychological subject, each person is a singular being, namely, he or she is not similar to any other person; he or she is unlike any other person. Singularity is the distinguishing mark of subjectivity, personhood, and selfhood. There is something substantial about each person that makes him or her dissimilar to any other being. As a biological creature, on the other hand, each person (self) is unique, namely, such a creature is not identical to any other biological creature. For instance, as a psychological subject the person called James Joyce was a singular being. In a substantial sense there was no other person like James Joyce; whereas as a human being, as a higher organism, James Joyce was unique, namely there was no other human being identical to James Joyce. Hence, James Joyce's brain, immune system, and fingerprints were unique; there was no other human being who could have had Joyce's brain, immune system, or fingerprints. They pertained exclusively to him, yet they shared many properties with other similar organs, systems, or tissues of other human beings. Thus, the brain, immune system, and fingerprints of each human being share some common properties, some similarity with those of other human beings.

Uniqueness implies irreplicability or unduplicability. Necessarily, there is no replica or duplicate of my fingerprints, immune system, or brain. My body as a whole is irreplicable or unduplicable. Even if two higher organisms, such as clones or "identical" twins, are supposed to share the same genes, these organisms are not identical or, at least, not strictly identical.

Singularity is what distinguishes us as psychological subjects from any objects. Subjectivity, personhood, and selfhood, as psychological traits, pertain to the category of singularity. Only psychological subjects are singular, whereas some objects can be unique but never singular.

Our psychological singularity has indications. Let me mention two of them. The first is the psychological anomalousness (*nomos* is "law" in ancient Greek, whereas "anomalous" means "being not subject to laws"): our mind, unlike our body, is not subject to the laws of nature, namely, to the laws of physics, chemistry, biochemistry, and biology. There is no physics, chemistry, biochemistry, or biology of the mind. As for psychology, it provides us with nothing that can be considered as a law of nature. Any psychological subject is exempt from such laws. Even if human ways of behavior, attitudes, reactions, and the like show some similarity, order, and structures, still there is no nomic necessity whatever that any of the human subjects should behave or react according to the expectation or predication that such similarity, order, and structures would imply. Human creativity, for instance, is not subject to any psychological law, structure, paradigm, or order and, thus, it can be entirely unexpected or unpredictable. Neither is human creativity subject to any rule. Unlike the psychological, the physical, whether chemical, biochemical, or biological, or strictly physical, is necessarily subject to the laws of nature. No individual body—

our immune system. Note that a novel discipline—psychoneuroimmunolgy—has gained great support recently thanks to brilliant Israeli researchers Professor Asya Rolls (Technion, Haifa, Faculty of Medicine) and her mentor Professor Michal Schwartz (Weizmann Institute, Rehovot).

an organism, for instance—is exempt from these laws. The bodily or the physical, unlike the psychological, is inescapably nomic, namely subject to laws.

The second indication of the psychological singularity is epistemic private accessibility. In principle, we have public access to any object, body, or organism. For instance, our brain is publically accessible directly or indirectly (by means of brain imaging technology), whereas there is no such access to our mind. Only I have access to my thoughts, feelings, and emotions. If I do not report or "reveal" them to other persons, no other person except me has epistemic access to them. My mind constitutes an inner reality, to which no access from without is possible. As for the future, I see no serious reason to believe that brain imaging, for example, will allow us access to the mind of any person (Gilead 2015b). Obviously, the way I experience, feel, or think about anything is singular, and nobody else can experience, feel, or think what is going on in my mind as I experience, feel, or think it. But private accessibility is much more than that, for it means that nobody else can have epistemic access to my mind. Even if my behavior, expressions, or reactions convey the impression to another person what my psychological state is, that person still has no epistemic access to my mind. Revealing my mind to others is simply a phrase, which does not indicate any possibility for epistemic access from without to my mind.

Our ordinary experience provides us with many solid reasons to consider mind and body as distinct and yet as inseparable. Thus, we have many good reasons to maintain both psychophysical irreducibility (namely, the mind is irreducible to the body and vice versa) and psychophysical unity, which means that mind and body are inseparable and that there is a full correspondence between them. In other words, psychophysical unity means that our body embodies our mind.[2]

Maintaining both these two psychophysical assumptions means that our psychological singularity necessarily corresponds to our biological uniqueness and vice versa. The singularity of the mind is the uniqueness of the body, or the singularity of the mind is embodied as the uniqueness of the body. As singularity is ascribable only to psychological subjects, on the grounds of the psychophysical unity or inseparability, each person as an organism is unique, whereas, as a psychological subject, this person is singular. Psychological singularity, which is anomalous, and biological uniqueness, which is yet nomic (i.e. subject to laws), are inseparable or united. In every case in which an individual organism is unique, namely, unduplicable or irreplicable, psychological traits should be ascribed to it as an embodied psychological subject. The uniqueness of a person as an organism reflects his or her singularity as a psychological subject. Unique organisms are thus subject to psychological and cognitive traits or, in other words, mind necessarily pertains to such organisms to the extent that each one of them is induplicable or irreplicable.

Because of the psychophysical inseparability or unity, our private accessibility to our mind corresponds to a parallel state concerning our immune system—as we

---

[2] Or, our mind is embodied in our body. In the possibilist terms which panenmentalism endorses, the body is the actuality of our mind, which, in turn, is a singular pure possibility. Hence, the mind is actualized in our body.

shall see below, the immune system of each one of us has a sort of private or individual accessibility, as it has access only to the body to which this system pertains.

Persons are systematic and coherent complexes of interconnected and interdependent heterogeneous parts, all sharing one and the same psychological reality, privately accessible. Hence, persons can be embodied only as higher, multicellular organisms of a special kind, capable of self-awareness and of relating to other persons, each of whom has a unique immune system.[3] Each such organism is a unique, irreplicable biological individual.

All the aforementioned psychophysical assumptions are undoubtedly subject to debates, criticisms, and questioning. My approach to the psychophysical problem is not only a new kind of possibilism, it also opposes externalism, naturalism, and physicalism (or materialism). As metaphysical, such controversies will not end, and each side has its strong supporters as well as its opponents. Recently, Saul Kripke, Frank Jackson, and David Chalmers, to mention only three, have suggested various very solid reasons and arguments against physicalism. Equally, there are some enlightening approaches subscribing to physicalism and suggesting counterarguments. The same holds true for externalism (like physicalism, there are various kinds of externalism) which I also oppose and whose classical adherents were Quine and Donald Davidson. Of course, there are strong supporters of naturalism as well as no less strong opponents. The issue of psychological private accessibility is also controversial, and naturalist and externalists, such as Donald Davidson, have generally denied its plausibility (as did Wittgenstein in *Philosophical Investigations*). All the more, the problem of personal identity has been very much with us, beginning with David Hume and going on to Derek Parfit and many others (see Gilead 2009). Finally, psychological anomalousness is also philosophically controversial and yet, undoubtedly, it is perfectly compatible with the singularity of each person as a psychological subject.

I would take issue with any of these controversies, but this Chapter is not the proper place to defend my metaphysical view as challenging other views. Let me say only this: the reducibility of mind to body, which the reductionist physicalism or materialism supports (is there any genuine physicalism that in fact does not support a reduction of the mind to the body?) is entirely incompatible with the widely acknowledged distinction between mind and body, of which all of us, whether physicalist-materialist or otherwise, should be well aware. Reduction should be considered as unacceptable whenever its price is too high, and any psychophysical

---

[3] For a general definition of organism—"a functionally integrated living thing, highly organized, and made of interdependent parts"—see Pradeu 2013, p. 79. Pradeu "makes clear why taking immunity into account sheds light on the individuation of every multicellular organism" (ibid.). Although most of his paper focuses on the organism as a unified and cohesive multicellular individual, he devotes some of it to show how immunology can be useful also to better understand biological individuals other than multicellular organisms. I, however, do not apply the concept of personal individuality to every multicellular organism, let alone to other kinds of organism (whether unicellular or superorganisms, such as bees, ants, and termites), but only to those who are endowed with highly functional brains that embody self-awareness and the capability of relating to other persons as subjects.

reduction ignores major psychophysical differences, varieties, and richness and renders our experience much shallower, poorer, and narrower than it really is. Such reductions inescapably lead to psychophysical poverty, which is susceptible to strong doubts and criticism. As for the psychophysical *unity*, dualists (to begin with Descartes), Spinoza, or any physicalist, *mutatis mutandis*, have acknowledged it. We are all well aware of the indisputable fact that there is a very strong connection between our mind and our body despite the differences between them. Finally, as for psychological private accessibility, despite the cliché "a penny for your thoughts," we have no epistemic access to the mind of another person. If we believe otherwise, it is only because of a category mistake—the access that we really have is to the intersubjective (or, in case of intimacy, the interpersonal) implications or reflections of what occurs in the other person's mind, whereas his or her mind per se, intrinsically, is beyond such access. Thus, the access we have to the intersubjective implications or reflections of this or that mind is only relational and by no means intrinsic. What is going on in a mind reflects on our intersubjective reality to which we have an epistemic access as we, as psychological subjects, share this reality (it is "inter us"). As I have just said, externalists and naturalists do not accept the idea of private accessibility, but I know of no externalist or naturalist who can offer a penny or even much more for any of my thoughts unless I informed him or her *about* them (for extensive arguments to defend psychological private accessibility consider Gilead 2003, pp. 43–75, 2009, 2011).

## 15.2 Applying Psychological Terms to Immunology: Breznits's Contribution

It was only in June 2013, while reading the psychologist Shlomo Breznitz's autobiography (Breznitz 2012, especially pp. 123–132), that I have become acquainted with his inspiring paper, "Immunoalienation: A Behavioral Analysis of the Immune System" (Breznitz 2001).[4] Having read this paper, it dawned on me that the aforementioned psychophysical assumptions may shed new light on some recent novelties in immunology.

Immunoalienation is the process in which the immune system deviates from its initial status. When the immune system recognizes factors of the organism ("self-factors") as alien, autoimmune reaction may occur. The main thesis of Breznitz's paper is about the possible incongruence within the context of the warning immune system of two separate *psychological* concepts—objective threat, signifying real danger, and threat "as defined by the appraiser," which, in this case, is the immune system (Breznitz 2001, p. 88). Having read Breznitz's paper, I have turned to some more recent developments in immunology. In this Chapter, I would like to show that

---

[4]Which Breznitz considers as the most important work he has ever made, expecting yet for acknowledgement (Breznitz 2012, p. 132).

the fascinating ideas that Breznitz and some immunologists share can be nicely interpreted in the terms of my psychophysical assumptions. In other words, his paper suggests a novel example in which our immune system, which is a biological entity, is also subject to psychological traits, in addition to chemical and biological properties. Though a great deal of the immune system is subject to physical, chemical, biochemical, or biological properties, it is still equally subject to psychological traits as they are embodied in that system. No wonder that psychological states strongly, sometimes even quite dramatically, reflect on our immunity and the behavior of our immune system.

One of the bold ideas in Breznitz's paper is that no error or mistake is involved in using psychological terms in immunology, for the psychological mechanism of false alarm ("Cry wolf") describes precisely and explains an undeniable immunological phenomenon. In fact, Breznitz demonstrates how psychological mechanism is biologically or immunologically applied, i.e. embodied. In the terms of my psychophysical assumptions, an immunological mechanism, in this case, implements, embodies, or actualizes a psychological trait or state of a special kind of reaction to threats. Below, I will say more about Breznitz's contribution.

The terms "self," "nonself," or "protected and unprotected self" frequently appear in Breznitz's paper. The use of such terms in immunology raises some philosophical and scientific problems. For instance, even though Anne-Marie Moulin and Alfred I. Tauber consider these terms as metaphors, they did not ignore their indispensability and fruitfulness for immunology (Tauber 1997). Thomas Pradeu points out the ineradicable alleged imprecision involved in these terms while applied to immunology. Such imprecision, he believes, is intolerable insofar as the exact sciences are concerned (Pradeu 2012, p. 129). Moreover, he believes that "the self-nonself theory did not allow for the full understanding of modern immunology's experimental data" (ibid., p. 130).

As early as 1974, Niels K. Jerne applied cognitive terms, such as recognition, learning, and memory, to the immune system (Jerne 1974), which he found analogous to the central nervous system.[5] Furthermore, his Nobel lecture in physiology or medicine of 1984, referring to Noam Chomsky's generative grammar, made an analogy between linguistics and immunology, between the description of language and that of the immune system (Jerne 1984a). Jerne says: "I find it astonishing that the immune system embodies a degree of complexity which suggests some more or less superficial though striking analogies with human language and that this *cognitive system* has evolved and functions without assistance of the brain" (Jerne 1984a, p. 223; italics added). To ascribe cognitive functions to the brain is undoubtedly justified, but what allows us to ascribe such functions to the immune system? In what follows I will show, on the grounds of my psychophysical assumptions, why we are allowed to apply cognitive and psychological terms to our immune system.

---

[5] Jerne's immunology endorsed the view that "the immune system (like the brain) reflects first ourselves, then produces a reflection of this reflection, and ... subsequently it reflects the outside world ... The mirror images of the outside world, however, do not have permanency in the genome. Every individual must start with self" (Jerne 1984b, pp. 19–20).

## 15.3   A Comment on the "Cognition" and "Individuation" of Bacteria, Social Amoeba, and Social Insects

Pamela Lyon's biogenic approach, "asking psychological questions as if they were biological questions" (Lyon 2006), is a tantalizing view according to which cognitive capability should be ascribed even to bacteria: "A bacterium may not remember much for long, but it must remember, if only for a few seconds—which may be, relatively speaking, a long time for a microbe" (ibid., which shows the author's pretty sense of humor).[6] Despite my appreciation and the significant merits of Lyon's biogenic approach, I do not subscribe to this approach but prefer quite a different one instead, considering individuality, uniqueness, and singularity and especially concerning the individuating function of the immune system. Against the background of the current Chapter and the aforementioned psychophysical assumptions, I would not ascribe cognition, let alone psychological traits, to bacteria unless in a highly metaphorical sense. Undoubtedly, at least to the extent that scientific approaches are concerned, the case appears that we are not allowed to attribute phenomenal qualities (*qualia*), even in a highly metaphorical sense, to bacteria. As for knowledge and intentions, to ascribe them to bacteria appears to me quite farfetched, as the *events* in which bacteria are involved should not be considered, under philosophical scrutiny, as *actions*. Why, according to Lyon's approach, do the reactions of an organism to its environment result in constituting the organism's "cognitive reality"? It is because Lyon defines knowing as "effective action." I can quite easily imagine an effective event, process, change, or even "behavior" (such as atomic, molecular, or mechanical "behavior") which should not be called "action" (unless by question begging) and which is involved with no knowledge or cognition at all (let alone consciousness). What adapts lower or simpler organisms to their environment would not be knowledge at all. Why should we subject evolutional changes of such organisms to "cognition" or "knowledge"? At least to the extent that these organisms are concerned, evolution implies events or processes, not actions. In general, the survival of the fittest does not necessarily depend on whatsoever action, cognition, knowledge, or intention of the creatures involved. I do not see how a theory of action can be applied to the evolution of simple organisms such as bacteria.

As for similar interesting contributions,[7] they should be discussed in a similar vein. Unlike bacteria, "social amoeba," bees, or ants, each higher organism, even those socialized in herds, has an individuality of its own. I understand the concept "higher organism" to apply only to organisms that are substantial individuals, each of which has a unique immune system of its own. We should not consider the herd as an individual unified organism. In contrast, we may consider colonies of bacteria or those of ants, bees, and other social insects as individual unified organisms. Each

---

[6] Cf: "recent developments in microbiology undermine standard arguments against bacterial cognition and a closer look at bacterial behavior would reward cognitive scientists" (Lyon 2007, p. 823).

[7] Such as Chen et al. 2007, Ugelvig and Cremer 2007, and Marraffini and Sontheimer 2010 (assuming a discrimination between "self" and "nonself" on p. 186).

of such colonies may be analogous to a multicellular individual. We may compare each ant or bee to a cell in a higher organism but we should ascribe non-metaphorical cognitive and psychological terms only to individuals which (or who) are higher organisms embodying cognitive and psychological traits. We should, of course, not ignore ecology, and we should treat our body as an individual ecological system (a conception which rendered possible the first successful kidney transplantation in humans, as Jean Hamburger reported about it). It is clear that in the ecological system of our body bacteria have a role in our immunity (see, for instance, Costello et al. 2012). Nevertheless, none of this diminishes even slightly the substantial individuality and relative separation of any higher organism and its immune system.

## 15.4   Irun Cohen's Novel Cognitive Paradigm for Immunology

Challenging the incompleteness of Burnet's clonal selection paradigm, Irun Cohen suggests a fascinating idea, especially from the perspective of my psychophysical assumptions. Cohen suggests a "cognitive paradigm" for immunology (Cohen 1992a), according to which the "immune system must *know* to focus on particular antigens and how to *evaluate* their context before it actually encounter the antigens" (ibid., p. 444; italics added).[8] Cohen defines cognitive paradigms as "founded on the idea that any system which collects and processes information will do its job most efficiently by having an internal representation of its subject. ... in a sense, a cognitive system is a one that *knows* what it should be looking for ... their internal organization endows them with a kind of *intentionality*" (Cohen 1992a, p. 443; italics added).[9] Even though "the intentions ... are ours, not nature's" (Cohen 2000, p. 50) and the cognition in discussion does not involve consciousness (ibid., p. 92), the intentionality he ascribed to the immune system is not metaphorical; it simply does not reflect any consciousness or teleological assumption (ibid.).

Cognitive systems differ strategically from other systems in the following capabilities: (1) exercising options, namely, making choices or decisions (given that there are also unconscious decisions[10]); (2) containing within them internal images

---

[8] Or, "the healthy immune system usually can be fooled only once. ...The system can *learn to interpret* context" (Cohen 1992b, p. 491; italics added).

[9] In contrast, philosopher Peter Malender, criticizing Edward Levy and Mohan Matthen, argues that immunology does not and needs no intentional explanation (Melander 1993). See Matten and Levy 1984.

[10] Cohen 2000, p. 68. Moreover, these decisions are associated with a will: "instead of passively receiving what the environment imposes, the cognitive system exerts its *will* ... in *choosing among alternatives*. ... cognitive systems are more resourceful: not only do they choose, they seek" (Cohen 2000, p. 69; italics added). Nevertheless, the choice under discussion is deterministic and "is not dependent on any self-reflective consciousness or mystically free will" (Cohen 2000, p. 182).

of the environment; and (3) self-organizing—using experience to build and update their internal structures and images (Cohen 2000, p. 64). Some of these images are *abstract* as they are not made of matter but created by processes (ibid., pp. 174–175; for instance the doctor, checking the patient's white blood cell count, consciously diagnoses an infection; similarly, the immune system unconsciously diagnoses or recognizes the state of some tissue as infected). Furthermore, the abovementioned capabilities "in concert make it possible for cognitive creatures to interact with the world in a way that supersedes the confines of evolutionary genetics. Cognition will turn out to be a form of meta-genetic adaptation. *Cognition, as it proceeds, creates individuality*" (ibid; italics added).[11] As we will see, on the grounds of my psycho-physical assumptions, individuality plays a crucial role in subjecting the immune system to cognitive and psychological traits.

Cohen shows in detail "how antigen selection operates through internal images of infection and of the self (the immunological homunculus). These images in part are encoded in the germ-line, refined in the thymus and primed by mother" (Cohen 1992a, p. 444).[12] Cohen mentions a threefold set of primary internal images, which constitute "a reference point that define the *intentionality* of the immune system: which antigens it should seek out and *remember*" (Cohen 1992b, p. 492; italics added), or "the germ-line elements … the B cells, and the T cells each analyze different features of the antigenic entity and extract the special *information they intend to see* …" (ibid; italics added).[13] He also mentions "apparatus of processing and *presentation*" (Cohen 1992b, p. 494; italics added) and "internal images" allowing "the system to encode the essential fragments of the antigenic world to the system specifications and utility" (ibid.). Thus, "contrary to the expectation of clonal selection, the germ-line effectively encodes a primitive internal image of bacteria, viruses, and the context of inflammation" (Cohen 1992b, p. 490). In addition to the term "internal image," Cohen refers to "primitive information arms cells with the capacity to *recognize* and respond to invaders" (ibid.; italics added). The daring idea of an immunological homunculus is about "an internal image of the self acquired by early recognition of self antigens, both in the thymus and in the periphery" (1992b,

---

[11] Cf.: "Cognition enriches the diversity of existence. … Cognitive creature, in contrast to non-cognitive creatures [such as bacteria or trees], learn individually and diversify as individuals, and not only as species" (Cohen 2000, p. 93).

[12] Of course, the immunological homunculus is not some "little man" sitting outside of the immune system and "rules" it; the homunculus is, instead, "the characteristic organization of autoimmunity itself" (Cohen 1992b, p. 493) or "a shorthand designation for the images of the body that self-organize in … the immune system" (Cohen 2000, p. 205).

[13] Though the information mentioned is not in a strictly psychological sense but rather the one used in Claude Shannon's information theory, consider, yet, the following analogy: "the eyes, organs designed for receiving information, also serve as organs designed for transmitting information [in the cognitive and psychological sense]. … there is a principle of biological signaling here, … it working at the molecular level in the immune system" (Cohen 2000, p. 77). And a little bit later, discussing information, Cohen writes: "Meaning is what information does. Indeed, the combination of information [now in a strictly psychological sense] with affect, which generates meaning, gives rise to behavior that feeds back to influence by cognitive creature's world" (ibid., p. 78).

p. 492). It is the immune system's *representation* of the body. Similar to the neurological homunculus, the immunological homunculus contains a "picture of the individual own body" (Cohen 1992a, p. 443). With Cohen, "the individuality of the mind arrives with the individuality of the brain" (Cohen 2000, p. 4), the immune system "defines the material components that make up the self," it is the "guardian of our chemical individuality," and it "establishes the molecular borders of each person" (ibid., p. 5). Note that Cohen's use of the attribute cognitive does not imply consciousness and the term "intentionality" in his works is devoid of personality (Cohen 1992a, p.443), yet it is certainly about individuality and uniqueness. Moreover, the affinity of Cohen's cognitive approach to the immune system and of his approach to the brain is closer to the way in which Gerald Edelman describes the central nervous system (Cohen 2000, p. 189).[14] This affinity, too, supports the application of psychological or cognitive terms to the immune system.

Uri Hershberg, one of the disciples and followers of Irun Cohen, claims in an interview with a philosophical journal that the line between biological systems and cognitive systems is a fake one, as there is no categorical difference between these two kinds of systems; the difference is, instead, of a "mental extension" (Hershberg 2012, p. 29). Thus,

> Even a single cell that needs to act in the world does not do it like a machine. It acts with signals, with meanings. I am not saying that cells have abstract thoughts the way we do. They do not have high cognitive potential. But the way we manage to manipulate those is that we have senses. Even a single cell organism does not really have sensors. They have senses. (Hershberg, ibid.)

As for the term "self," Hershberg admits that it is "a philosophical concept that stands on individuality and what individuality means or how important it is" (ibid., p. 28). Yet, he states, without such a concept immunology cannot do their work, unless immunologists lie to themselves (following ibid., p. 36). The understanding of the immune system implies understanding of biological individuality[15] or uniqueness. Moreover, as the current Chapter attempts to show, a philosophical view is essential for immunology.

I consider Cohen's cognitive paradigm for immunology as most interesting from the perspective of my psychophysical assumptions, first of all for its use of psychological or cognitive terms literally and not metaphorically. What makes it even more attractive for my analysis is its aim to explain how the immune system "maintains and protects the individual" (Cohen 2000, p. xix). As I have mentioned above,

---

[14] For a panenmentalist critique of the emergentist psychophysical view of the mind in general and of that of Edelman in particular, consult Gilead 1999, pp. 12, 143, 144, and 161. Cohen, too, is a Darwinian emergentist. Yet, the divergences in psychophysical views should not hinder us from accepting much of Cohen's cognitive conception of the immune system.

[15] Which is a separate important issue in the current philosophy of biology. I do not discuss this issue in the present Chapter. I refer to the biological individuality, as a multicellular organism's individuality, only to the extent that the immune system is concerned. Note that Pradeu convincingly demonstrates that the immune system plays a crucial role in the emergence and maintenance of individuality and, thus, in defining biological individuality (Pradeu 2013).

Cohen's conceptual framework "deals with the way the immune system relates to the individual body and defines its individuality" (ibid.; cf. p. 244–245: "tale of cognitive individuality" and "the story of the self"). In the terms of my psychophysical assumptions, the immune system in Cohen's cognitive view relates to the unique identity of each human body which, in these terms, embodies the singularity of a person (at least as far as human beings are concerned). As Cohen's analysis shows, following the sublime Talmudic idea that "Adam was created as a singularity," "each individual fashions a unique world out of his or her unique somatic experience. Therefore, no individual is redundant, ever" (ibid., p. 244). Cohen's analysis has "added the cognitive immune system to the armor of individuality. ... The tale of Adam, like the message of this book, is a tale of cognitive individuality" (ibid.). This is beautifully compatible with my independent view of singularity (especially in Gilead 2003 and, to begin with, Gilead 1999).

Contrary to Tauber and others, Breznitz, Cohen, Polly Matzinger, and other immunologists[16] use the abovementioned indispensible cognitive and psychological terms in a literal sense. In the terms of my psychophysical assumptions, they recognize, in fact though sometimes unknowingly to this or that extent, that our immune system embodies or implements *unconscious* psychological traits.

Cohen ends his sequel paper thus: "The cognitive paradigm is an immunologist's paradigm of the immune system's paradigm of the molecular world" (Cohen 1992b, p. 494), which means that his paradigm represents the paradigm according to which the immune system consists. Thus, Cohen commits his paradigm to, at least some, correspondence with the immune reality, which means that in fact he considers this paradigm as not only useful but also as true. My view explains this better, I believe, for Cohen does not show what makes the immune reality to correspond to his or other's paradigm. Speaking about usefulness, Cohen's view can be considered as compatible with Tauber's view about "self" and "nonself" as useful metaphors, whereas Cohen, in fact, suggests much more than that. On the basis of my psychophysical assumptions, he would be entirely allowed to use psychological or cognitive terms literally in analyzing the immune system. Furthermore, Cohen has taught us brilliantly that borrowing ideas and insights from art can contribute to immunology (Cohen 1994).

---

[16] Cohen mentions that F. J. Varela, Antonio Coutinho, and others, too, "have called attention to the cognitive properties of recognition, learning, and memory as fundamental to immune behavior" (Cohen 1992a, p. 444).

## 15.5    Tauber's Criticism of Cohen and of Other Immunological Cognitivists

Tauber refers to Cohen's immunological homunculus and to what he calls Cohen's "cognitive metaphor" (Tauber 1997, pp. 178–182). Tauber ascribes the theories of both Matzinger and Cohen to the "contextualist scheme" (Tauber 1998). According to Tauber, Cohen made a critical theoretical turn regarding the entire notion of self-hood (ibid., p. 466). As I see it, Tauber's recent challenge of the immune self or individuality (Tauber 2012), concerning especially the symbionts hosted in our body and which take part in the immune system, does not affect Cohen's idea of the immunological homunculus in particular and the immune self in general. Given the fact, which Tauber 2012 mentions, that the immune system does not function properly if the symbiotic microbes are not residing, for instance, within the gut, why should the immune system not *mobilize* such organisms for the sake of the body's maintenance and immunity? The hologenome theory of evolution, to which Tauber 2012 refers, is possibly adequate for understanding the evolution of coral reefs, for instance, but not necessarily for that of human beings. This may not affect at all the understanding of the human individual organism (or even of the higher vertebrates) as the object of natural selection and of immunology as well. Is the hologenome theory of the evolution of the immune system valid for higher vertebrates in general and for human beings in particular? No human being is simply a "society of cells." Rather, each human being is a singular individual, a self. Contrary to Tauber, I attempt to demonstrate that self, recognition, memory, and learning, as correctly applied to the immune system, are not metaphors at all. Note that Tauber 2012, contrary to Cohen, entirely ignores the unconscious nature of these terms when applied to the immune system.

In 2013, Tauber suggests replacing the representational cognitivist approaches to immunology with "ecological" approaches. Tauber states:

> While a representational model has dominated immune theorizing, recent research supports the utility of an "ecological" orientation, which reflects the growing interests in systems biology, where the organism becomes a "node" in an ecological network (Tauber 2012). This approach explores integration of functions (e.g., development, metabolism, immunity) and thereby emphasizes inter-connections, regulative dynamics, and organizational structures of holistic constructs, where *individuals* become subsumed to *relationships* of various kinds. (Tauber 2013, p. 241)

In a striking contrast, Cohen, Breznitz, and my philosophical approach consider the individual, especially the human individual, by no means as a "node" in an ecological network, which may be more suitable for bacteria, ants, and the like but not for higher organisms, especially human beings. Such individuals are not "subsumed to relationships of various kinds" but, on the contrary—the relationships are subsumed to the individuals. What comes first is the immune system of the individual—not the external ecological system. Human society, per se, has no immune system; each individual owns it, and the first function of this system is to ensure the existence of the individual organism and to maintain its integrity and health. Actions and agency come later; they are secondary. First the system has to perceive and to learn, and only then to act or to decide to remain passive and not to react. Tauber replaces identity by agency, representation and learning—by action, whereas the case appears to be otherwise—our immune system aims primarily at individual being and existence and only secondly at agency. The individuating function of our immune system should not be ignored.

Tauber complains that "notions of selfhood still undergird various contextual theories ...., where a "homunculus" (Cohen 1992b) or even a Bumetian-inspired concept of selfhood ... obscures Jerne's crucial insight that the self-other distinction as a metaphorical extension of human personal identity distorts the character of immune perception" (Tuaber 2013, p. 249, n 9). Again, immunological cognitivists such as Irun Cohen and Shlomo Breznitz make a legitimate literal use of psychological and cognitive terms in describing and analyzing our immune system. Moreover, I have not found even a shred of evidence that Cohen relies upon the problematic, even mistaken, analogue of the homunculus in the "Cartesian Theater." On the contrary, the homunculus that Cohen mentions reflects a fruitful neurological concept, to which the immunological homunculus is similar (Cohen 1992a, p. 443),[17] rather than anything of the "Cartesian homunculus." Such cognitivism does not pertain at all to the "philosophical infrastructure" that guides Cohen and other immunological cognitivists. Cohen never confused the immune system with the scientist as a person. Unlike us, the immune system, as a cognitive system, observes, perceives, reflects, considers, chooses, and decides about anything *only unconsciously*. This makes a major difference between scientists and the cognitive system as an *agency* of a self, protecting it and maintaining its integrity and health. Hence, the following criticism by Tauber has no solid ground:

> If immune selfhood reflects an underlying conception of a homunculus discerning itself from the other through a *cognitive* faculty (e.g., Cohen 1992b), then the same issues confronting current representational philosophies of mind lie latent in contemporary immune theory. Discerning those issues raises a new dimension in the critique of the immune self. Simply, with the invocation of agency, the weakness of the "Cartesian Theater" in the immune setting is apparent: The immune system does not reflect; it perceives without the interposition of an agent reviewing its findings. The *scientist* observes and constructs the self and its cognition. The phenomenon and the judgment of that phenomenon have distinctive epistemological characteristics that must be maintained. Simply, a sympathetic fallacy is committed when the immune system becomes an immune *self*. (Tauber 2013, p. 259)

---

[17] Both the neurological homunculus and the analogue "immunological homunculus" have recently proven to be quite fruitful (see, for instance, Poletaev 2012; Poletaev and Osipenko 2003; Poletaev et al. 2008, 2012; Gonzales and Lange 2007; and Zingrone et al. 2010). Following Irun Cohen, Alexander Poletaev coined the term "Immunculus," which is also the name of the scientific institute in Moscow, which Poletaev heads.

## 15.6    Biological Agency, Metaphors and Biological Individuality: Comments on the Views of Wilson and Dennett

At this point, we have to pay attention to Robert A. Wilson's arguments concerning the "metaphoric" use of cognitive and agency terms in biology (Wilson 2005). Wilson starts his book with the following striking question: "What are the agents of life?" (Wilson 2005, p. 3). Is this phrase "the agents of life" a metaphor? Such appears not to be the case. *Literary speaking*, there *are* agents of life and life is existence as well as activity (agency). In this sense, even a single ant is an agent of life, yet not an individual one, as it vitally depends upon its colony of which it is only a member or a part and in which it functions like a cell of a more complex organism. At least, from Wilson's viewpoint, too, human actions are not only biological events, as, in my terms, they are actualities of a human agent who is a psychical and cognitive subject and not only a biological one; these are human *actions* and not human *events*. Such events pertain only to the body and have merely physical or material significance (for instance, a patient's involuntary movements due to neural states are events and not actions though they are human). Thus, at least when humans are concerned, from a *biological* viewpoint, they are undoubtedly *agents* (in my terms—actualities of agents) in the full sense of the word. Moreover, in any case in which *information and representations (or images)* are concerned, activity (agency) is also biological, and the *literal* sense of the word "activity" is indispensable. For this reason, too, when brain studies are concerned, the term "cerebral homunculus" is not an anthropomorphic metaphor, or a metaphor at all; it is a biologically literal sense. And, as far as Irun Cohen's (and others') "immunological homunculus" is concerned, it is an immunological *literary* concept, by no means a metaphorical one. There are biological information, cognition, and images; and there are biological agents in the full sense of the term "agency" or "activity" and simply in the literal senses of these terms. To treat such terms as metaphors ends in rendering them meaningless or senseless. This would be so heavy a price that biology cannot afford it. Such terms are biologically indispensable and they are irreplaceable by other literal terms. Hence, they do not function metaphorically, without underestimating the indispensability of metaphors in rigorous scientific language.

Referring to Richard Lewontin's view that metaphors are indispensable for the languages of sciences and Tauber's view on metaphors in immunology, Wilson writes:

> In the biological and social sciences we often metaphorically extend our conception of mind from our paradigmatic individuals, human agents, to things that do not have minds. This use of the cognitive metaphor gains purchase across various biological hierarchies. In characterizing biological agents that are smaller than organisms, and typically a part of them, the cognitive metaphor is manifest in Richard Dawkins's metaphor of the selfish gene; in cell biology, where cells are described as recognizing, remembering, preferring, and seeking certain other cells or molecules; and in immunology, where the immune system is conceptualized as distinguishing self from nonself. (Wilson 2005, p. 42)

The viewpoint that I take in this Chapter is quite different, *mutatis mutandis*, from those of Wilson, Lewontin, and Tauber: because the body is an actuality of the mind, when minded organisms are concerned—the organism as a whole and *also its parts, including the microscopic ones*, consists of actualizations of a mind including its tiniest "parts" or "members." My psychophysical approach is holistic, namely, mind and body are inseparable and we cannot identify and understand the parts or the members but only against the background of the organism as a whole. Thus, when it comes to minded beings, the organism is literary characterized by cognitive and psychical or psychological terms. Also, the minded being as a biological agent as well as its microscopic parts or members together is an embodied or actualized psychical subject. As a result, human genes or those of a minded being are entitled to be literally characterized as selfish. In the cases of minded organisms, cognitive and psychical terms are *literary* indispensable.

Hence, Wilson's above-cited passage is not valid, I think, for terms such as the cerebral homunculus or the immunological homunculus as taking part in actualities or the actuality (the body) of any psychical subject-agent. Yet, Wilson writes, justifiably: "The cognitive metaphor is operative whenever psychological terms are used to describe actions or behaviors of nonpsychological agents, or to explain actions or behaviors not caused by psychological states" (Wilson 2005, p. 75). He is quite right not to apply such terms on microorganisms, ants, and other non-psychical or non-cognitive factors that are independent of the human body or of that of higher animals.

Wilson writes: "living things are agents" (op. cit., p. 4), which are individuals (ibid.). Moreover, "we think of life as we think of the mind—as tied to and delimited by agents" (ibid.). And, indeed, "this raises the question of just how closely life and mind are related" (ibid.). Undoubtedly, all this is very intricate and complicated. As you will see below, panenmentalism has tried its best to answer these questions with clear and explicit answers.

I find great merit in the following:

> The study of life and mind has been compartmentalized into, respectively, the biological and the cognitive sciences. This sort of disciplining of the domains of life and mind, however, was contingent rather than inevitable. While it has created the opportunity for deep insights into both life and mind, it has also produced its own blind spots. One of these concerns the role and conception of agents in thinking about life and mind. (op. cit., p. 5)

In this Chapter, I try to overcome this dispensable compartmentalization and to show how psychical singularity is embodied, better actualized, as biological uniqueness. This actualization is vital for understanding how our immune system operates.

Suppose that a critic would challenge me, using the following argument: cognitive and psychological metaphors are effective ways to model behavior, better agency, but to transfer this to the acts of the immune system, would simply be a metaphoric use, entirely groundless, from a viewpoint of Daniel C. Dennett, for instance. But Dennett is a reductionist, whereas my approach is very much against any sort of psychophysical reduction. Do I model biological phenomena according to human experience? Not at all. On the contrary, I point out an inseparable connection between singularity, which is psychical, and uniqueness, which is bio-logical or bodily. I refer to a metaphysical or ontological correlation or compatibil-ity, not to any metaphoric use of terms, and I do so by avoiding any psychophysical reductionism. Information and representation should be termed in cognitive terms, otherwise they would lose their meaning and sense. Thus, immunological or cere-bral homunculus is not a metaphor and it is biologically indispensable in its literal, rigorous sense.

When it comes to a radical attack on any application of cognitive and psychical terms on our brain, Daniel C. Dennett certainly comes to mind. Let us consider the following

> It is literally child's play to imagine the stream of consciousness of an "inanimate" thing. Children do it all the time. Not only do teddy bears have inner lives, but so does the Little Engine That Could. … Children's literature (to say nothing of television) is chock full of opportunities to imagine the conscious lives of such mere things. … It's obvious that no teddy bear is conscious, but it's really not obvious that no robot could be. What is obvious is just that it's hard to imagine how they could be. (Dennett 1991, p. 432)

Undoubtedly, children believe in monstrosities, witches, and magic beings or acts, and are under many illusions. To ascribe thinking or consciousness to machines may still be an illusion, no less illusory than ascribing it to teddy bears. At most, Dennett shows that a theory or belief such as his is possible for imagination, but no more. Illusions are possible, as other mistakes are possible, but this proves nothing about their veridicality. Dennett draws a comparison between "a bunch of silicon chips" and the human brain (1991, p. 433), but all the difference in the world lies between a machine, whether virtual or actual, and an organism or a part of it. Organisms endowed with a brain and nervous system can actualize thinking and consciousness, but no machine can. No machine, say, a robot or a computer, can think or be con-scious, not because to imagine how it could is sometimes difficult, but for the mani-fest reason that with my approach "a thinking machine" and "a conscious machine" are contradictions in terms or in concepts, unless we wrongly assume that con-sciousness and thinking are not necessarily psychical (Gilead 1999, pp. 137–158). Contradictions in terms are possible, albeit not logically, but they are still irrational or untenable, just as beliefs in magic of any kind are possible yet irrational and illusory.

Suppose, for a moment, that Dennett is right in saying about brain cells: "*not a single one of the cells that compose you knows who you are or cares*. … The indi-vidual cells that compose you are alive, but we now understand life well enough to appreciate that each cell is a mindless mechanism, a largely autonomous microro-bot, no more conscious than a yeast cell" (Dennett 2005, p. 2). First, this is a dog-matic reductionist view that should not be taken seriously in the biology of human beings. As an organism, each one of us is not just the sum of all its cells, for each cell cannot exist or be known and understood, unless it is considered as a part of the whole, which ontologically and epistemologically is prior to the cells. None of these cells would have existed, had it not been an actuality of a psychical subject who is a human being. All the cells and other members of our body are actualities of a psychical and cognitive, minded subject or agent. Not to see that simple fact is to be totally blind to the psychophysical unity and its meaning, to be a dogmatic materialist or physicalist who is completely blind to some undeniable basic facts about our life, or at least to dogmatically deny them. Moreover, Dennett appears to be completely blind to the complexity which is inseparable from organisms especially from human organisms; instead, he follows a very simple-minded reductionism. Second and finally, the unhappy example of a yeast cell is quite misleading. Erez Braun's novel discovery (Braun 2015) is that yeast cells are by no means such microrobots and, in fact, they are not subject to determinism; their self-regulation and free, spontaneous reactions to environmental stress is not subject to reductionist explanation such as that in Dennett's example.

As for individual organisms, biologically speaking, an individual is an organism that *to some extent* is independent on other organisms and the environment as well. We are not entitled to consider an ant or a cell in our body as a biological individual, since there is no existential independence that might be ascribed to any of them. No ant can live for a long while outside its colony. Equally, no cell can live outside the body unless it is fed by a special substratum. To the extent that the immune system is concerned, biological individuality is defined by the accessibility of the immune system of the individual organism in question. The immune system of an individual organism has access only to what is included within the boundaries of that organism (the body) and no access whatsoever to any other organism. Equally, the cognitive access of its mind is limited to that mind only; it is a private access. We have no cognitive access to any other mind. I say more of biological individuality in this Chapter.[18]

Wilson refers to the debate between individualism and externalism in biology (for instance, Wilson 2005, p. 31). This is a major question. On the background of this Chapter as a whole, I prefer biological individualism to biological externalism, but I do not ignore ecological views such as Spinoza's nor the Gaia hypothesis by James Lovelock and others. Such views do not consider this or that organism as a real or substantial individual, for each one of them is just a mode, member, or node (to take a term that Tauber uses) of a comprehensive living system. In contrast, at this point, I take a somewhat Aristotelian stance: each organism is an individual, primary substance, namely, despite its dependence on other organisms and on the environment, each of which enjoys some degree of independence, autonomy, and self-regulation.

---

[18] In contrast, compare the following: "individualism should be rejected. ... there are contrasting externalist views within the cognitive sciences worth developing further. These require rethinking many concepts central to the philosophy of mind and cognitive science, such as physicalism, computation, and representation" (Wilson 2005, p. 26). He appears to adopt this stance in his view of biological externalism. The same holds true for Tauber's recent view (2012 and 2013). My approach is very much against externalism, physicalism, and computation (according to which our brain is a most complicated and powerful computer).

## 15.7    The Application of My Psychophysical Assumptions to Immunology

The case appears to be that the self-nonself model is not sufficient for understanding and explaining many of the immune phenomena, which are explained by the danger model in its evolution.[19] Nevertheless, we must not decide between the models. As the immune system is so comprehensive and diverse, it is possible that no model can exhaust it (see, for instance, Vance 2000, pp. 1727–1728).[20] In any event, this does not mean that self-nonself factors should not be taken into consideration, either. To the extent that Breznitz's novel idea concerning danger and the "Cry wolf" phenomenon is concerned, the danger model is even more fruitful and compatible with the relevant immunological phenomena than the self-nonself model alone. Whichever model one may choose, we still have to apply terms such as "danger," "foreignness" (or "alienation"), and "self" to the immune system, literally and not in their metaphorical sense. I will now try to establish this literal application on the basis of my psychophysical assumptions.

From this viewpoint, our body, as a biological entity, embodies our mind, which is a psychological, singular being. The singularity of each mind is the singularity of a person, a psychological subject. According to my psychophysical assumptions, we ascribe singularity only to psychological subjects, whereas as organisms, which are subject to the laws of physics, chemistry, and biology, they share some similarities. The singularity of each one of them corresponds to the biological uniqueness of each.[21] Thus, my genetic signature, my genetic self, is unique as are my finger prints and my immunological system. It is a received view that selfhood is a psychological matter, not simply a physical or biological one, whereas uniqueness is a biological trait of higher vertebrates or mammals. Thus, the question of "alienation" or "foreignness" is crucial when it comes to the biology of higher vertebrates, especially human beings. As our immunological system reflects the uniqueness of each one of us as a biological creature, as a human being, we are allowed to use the

---

[19] The creator of the danger model and theory in immunology is Polly Matzinger. See, for instance, the following: "For many years immunologists have been well served by the viewpoint that the immune system's primary goal is to discriminate between self and non-self. I believe that it is time to change viewpoints and, in this essay, I discuss the possibility that the immune system does not care about self and non-self, that its primary driving force is the need to detect and protect against danger, and that it does not do the job alone, but receives positive and negative communications from an extended network of other bodily tissues" (Matzinger 1994, p. 991). Cf. Matzinger 2002a, b, 2007, 2011, 2012.

[20] Thus, Vance's concluding suggestion is quite reasonable: "Why not, rather, think of the immune system as a much more diverse collection of mechanisms and processes that have been cobbled together during the course of evolution? If I am right, it will be the details of these mechanisms and their interactions that will ultimately be of interest to immunologists, and not whether they conform to one 'paradigm' or another" (Vance 2000, p. 1728).

[21] Biological uniqueness is one of the major issues with which Pradeu 2012, too, deals with, raising the question, "What makes a living thing different from all other living things, including those that belong to the same species?" (Pradeu 2012, p. 2).

precise language of "foreignness" and "alienation," and "self-nonself." There is a strong, necessary, or inseparable connection between our mind and our body and, as a matter of fact, this connection is the psychophysical *unity*. This unity should be distinguished from identity, as mind and body are not identical or reducible one to the other, although their unity is inseparable. *There is, thus, an inseparability of personhood or singularity and biological uniqueness.* The immune system of a higher organism guards it against what endangers or threatens its uniqueness. Because the immune system, as part and parcel of our body, implements or embodies some psychological traits as biological actualities, we are allowed, contrary to Tauber, Pradeu, and others, to attribute memory,[22] recognition, person-nonperson, meaning,[23] and other psychological traits to our immune system, not in a metaphoric,[24] in however useful or fruitful way, but literally, as long as we are aware of the fact that we are discussing individual organisms, each of which is unique, and not psychological subjects each of whom is singular. We are thus discussing

---

[22] Cf.: "'Memory,' whether in the form of differentiated B cells or sensitized T lymphocytes, is an essential component of the immune reaction. … This 'positive
recency effect' (to borrow an expression from the *psychology of learning*) introduces a systematic *bias* into the cell population that composes the immune system" (Breznitz 2001, p. 89; italics added). Cf. Cohen: "Memory is another cognitive concept whose mechanism is clearer in the immune system than it is in the brain" (Cohen 2000, p. 186). As for biases, see those concerning the inner images in the cognition systems of our body according to Cohen (2000, pp. 74 and 198).

[23] For instance, Cohen 1992a, p. 442: "To rescue a signal from noise is not sufficient for fitness; *we* have to *know* the context in which the signal arrives. Context bestows *meaning*. The context *tells us* if the gun we *see* [cf. ibid.: "*what the system can see*"] is likely to be a toy or a weapon, if it is theater or murder. A processed peptide presented in the pocket of an MHC molecule may constitute an antigenic epitope for a T cell, but fitness cannot be promoted without more information. … Appreciation of context is the beginning of *wisdom*" (all italics added). With my interpretation, the italicized psychological or cognitive terms are mentioned in this text not as metaphors but literally with the stipulation that they do not involve consciousness and that meaning is simply "the impact of information as an outcome of interaction" (Cohen 2000, p. 98). For more about immune information, meaning, and chemical language, see Atlan and Cohen 1998. Note that Cohen discusses a "molecular dialogue" whose character "can be analyzed by exploring five attributes of *linguistic* communication: *abstraction*, combinatorial signal, *semantics*, *syntax*, and context" (Cohen 2000, p. 183; italics added). The term "abstraction" is in place here, for the immune system reacts to a processed peptide serving as a *representation* of an infection agent that is not in presence (ibid.). As a cognitive system, "the immune system recognizes not entities but *signs* of entities. Just as a spoken word is both a physical reality and an abstraction [namely, it is a symbol], a molecule may function as a physical abstraction" (ibid'.). See more about the immune semantic and syntax, language and dialogue (ibid., pp. 184–185). In sum, Cohen believes that "the immune language might share strategic structures with natural language worthy of study; they both are concerned with generating meaning out of information" (Cohen 2000, p. 185). Cf. Jerne 1984a.

[24] Unlike the following, concerning the danger theory: "Those who insist on experimental verification miss the point of these theories, which are essentially metaphorical generalizations, far abstracted from the gritty but testable details of immunity" (Vance 2000, p. 1727). Endorsing the idea of saving possibilities, I prefer instead Carolyn Strange's view as, while referring to Matzinger's novelty, she reminds the reader: "Upon breaking free of old assumptions, researchers can begin to consider new possibilities that had been quite literally unthinkable. When dogma predicts that an experiment would not work, it rarely is performed" (Strange 1995, p. 665).

psychological traits as biologically embodied under spatiotemporal and causal circumstances and as subject to the laws of physics, chemistry, and biology.

In light of such an explanation, Breznitz's view makes much sense. It was a brilliant idea to demonstrate that psychological patterns are implemented or embodied by the immune system. I believe that, in the future, immunologists may gain great benefit from learning more and more from psychologists in understanding recognition, memory, commitment,[25] tolerance, decision,[26] defense mechanisms, guardianship, protection, treating and mistreating threats and danger, containing or holding, associating,[27] responding or answering,[28] education (Matzinger 2012, p. 316), dialogue or conversation (ibid., op. cit., p. 317), and the like, all of which have served immunologists quite fruitfully, not as merely metaphors, however useful and even indispensable. Such terms shed light on immunological phenomena. Such is the case because the identity of some of the biological embodiments or implementations taking part in our body rests inescapably on the psychological, on the singularity of each one of us.

Our era is facing an imminent danger to the independence of psychology, which many brain scientists as well as psychologists attempt to reduce to the research into the functions of the brain. This most dangerous trend may endanger not only the existence of psychology as a substantive, independent science, irreducible to biology, neuroscience, computer science, and the like, but, even more, it may endanger biology as a wide-ranging and promising science and render it less abundant and much less promising. Psychology may save many fruitful possibilities for biology. Breznitz has made a pioneering step toward this aim, but so far this step appears to have no major continuation. It is an annoying symptom of the intellectual poverty of

---

[25] Which Breznitz 2001, p. 90 mentions.

[26] Breznitz refers to "a 'decision' must be made whether to effect an immune response against them or to tolerate them. The most typical instance of the need to deal with what psychologists call ambivalence is in the context of an immunological *cross-reaction*. … In this respect, not unlike in the case of the psychology of conflict, subsequent reactions follow prior commitments" (Breznitz 2001, p. 91). It appears that Breznitz hesitates between his bold application of psychological terms to immunology and the dogmatic reluctance to do so, for there are cases in which he does not mark psychological terms with inverted commas (in this citation, ambivalence, tolerance, and commitments), whereas in other cases he uses them (in this citation, "decision;" and, while Clarke and Playfair mention the term "decision" with no inverted commas, in the citation from their paper, Breznitz, nevertheless, adds them in his comment to the cited term). As far as I can see, he is not consistent in this matter. From my viewpoint, he is consistently allowed to dispense with the inverted commas in each of these cases. So does Matzinger (except for "self" and "nonself") by and large in her papers known to me.

[27] Cf., for instance: "If we give the body a molecule at the time that something else has caused damage, the immune system will *associate* that new molecule with the (unassociated) damage, and respond to it" (Matzinger 2012, p. 311; italics added). Cf. "cognitive systems make associations" (Cohen 2000, p. 69).

[28] "The first question the immune system needs to *ask* when faced with something" is "'Do I *respond* or not?' However, once *you respond*, there is a second question to *ask*, 'What kind of response do I make?' … How does the immune system *determine* what kind of response it is going to make?" (ibid., p. 313; italics added).

our era. Applying the "psychological properties of warning systems, and particularly problems of credibility and false alarms" to the "lawfulness discovered in biological warning systems" (Breznitz 2001, pp. 86–87) is a most promising and fruitful idea. Curiously enough, the danger immunologists, such as Matzinger,[29] often discuss the stress of cells and tissues,[30] whereas though Breznitz is a celebrated specialist in the psychological research of the phenomenon of stress, he does not mention the term "stress" or "distress" in his paper on immunoalienation. Nevertheless, he explicitly applies his false alarm theory ("Cry wolf") to immunology. The association between Matzinger's danger theory and the false alarm concerning the immune system has been suggested in the literature,[31] though not by Matzinger herself. The danger theory has an interesting application to artificial immune systems in computer science and engineering, also associated with the problem of false alarm (Dasgupta et al. 2010).

If Breznitz were familiar with Matzinger's theory, according to which not only the immune system takes part in protecting our body against danger and threats but our tissues and members also take part in it, the following problem would have been solved according to his own theory:

> From the viewpoint of the behavioral analysis of warning systems, a central weakness of the immune system stems from the fact that it has a virtual monopoly over the protection of the organism of which it is a part. This precludes corrective feedback about its effectiveness. The history of human warning systems abounds in examples in which the absence of extrinsic information (that is, from sources other than the warning system itself) can lead to major mistakes. The monopoly over information ensures the unchecked growth of biases. (Breznitz 2001, p. 92)

In Matzinger's theory, there is always room enough for extrinsic information, which the immune system processes. Thus, much like in human behavior, in cases where there is extrinsic information, the chance of making mistakes, especially serious ones, is minimized. Were human intelligence agencies as sophisticated as our immune system, they would undoubtedly have made fewer mistakes in processing information. No act of espionage—however sophisticated—can do the job as efficiently as our immune system.

---

[29] The immune system selects as dangerous anything that causes cell stress or necrotic cell death (Matzinger 1994, pp. 1023 and 1037). Cf. Vance 2000, p. 1727. Note that "while it might be agreed that the immune system can sense endogenous signs of distress (danger), the question is whether immunology should be limited, a priori, to the study of these endogenous signs—especially as the molecular mechanisms that the immune system uses to detect 'strangers' is becoming increasingly understood" (Vence 2000, ibid.). This may make Breznitz's immunoalienation even more attractive. After one year of the publication of his paper, Matzinger had published a very short outline of her novel theory in the very same journal (Matzinger 2002b)!

[30] Cohen's immunological theory explains how the immune system detects and focuses on self stress proteins and reacts to them (Cohen 2007, p. 571). Stress proteins "can provide the … immune system with crucial information about the local state (stressed or un-stressed) of the tissue" (ibid.). Breznitz's psychological theory of stress may shed more light on this important aspect of the immune system.

[31] For instance, Pittman and Kubes 2013, p. 320; and Mills 2012.

As for Breznitz's application of the psychological distinction between minimal and difference thresholds of the immune system, it may be found useful in understanding, within the danger model, why monoclonal processes (whose evolvement begins with one single cell), such as monoclonal tumors or pregnancy, are not attacked by our immune system (Breznitz 2001, p. 94): because of subthreshold kinetics the slower rate of growth increases the chance of avoiding detection by the immune system.[32]

The false alarm phenomenon, which Breznitz explored as a psychologist and to which he devoted a whole book (Breznitz 1984), is not valid for fully automated warning systems; it is rather valid for an entirely different kind of alarm systems in which decision-making or the evaluation of alternatives and learning from previous experiences take place (Breznitz 2001, p. 87).[33] Such, obviously, is our psychological alarm system. Only in such, non-fully automated warning systems, may the phenomenon of false alarm based on previous experiences occur. The gist of the matter is that Breznitz finds evidence for the existence of the false alarm phenomenon not only in human behavior but also in the behavior of the immune system (which is well compatible with Matzinger's danger model). In this system, too, the "false alarm phenomenon" is not a metaphor but its sense is fully literal. In other words, or in the terms of my psychophysical assumptions, our immune system *implements or embodies* processes of decision-making, evaluations of alternatives, and learning. The difference between this biological embodiment and that of our similar psychological processes of *conscious* decision-making, choosing and deciding between alternatives, and of learning is quite clear. Our immune system needs no consciousness to perform its tasks. While our immune system makes some choices and learns from previous experience, there is no deliberation which is subject to consciousness.[34] The phenomena of detection, recognition, processing information, and alienation are also literally valid for our immune system to the extent that none of them requires consciousness. This holds true for alarms, threats, signals, memory, inter-cell communication, stress, and distress, as these terms are used in immunology.

---

[32] Cf. Matzinger's explanation: "An early growing tumor is a healthy tissue not sending alarm signals, and therefore is constantly inducing tolerance to itself" (Matzinger 2012, p. 316).

[33] Cf. "T cells and B cells are central to the cognitive enterprise because they can learn from experience" (Cohen 2000, p. 107).

[34] Yet consider the following as regards *unconscious* deliberation in the immune system: "Co-respondence can even help some of the logic of immune anatomy. Lymph nodes can be viewed as the courts where immune agents can gather to present their findings for communal co-respondence *in camera*, secluded from distractions in the tissue arena of action. Once their mutual deliberations lead to a joint immune decision, the agent can exit the lymph node to return to the blood for delivery of their verdict to the tissues. Lymph nodes are not only district offices ..., they can function as local, ad hoc, brains" (Cohen 2000, pp. 161–162). Here, too, Cohen makes some analogies between the brain and the immune system. When it comes to cognitive and psychological terms, these analogies make much sense, for undoubtedly the brain embodies psychological traits and states. The difference is that while the brain can also embody conscious psychological traits, the immune system cannot do so.

To show that the psychological terms and the psychological traits that have been ascribed above to the immune system cannot be ascribed to inanimate systems, such as computers or any software (which are fully automated systems),[35] we have to refer again to the issues of biological identity, uniqueness, and individuation to the extent that the immune system is concerned. No immune system of a higher individual multicellular organism can be duplicated or replicated, for each such a system is unique, bearing its single identity. Unconscious decisions, recognition, a sense of danger or threat, and all the other abovementioned psychological traits are applicable to the immune system of higher organisms but never to inanimate beings, none of which is unique, irreplicable, or unduplicable. Any hardware or software is replicable or duplicable; no hardware or software is unique in principle. It is the nature of such devices that they are duplicable or replicable. In contrast, my finger prints are unique and there cannot be any other person who has the same finger prints. My immune system is unique; nobody else can share it with me. This holds true also for so-called "identical" twins.

As for the individuation function that Pardeu ascribes to any immune system (Pradeu 2013, pp. 90–92), including that of unicellular "organisms" and "superorganisms," any biological individuality ascribed to such a system does not entail subjecting it literally, and not only in a metaphorical sense, to psychological traits or to cognitive and psychological terms. Only the immune system of an organism that is a biological embodiment of a person, namely, of a singular subject, is allowed to be subject non-metaphorically but rather literally to psychological traits and to cognitive and psychological terms.

Psychological subjects, each of which is singular, can be thus embodied only as organisms that have a higher biological individuality and identity. When we ascribe psychological traits to inanimate beings, such as computers, we are dealing only with metaphors, not with the literal sense of these terms. Thus, we ascribe "(artificial) intelligence," "memory," "information processing," "thinking," "decisions," or "intentions" to computers metaphorically only. We borrow the terms from the concepts pertaining to our mind and metaphorically apply them to human-made machines. The intelligence, memory, meaning, thinking, deciding, and intending in the case of these machines pertain literally only to the human beings who planned and produced the hardware or the software. Still, consciousness is not a necessary condition for applying these terms literally. Thus, though our immune system is not a conscious system, it implements and performs unconscious psychological processes, and we are entitled to ascribe to it not only beauty but also admirable wisdom.

---

[35] Even though Cohen argues that the immune system uses a computational strategy to carry out its function and that reframing our view of this system in computational terms is worth our while, he, nevertheless, clearly emphasizes that there is "a fundamental difference between the computations performed by the immune system and those done by a computer. A Turing machine is not modified by either its input or its output; it simply functions according to a preset program. The immune system of every individual, in contrast to a Turing machine, is self-organizing; it learns from experience; it has memory" (Cohen 2007, p. 570). Cf. Hershberg 2012, p. 29.

I have argued above that psychological subjects, who are singular beings, are embodied as organisms, each one of which has a unique identity. When it comes to our immune system, it is inevitable to emphasize that biological uniqueness and identity are indispensable, hence we cannot dispense with the self-nonself distinction.

The self-nonself distinction (in any of its variations, including Cohen's immunological homunculus) is inevitable in understanding and explaining our immune processes. Insofar as we must apply to them the abovementioned cognitive or psychological terms and psychological traits, the distinction self-nonself should be valid for the immune system. Of course, the danger theory is equally indispensable, but it is not sufficient to justify the claim that only some biological embodiments which have a biological individuality and uniqueness can be of singular, psychological subjects, which require the self-nonself distinction. Note that Matzinger, although following consistently the danger theory/model, does not dispense with the self-nonself distinction entirely. After all, the damage in question is of a multicellular individual creature not of a collective of whatsoever kind. There is no immune system of a species or a genus. The immune system protects biological individuals.[36] When it comes to human individuals the term "self" is entirely in place.

The singularity of each mind implies private accessibility, which means that each person has exclusive access, first of all epistemic, to his or her mind (Gilead 2003, pp. 43–75, 2009, 2011). No one else, whether omniscient or not, has any access to any other mind. I have no access to the thoughts, emotions, feelings, sensations, and volitions of any other person. Equally, my immune system cannot perceive and respond to the signals pertaining or addressed to the immune system of other person. Following the approach that Irun Cohen suggests, my immune system can defend and maintain only my body; it has no access to the body of any other person.[37] *The psychological accessibility and the immune accessibility are thus entirely compatible and they are, like the psychophysical unity as a whole, inseparable.* Contrary to the view of Tauber and others, at least the borders of the self, physically

---

[36] For a contrary example see Cohen 2000, pp. 251–252, as there are circumstances in which the immune system kills the individual for the benefit of the species. How is this compatible with Cohen's leitmotiv that the story of the immune system is the "tale of cognitive individuality" and "the story of the self"? He himself raises the question and tries to answer it: "Evolution is supposed to work on individual survival …, not on group survival. How could a species have ever evolved an immune program to kill the individual? … Just note that the immune system is a contractor of apoptosis for sick cells …, and sick individual, too, when the need arises" (Cohen 2000, p. 252). Nevertheless, sick cells are not individuals, at least not in the higher or complex sense that human beings undoubtedly are. As for the issue of social immunity, consider also the papers mentioned in a footnote above.

[37] Only the case of the fetus is different to some extent, and the immune system of the mother takes part in its immunity. Yet this does not allow her access to the mind of the fetus as they are two distinct psychological subjects. Furthermore, breastfeeding shares, as does blood transfusion to some extent, some of the antibodies of one person with the other but all such antibodies are mobilized to the service of the unique immune system of the other, receiving individual.

and psychologically, are well defined, and at least from this respect, the immune self is quite clear and sound. Moreover, the private accessibility concerning our mind, on the one hand, and our immune system, on the other, is more than enough to safeguard the idea of the self in general and the immune self in particular.[38]

## 15.8 Immunology and Panenmentalism: Why Does Our Immune System Actualize Psychical and Cognitive Possibilities?

Let me put my approach to the above in panenmentalist terms.

According to panenmentalism, every actuality is physical. The combination actual-physical requires some explanation. Panenmentalism begins its investigation in dealing with the psychophysical problem. It offers a novel way of dealing with this time-honored problem—the body, as a physical entity, is the actuality of the psychical pure possibility that is the mind. In this way, the psychophysical unity and the psychophysical irreducibility (namely, the mind is irreducible to the body and vice versa) are both well kept. Furthermore, the conditions, especially the spatio-temporal and causal ones, to which the actual and the physical are subject, are the same. Hence, the only possible actualization of any pure possibility, psychical or otherwise, is physical. By "physical" I mean strictly physical, chemical, biochemical, or biological.

Each individual actual existent—actuality—is an actualization of an individual pure possibility. Such an individual pure possibility serves as the identity of the relevant actuality, thus it is the pure possibility-identity of that actuality. Indeed, there is no entity without identity, yet, unlike Quine's precept (Quine 1969, p. 23), the identity in discussion is individual pure possibility. No two individual pure possibilities can be identical otherwise they would have been one and not two. As exempt from any spatiotemporality, individual pure possibilities are inescapably subject to the metaphysical principle of the identity of the indiscernibles. Individual pure possibilities, however, can be similar unless they are psychical possibilities pertaining to different psychical subjects, i.e., persons. Each psychical possibility is singular, namely, it cannot be similar to the psychical possibilities of other persons. Such is the nature only of psychical possibilities, which distinguishes them from all the other individual pure possibilities. The psychical is a singular section of the purely possible as a whole. Panenmentalism endorses psychical anomalousness,

---

[38] This view may shed a surprising light on the following ideas of Pradeu concerning the immune system: "The immune system offers a *principle of inclusion* ..., because it establishes what is rejected and what is not rejected by an organism. In so doing, the immune system determines which constituents stick together and thus are parts of one and the same organism. In addition to this exclusion-inclusion mechanism, the immune system is truly 'systemic' in the sense that ... it exerts its activity everywhere in the organism, insuring the unity and the cohesiveness of the organism as a whole" (Pradeu 2013, pp. 79–80).

namely, the psychical is anomalous—it is not subject to laws—whereas the actual-physical is necessarily subject to the laws of nature.

The singularity of each person, as a psychical subject, is a central idea in panenmentalism (to begin with Gilead 1999 and as elaborated in Gilead 2003). Since the actualization of any pure possibility, singular or not, is subject to some similarity according to the laws of physics, chemistry, biochemistry, and biology, psychical singularity is actualized merely as physical *uniqueness*. For instance, each human being is a singular person, sharing no similarity with any other person, whereas each person has, for instance, a unique brain, a unique immune system, unique finger prints, and the like. Any finger prints share some common properties with the finger prints of other persons, but each person has unique finger prints, which pertain to him or her alone. The same holds for our immune system—the immune systems of all human beings share some similarities, yet each human being has a unique immune system that cannot be shared with other human beings, even if the other human being is an "identical" twin.

Persons are systematic and coherent complexes of interconnected and interdependent heterogeneous parts, all sharing one and the same psychical reality, privately accessible. Hence, persons can be actualized only as higher, multicellular organisms of a special kind, capable of self-awareness and of relating to other persons, each of whom has a unique immune system. Each such organism is a unique, irreplicable biological individual. The biological individuality mentioned and discussed in this Chapter is thus only of persons or "selves" actualized as such organisms.

## 15.9   Conclusions

On the grounds of my psychophysical assumptions, the unity or inseparability of the self, as a psychological subject, and the unique immune system of each individual, the immune self, is well kept.

One of the lessons that this Chapter suggests is that immunologists may benefit significantly from borrowing insights and ideas suggested by metaphysicians, philosophers, psychologists, linguists, cognitive scientists, and artists, and vice versa.

I have demonstrated above that applying psychological traits and cognitive or psychological terms literally to our immune system, in any of the different immunological studies discussed above, is perfectly legitimate, at least in the light of my psychophysical assumptions.

# References

Abboud, S., et al. (2015). A number-form area in the blind. *Nature Communications, 6*(6026), 1–9.

Adams, R. M. (1979). Primitive thisness and primitive identity. *The Journal of Philosophy, 76*, 5–26.

Ader, R., Felten, D. L., & Cohen, N. (Eds.). (2001). *Psychoneuroimmunology*. San Diego: Academic.

Aerts, D. (2010). A potentiality and conceptuality interpretation of quantum physics. *Philosophica, 83*, 15–52.

Aharonov, Y., & Gruss, E. (2005). *Two-time interpretation of quantum mechanics*, arXiv:quant-ph/0507269v1 28 July 2005.

Aharonov, Y., & Tollaksen, J. (2007). *New insights on time-symmetry in quantum mechanics*, arXiv:0706.1232v1 [quant-ph] 8 June 2007.

Aharonov, Y., Popescu, S., & Tollaksen, J. (2010). A time-symmetric formulation of quantum mechanics. *Physics Today, 63*(11), 27–32.

Aharonov, Y., Cohen, E., Grossman, D., & Elitzur, A. C. (2013). Can weak measurement lend empirical support to quantum retrocausality? *DPJ Web of Conferences, 58*, 01015. http://www.cpj-conferences.org

Aharonov, Y., Cohen, E., Gruss, E., & Landsberger, T. (2014). *Measurement and collapse within the two-state-vector formalism*, arXiv:1406.6382v1 [quant-ph] 24 June 2014.

Aharonov, Y., Cohen, E., & Shushi, T. (2016). Accommodating retrocausality with free will. *Quanta* 5:1, pp. 53–59. *arXiv:quant-ph/0006070v2* 4 Aug 2000.

Anderson, C. D. (1936, December 12). The production and properties of positrons. *Nobel Lecture*. http://www.nobelprize.org/nobel_prizes/physics/laureates/1936/anderson-lecture.pdf, pp. 365–376

Armstrong, D. M. (2004). *Truth and Truthmakers*. Cambridge: Cambridge University Press.

Armstrong, D. M. (2010). *Sketch for a systematic metaphysics*. Oxford: Oxford University Press.

Arnold, F. H. (1998). Design by directed evolution. *Accounts of Chemical Research, 31*, 125–131.

Arnold, F. H. (2001). Combinatorial and computational challenges for biocatalyst design. *Nature, 109*, 253–257.

Arstila, V., & Scott, F. (June 2011). Brain reading and mental privacy. *Trames, 15*(2), 204–212.

Atiyah, M. (1995). Address of the president. *Notes and Records of the Royal Society of London, 49*(1), 141–151.

Atlan, H., & Cohen, I. R. (1998). Immune information, self-organization, and meaning. *International Immunology, 10*, 711–717.

Ayer, A. J. (1971). *The problem of knowledge*. Harmondsworth: Penguin Books.

Azzouni, J. (1994). *Metaphysical myths, mathematical practice: The ontology and epistemology of the exact sciences*. Cambridge: Cambridge University Press.

A. Gilead, *The Panenmentalist Philosophy of Science*, Synthese Library 424,
https://doi.org/10.1007/978-3-030-41124-4

Azzouni, J. (2000). Stipulation, Logic, and Ontological Independence. *Philosophia Mathematica, 8*, 225–243.

Balaguer, M. (2011). Fictionalism in the Philosophy of Mathematics. In E. N. Zalta (Ed.), *The Stanford Encyclopedia of Philosophy (Fall 2011 Edition)*. http://plato.stanford.edu/archives/fall2011/entries/fictionalism-mathematics/

Bangu, S. (2008). Reifying mathematics? Prediction and symmetry classification. *Studies in History and Philosophy of Modern Physics, 39*, 239–258.

Barbieri, R., Hall, L. J., & Rychkov, V. S. (2006) Improved naturalness with a heavy Higgs: An alternative road to LHC physics. *Physical Review D* 74 (arXiv:hep-ph/0603188).

Bell, J. S. (1987). *Speakable and unspeakable in quantum mechanics*. Cambridge: Cambridge University Press.

Bemporad, J. R. (1991). Dementia praecox as a failure of Neoteny. *Theoretical Medicine and Bioethics, 12*(1), 45–51.

Bennett, K. (2005). Two axes of Actualism. *The Philosophical Review, 114*, 297–326.

Bennett, K. (2006). Proxy 'Actualism'. *Philosophical Studies, 129*, 263–294.

Bigaj, T. (2012). Ungrounded dispositions in quantum mechanics. *Foundations of Science, 17*, 205–221.

Bigelow, J. (1988). *The reality of numbers: A Physicalist's philosophy of mathematics*. Oxford: Oxford University Press.

Bigelow, J., & Pargetter, R. (1990). *Science and necessity*. Cambridge: Cambridge University Press.

Bindi, L., et al. (2009, June 5). Natural quasicrystals. *Science, 324*, 1306–1309. https://doi.org/10.1126/science.1170827

Bindi, L., et al. (2011). Icosahedrite, $Al_{63}Cu_{24}Fe_{13}$, the first natural quasicrystal. *American Mineralogist, 96*, 928–931.

Bindi, L. et al., (2012, January). Evidence for the extraterrestrial origin of a natural quasicrystal. *Proceedings of the National Academy*. www.pnas.org/cgi/doi/10.1073/pnas.1111115109

Bjerring, J. C. (2014). On counterpossibles. *Philosophical Studies, 168*, 327–353.

Bjorklund, D. F. (1997). The role of immaturity in human development. *Psychological Bulletin, 122*(2), 153–169.

Black, M. (1952). The identity of Indiscernibles. *Mind, 61*, 153–164.

Blass, T. (1999). The Milgram paradigm after 35 years: Some things we now know about obedience to authority. *Journal of Applied Social Psychology, 29*, 955–978.

Bles, M., & Haynes, J.-D. (2008). Detecting concealed information using brain-imaging technology. *Neurocase: The Neural Basis of Cognition, 14*(1), 82–92.

Blood, C. (2011). No evidence for particles. *arXiv:0807.3930v2 [quant-ph]* (the first version appeared in 2008).

Bloom, J. D., Labthavikul, S. T., Otey, C. R., & Arnold, F. (2006). Protein stability promotes evolvability. *PNAS, 103*(15), 5869–5874.

Brading, K., & Castellani, E. (2008). Symmetry and symmetry breaking. In E. N. Zalta (Ed.), *The Stanford Encyclopedia of Philosophy*. http://plato.stanford.edu/archives/fall2008/entries/symmetry-breaking/

Bramlett, D. C., & Drake, C. T. (2013). A history of mathematical proof: Ancient to the computer age. *Journal of Mathematical Sciences & Mathematical Education, 8*, 20–33.

Brandt, R. (2005). Comments on the question of the discovery of element 112 as early as 1971. *Kerntechnik, 70*(3), 170–172.

Braun, E. (2015). The unforeseen challenge: From genotype-to-phenotype in cell populations. *Reports on Progress in Physics, 78*, 1–51.

Breznitz, S. (1984). *Cry wolf: The psychology of false alarms*. Hillsdale: Laurence Erlbaum Associates.

Breznitz, S. (2001). Immunoalienation: A behavioral analysis of the immune system. *Annals of the New York Academy of Science, 935*, 86–97. https://doi.org/10.1111/j.1749-6632.2001.tb03473.x

Breznitz, S. (2012). *The Tapestry of life* (in Hebrew). Tel-Aviv: Hakibbutz Hameuchad.

Brumfiel, G. (2008, May 1). The heaviest element yet? *Nature*, published online https://doi.org/10.1038/news.2008.794

Brustad, E. M., & Arnold, F. H. (2011). Optimizing non-natural protein function with directed evolution. *Current Opinion in Chemical Biology, 15*, 201–210.

Bub, J. (1997). *Interpreting The Quantum World.* Cambridge: Cambridge University Press.

Bueno, O. (2014). Constructive empiricism, partial structures, and the modal interpretation of quantum mechanics. *Quanta, 3*(1), 1–15.

Bufill, E., Agusti, J., & Blesa, R. (2011). Human Neoteny revisited: The case of synaptic plasticity. *American Journal of Human Biology, 23*, 729–739.

Burgess, J. P. (2005). Review of Charles Chihara's. *A structural account of mathematics. Philosophia Mathematica, 13*, 78–113.

Campbell, K. (1983). Abstract particulars and the philosophy of mind. *Australasian Journal of Philosophy, 61*, 129–141.

Campos, D. G. (2009). Imagination, concentration, and generalization: Peirce on the reasoning abilities of the mathematician. *Transactions of the Charles Peirce Society, 45*(2), 136–156.

Cartwright, N. (1983). *How the laws of nature lie.* Oxford: Oxford University Press.

Cartwright, J. H. E., & Mackay, A. (2012). Beyond crystals: The dialectic of materials and information. *Philos Trans A Math Phys Eng Sci, 370*, 2807–2822.

Casullo, A. (1982). Particulars, substrata, and the identity of Indiscernibles. *Philosophy of Science, 49*, 591–603.

CERN. (1983). *Achievements with antimatter.* http://cern-discoveries.web.cern.ch/cern-discoveries/Courier/HeavyLight/Heavylight.html

Changeux, J.-P., & Connes, A. (1989). *Conversations on mind, matter, and mathematics* (M. B. DeBevoise, Ed. & Trans.). Princeton: Princeton University Press.

Chen, G., et al. (2007). Immune-like phagocyte activity in the social Amoeba. *Science, 317*, 678–681.

Cheng, T. (2015). Obstacles to testing Molyneux's question empirically. *i-Perception, 6*, 1–5.

Cheselden, W. (1727–1728). An account of some observations made by a young gentleman, who was born blind, or lost his sight so early, that he had no remembrance of ever having seen, and was couched between 13 and 14 years of age. *Philosophical Transactions, 35*, 447–450.

Cocchiarella, N. B. (2007). *Formal ontology and conceptual realism.* Dordrecht: Springer.

Cohen, I. R. (1992a). The cognitive principle challenges clonal selection. *Immunology Today, 13*, 441–444.

Cohen, I. R. (1992b). The cognitive paradigm and immunological homunculus. *Immunology Today, 13*, 490–494.

Cohen, I. R. (1994). Kadishman's tree, Escher's angels, and the immunological homunculus. In A. Coutinho & M. D. Kazatchkine (Eds.), *Autoimmunity: Physiology and disease* (pp. 7–18). New York: Wiley.

Cohen, I. R. (2000). *Tending Adam's garden: Evolving the cognitive immune self.* San Diego: Academic.

Cohen, I. R. (2007). Real and artificial immune systems: Computing the state of the body. *Nature Reviews/Immunology, 7*, 569–570.

Cohen, I. R. (2014). *A fresh look at evolution: Running with the Wolf of Entropy* (in preparation).

Coltheart, M. (2006a). What has functional neuroimaging told us about the mind (so far). *Cortex, 42*, 323–331.

Coltheart, M. (2006b). Perhaps functional neuroimaging has not told us anything about the mind (so far). *Cortex, 42*, 422–427.

Colyvan, M. (2011). Indispensability arguments in the philosophy of mathematics. In E. N. Zalta (Ed.), *The Stanford Encyclopedia of Philosophy* (Spring 2011 Edition). http://plato.stanford.edu/archives/spr2011/entries/mathphil-indis/

Contessa, G. (2010). Modal Truthmakers and two varieties of Actualism. *Synthese, 174*, 341–353.

Copenhaver, R. (2010). Thomas Reid on acquired perception. *Pacific Philosophical Quarterly, 91*, 285–312.

Costello, E. K., et al. (2012). The application of ecological theory toward an understanding of the human microbiome. *Science, 336*, 1255–1262.

Cowan, C., Reines, F., et al. (1956). Detection of the free neutrino: A confirmation. *Science, 124*, 103–104.

Cramer, J. G. (2015) *The transactional interpretation of quantum mechanics and quantum nonlocality.* arXiv: 1503.00039v1. https://arxiv.org/pdf/1503.00039.pdf

Cross, C. B. (1995). Max Black on the identity of Indiscernibles. *The Philosophical Quarterly, 45*, 350–360.

Dasgupta, D., Yu, D., & Nino, F. (2010). Recent advances in artificial immune systems: Models and applications. *Applied Soft Computing Journal.* https://doi.org/10.1016/j.asoc.2010.08.024.

Davidson, D. (1989). The Myth of the subjective. In M. Krausz (Ed.), *Relativism: Interpretation and confrontation* (pp. 159–172). Notre Dame: University of Notre Dame Press.

Davidson, D. (1991). Epistemology externalized. *Dialectica, 45*, 191–202.

Davidson, D. (1994). Knowing one's own mind. In Q. Cassam (Ed.), *Self-knowledge* (pp. 43–64). Oxford: Oxford University Press.

Davidson, D. (1996). Subjective, intersubjective, objective. In P. Coates (Ed.), *Current issues in idealism* (pp. 155–176). Bristol: Thoemmes Press.

Davidson, D. (2003). Quine's Externalism. *Grazer Philosophische Studien, 66*, 281–297.

Davis, P. J., & Hersh, R. (1998). *The mathematical experience.* New York: Martiner Books.

Day, M. V. (2012). *An introduction to proofs and the mathematical vernacular.* http://www.math.vt.edu/people/day/ProofsBook

Degenaar, M., & Lokhorst, G.-J. (2014). Molyneux's Problem. In E. N. Zalta (Ed.), *The Stanford Encyclopedia of Philosophy.* http://plato.stanford.edu/acrhieves/spr2014/entries/molyneux-problem/

Dellinger, F. (2010). Search for a superheavy nuclide with A = 292 and neutron-deficient thorium isotopes in natural Thorianite. *Nuclear Instruments and Methods in Physics Research B, 268*, 1287–1290.

Denkel, A. (1991). Principia Individuationis. *The Philosophical Quarterly, 41*, 212–228.

Dennett, D. C. (1987). *The intentional stance.* MIT-Bradford: Cambridge, MA.

Dennett, D. C. (1991). *Consciousness explained.* Boston: Little, Brown and Co..

Dennett, D. C. (2005). *Sweet dreams: Philosophical obstacles to a science of consciousness.* Cambridge, MA: MIT Press.

Descartes, R. (1984). *Meditations on First Philosophy.* In: *The Philosophical Writings of Descartes* (Vol. 2, J. Cottingham, R. Stoothoff & D. Murdoch, Trans.). Cambridge: Cambridge University Press.

Deutsch, D. (2004). It from Qubit. In J. D. Barrow, P. C. Davis, & C. L. Harper (Eds.), *Science and ultimate reality: Quantum theory, cosmology, and complexity* (pp. 90–102). Cambridge: Cambridge University Press.

Dieks, D. (2010). Quantum mechanics, chance and modality. *Philosophica, 83*, 117–137.

Dirac, P. (1928). The quantum theory of the electron. *Proceedings of the Royal Society of London A, 117*, 610–624.

Dirac, P. (1933). *Theory of electron and positrons: The nobel lecture* (pp. 320–325). http://www.nobelprize.org/nobel_prizes/physics/laureates/1933/dirac-lecture.pdf

Dixon, D. (1972). Some conditions of obedience and disobedience to authority. *Human Relations, 18*, 57–76.

Dorato, M. (2011). *Do dispositions and propensities have a role in the Ontology of Quantum Mechanics? Some Critical Remarks.* http://www.philsci-archive.pitt.edu/8482/1/drmadrid1final.pdf

Dretske, F. (1997). *Naturalizing the Mind.* Cambridge, MA: The MIT Press.

Du Sautoy, M. (2003). *The music of the primes.* New York: HarperCollins.

Du Sautoy, M. (2011). Exploring the mathematical library of babel. In J. Polkinghorne (Ed.), *Meaning in mathematics* (pp. 17–25). Oxford: Oxford University Press.

Dyson, F. J. (2002). *Random matrices, neutron capture levels, Quasicrystals and Zeta-functions Zero.* http://www.msri.org/realvideo/ln/msri/2002/rmt/dyson/1/

Earman, J. (2004). Laws, symmetry, and symmetry breaking: Invariance, conservation principles, and objectivity. *Philosophy of Science, 71*, 1227–1241.

Eddington, A. S. (1928). The deviation of stellar material from a perfect gas. *Monthly Notices of the Royal Astronomical Society, 88*, 352–369.

Einstein, A., & Rosen, N. (1935). The particle problem in the general theory of relativity. *Physical Review, 48*, 73–77. In short, ER.

Einstein, A., Podolsky, B., & Rosen, N. (1935). Can quantum-mechanical description of physical reality be considered complete? *Physical Review, 47*, 777–780. In short, EPR.

Emsley, J. (2011). *Nature building blocks: An A–Z guide to the elements.* Oxford: Oxford University Press.

Evans, G. (2002). "Molyneux's Question". Chapter 14. In A. Noë & E. Thompson (Eds.), *Vision and mind: Selected readings in the philosophy of perception* (pp. 319–350). Cambridge, MA: MIT Press/Bradford Book.

Farah, M. J. (January 2005). Neuroethics: The practical and the philosophical. *Trends in Cognitive Sciences, 9*(1), 34–40.

Farah, M. J., et al. (2009, January). Brain imagining and brain privacy: A realistic concern. *Journal of Cognitive Neuroscience, 21*(1), 119–127.

Feldman, F. (1973). Sortal Predicates. *Nous, 7*, 268–282.

Fermi, E. (1933). An attempt at a theory of 'Beta' ray emission. *Ricerca Scientifica, 4*, 491–495. Translated by L. Menthe, in http://www.zeroic.com/writing/neutrino.pdf

Field, H. (1980). *Science without numbers.* Princeton: Princeton University Press.

Fine, K. (2012). Counterfactuals without possible worlds. *The Journal of Philosophy, 109*(3), 221–246.

Fitch, G. W. (Ed.). (1996). Possibilism and Actualism. *Philosophical Studies, 84*, 107–320.

Folescu, M. (2016). Thinking about different nonexistents of the same kind: Reid's account of the imagination and its nonexistent object. *Philosophy and Phenomenological Research, 53*, 627–649.

Forrest, P. (1986). Ways worlds could be. *Australasian Journal of Philosophy, 64*, 1273–1285.

Fossum, E. R. (2016). The quanta image sensor: Every photon count. *Sensors, 16*(8), 1260–1285.

Foster, J. (1991). *The immaterial self: A defence of the Cartesian dualist conception of the mind.* London: Routledge.

Fraser, A., et al. (2006). Synesthesia: The prevalence of atypical cross-modal experiences. *Perception, 35*, 1024–1033.

French, S. (1995). Hacking away at the identity of Indiscernibles: Possible worlds and Einstein's principle of equivalence. *The Journal of Philosophy, 92*, 455–466.

French, S., & Redhead, M. (1988). Quantum physics and the identity of indiscernibles. *The British Journal for the Philosophy of Science, 39*, 233–246.

Freud, S. (1901). The psychopathology of everyday life. In *The Standard edition of the complete psychological works of Sigmund Freud.* (Trans. under the general editorship of James Strachey) London: Hogarth Press and the Institute of Psycho-Analysis [1953–1974]) (hereafter *SE*) vol. 6.

Freud, S. (1910). Five lectures on psycho-analysis. *SE* vol. 11, pp. 3 ff.

Freud, S. (1913). On beginning the treatment (Further Recommendations on the Technique of Psycho-Analysis, I). *SE* 12: p. 123 ff.

Freud, S. (1914). Remembering, Repeating, and Working-through (Further Recommendations on the Technique of Psycho-Analysis, II). in *SE*, pp.147 ff.

Freud, S. (1917). Introductory Lectures on Psychoanalysis, *SE* vol. 16, pp. 241 ff.

Freund, M. A., & Cocchiarella, N. B. (2008). *Modal logic: An introduction to its syntax and semantics.* Oxford: Oxford University Press.

Furka, Á. (1999). Combinatorial chemistry. *The Chemical Intelligencer, 5*(1), 22–27.

Furka, Á. (2007). *Combinatorial chemistry: Principles and techniques.* Budapest: The author's site http://members.iif.hu/furka.arpad/BookPDF.pdf

Gendler, T., & Hawthorne, J. (Eds.). (2002). *Conceivability and possibility.* Oxford: OUP.

Giaquinto, M. (1996). Non-analytic conceptual knowledge. *Mind, 105,* 249–286.

Giaquinto, M. (2007). *Visual thinking in mathematics: An epistemological study.* Oxford: Oxford University Press.

Gilead, A. (1985). Teleological time: A variation on a Kantian theme. *Review of Metaphysics, 38*(3), 529–562.

Gilead, A. (1999). *Saving possibilities: An essay in philosophical psychology* (Value Inquiry Book Series) (Vol. 80). Amsterdam: Rodopi.

Gilead, A. (2003). *Singularity and Other Possibilities: Panenmentalist Novelties* (Value Inquiry Book Series) (Vol. 139). Amsterdam: Rodopi.

Gilead, A. (2004a). Philosophical blindness: Between arguments and insights. *The Review of Metaphysics, 58,* 147–170.

Gilead, A. (2004b). How many pure possibilities are there? *Metaphysica, 5,* 85–103.

Gilead, A. (2005a). A Possibilist metaphysical reconsideration of the identity of indiscernibles and free will. *Metaphysica, 6,* 25–51.

Gilead, A. (2005b). Torture and singularity. *Public Affairs Quarterly, 19*(3), 163–176.

Gilead, A. (2009). *Necessity and truthful fictions: Panenmentalist observations* (Value Inquiry Book Series) (Vol. 202). Amsterdam: Rodopi.

Gilead, A. (2010). Actualist fallacies, from fax technology to lunar journeys. *Philosophy and Literature, 34,* 173–187.

Gilead, A. (2011). *The privacy of the psychical* (Value Inquiry Book Series) (Vol. 233). Amsterdam: Rodopi.

Gilead, A. (2013). Shechtman's three question marks: Impossibility, possibility, and quasicrystals. *Foundations of Chemistry, 15,* 209–224.

Gilead, A. (2014a). Pure possibilities and some striking scientific discoveries. *Foundations of Chemistry, 16,* 149–163.

Gilead, A. (2014b). We are not replicable: A challenge to Parfit's view. *International Philosophical Quarterly, 54,* 453–460.

Gilead, A. (2014c). Chain reactions, 'impossible' reactions, and panenmentalist possibilities. *Foundations of Chemistry, 16,* 201–214.

Gilead, A. (2015a). Self-Referentiality and two arguments refuting physicalism. *International Philosophical Quarterly, 55,* 471–477.

Gilead, A. (2015b). Can brain imaging breach our mental privacy? *Review of Philosophy and Psychology, 6,* 275–291.

Gilead, A. (2015c). Cruelty, singular individuality, and Peter the great. *Philosophia, 43,* 337–354.

Gilead, A. (2015d). Neoteny and the playground of pure possibilities. *International Journal of Humanities and Social Science, 5,* 30–39.

Gilead, A. (2015e). *The twilight of determinism: At least in biophysical novelties.* http://arxiv.org/ftp/arxiv/papers/1510/1510.04919.pdf

Gilead, A. (2016). Eka-elements as chemical pure possibilities. *Foundations of Chemistry, 18,* 1–16. https://doi.org/10.1007/s10698-016-9250-7.

Gilead, A. (2017). The philosophical significance of Alan Mackay's theoretical discovery of Quasicrystals. *Structural Chemistry, 28,* 249–256.

Gilead, A. (2018). Further light on the philosophical significance of Mackay's theoretical discovery of crystalline pure possibilities. *Foundations of Chemistry: Philosophical, Historical, Educational and Interdisciplinary Studies of Chemistry, 21*(3), 285–296.

Gillespie, R. J., & Hargittai, I. (2012). *The VSEPR model of molecular geometry.* New York: Dover.

Godfrey, L. R., & Sutherland, M. (1996). Paradox of Peramorphic Paedomorphosis: Heterochrony and human evolution. *American Journal of Physical Anthropology, 99,* 17–42.

Gonzales, G., & Lange, A. (2007). Cancer vaccine for hormone/growth factor immune deprivation: A feasible approach for cancer treatment. *Current Cancer Drug Targets, 7*, 229–241.

Gould, S. J. (1977). *Ontogeny and phylogeny*. Cambridge, MA: Harvard University Press.

Gregoire, T., Smith, D. R., & Wacker, J. G. (2004). What precision electroweak physics says about the SU(6)/Sp(6) little Higgs. *Physical Review D, 69*, 115008. (arXiv:hep-ph/0305275).

Gross, E. (2009, June 15). In search of the God particle. *PhysicaPlus: Online Magazine of the Israel Physical Society* 12. http://physicaplus.org.il/zope/home/1223030912/god_particle_en?curr_issue=1223030912

Gruss, E. (2000). *A suggestion for a teleological interpretation of quantum mechanics*. An MA Thesis instructed by Yakir Aharonov and Issachar Unna, Racah Institute of Physics, The Hebrew University, Jerusalem.

Hacking, I. (1975). The identity of indiscernibles. *The Journal of Philosophy, 72*, 249–256.

Hamilton, J., et al. (2010, April 9). Synthesis of a new element with atomic number Z=117. *Physical Review Letters, 104*.

Hargittai, I. (Ed.). (1992). *Fivefold Symmetry*. Singapore: World Scientific.

Hargittai, I. (1997). Quasicrystals discovery: A personal account. *The Chemical Intelligencer, 3*, 26–34.

Hargittai, I. (2010). Structures beyond crystals. *Journal of Molecular Structure, 976*, 81–86.

Hargittai, I. (2011a). 'There is no such animal (אין חיה כזו)'—Lessons of a discovery. *Structural Chemistry, 22*, 745–748.

Hargittai, I. (2011b). Geometry and models in chemistry. *Structural Chemistry, 22*, 3–10.

Hargittai, I. (2011c). Stubbornness: 'Impossible' matter. In *Drive and Curiosity: What Fuels the Passion for Science* (pp. 155–172). Amherst: Prometheus Books.

Hargittai, I. (2011d). Dan Shechtman's Quasicrystal discovery in perspective. *Israel Journal of Chemistry, 51*, 1–9. https://doi.org/10.1002/ijch.201100137.

Hargittai, I. (2017). Generalizing crystallography: A tribute to Alan L. Mackay at 90. *Structural Chemistry, 28*, 1–16.

Hargittai, I., & Hargittai, M. (2000). In our own image. In *Personal symmetry in discovery* (Vol. 8, pp. 151–157). New York: Kluwer Academic/Plenum.

Hargittai, I., & Hargittai, M. (2003). *Candid science III: More conversations with famous chemists*. Magdolna Hargittai (Ed.). London: Imperial College Press.

Hargittai, I., & Hargittai, M. (2004). *Candid science IV: Conversations with famous physicists* (pp. 32–63). London: Imperial College Press.

Hargittai, I., & Hargittai, M. (2009). *Symmetry through the eyes of a chemist* (p. 16). Berlin: Springer.

Hargittai, I., & Hargittai, M. (2010). Structures beyond crystals. *Journal of Molecular Structure, 976*, 81–86.

Hargittai, I., & Hargittai, M. (2011a). Dan Shechtman's Quasicrystal discovery in perspective. *Israel Journal of Chemistry, 51*, 1145–1148.

Hargittai, I., & Hargittai, M. (2011b). *Drive and curiosity: What fuels the passions for science*. Amherst: Prometheus Books.

Heisenberg, W. (1959). *Physics and philosophy: The revolution in Modern Physics* (World Perspectives vol. 16). London: Unwin.

Held, R., et al. (2011). The newly sighted fail to match seen with felt. *Nature Neuroscience, 14*, 551–553.

Hellman, G. (1989). *Mathematics without numbers: Towards a modal-structural interpretation*. Oxford: Oxford University Press.

Hellman, G. (1993). Review of Charles Chihara's *constructibility and mathematical existence*. *Philosophia Mathematica, 1*, 75–88.

Hellman, G. (2003). Does category theory provide a framework for mathematical structuralism? *Philosophia Mathematica, 11*, 129–157.

Hersh, R. (1993). Proving and explaining. *Educational studies in Mathematics, 24*, 389–399.

Hersh, R. (1997). *What is mathematics, really?* New York: Oxford University Press.

Hershberg, U. (2012). Life as a meshwork of selves: Interview with Uri Hershberg. *Avant, 3*, 26–36.

Herzberg, R. D., et al. (2006). Nuclear isomers in superheavy elements as stepping stones towards the island of stability. *Nature, 442*, 896–899.

Hoffmann, R. (1995). *The same and not the same*. New York: Columbia University Press.

Hopkins, R. (2005). Thomas Reid on Molyneux's question. *Pacific Philosophical Quarterly, 89*, 340–364.

Howes, M. (1998). The self of philosophy and the self of immunology. *Perspectives in Biology and Medicine, 42*, 118 ff.

Howkes, K. (2014). Primate sociality to human cooperation. *Human Nature, 25*, 28–48.

Ingram, D. (2015). Platonism, alienation, and negativity. *Erkenntnis, 81*, 1273–1285.

Isaacson, W. (2007). *Einstein: His Life and Universe*. New York: Simon and Schuster.

Jacobs, J. D. (2010). A power theory of modality: Or, how I learned to stop worrying and reject possible worlds. *Philosophical Studies, 151*, 227–248.

Jacquette, D. (2008). Mathematical proofs and discovery *Reductio ad Absurdum*. *Informal Logic, 28*, 242–261.

Janeway, C. A., Goodnow, A. C. C., & Medzhitov, R. (1996). Immunological tolerance: Danger— Pathogen on the premises! *Current Biology, 6*, 519–522.

Janeway, C. A., Goodnow, A. C. C., & Medzhitov, R. (2001). How the immune system works to protect the host from infection: A personal view. *Proceedings of the National Academy of Science of the USA, 98*, 7461–7468.

Jerne, N. K. (1974). Towards a network theory of the immune system. *Annals of Institute Pasteur/ Immunology, 125C*, 373–389.

Jerne, N. K. (1984a). The generative grammar of the immune system. In J. Lindsten (Ed.), *Nobel Lectures: Physiology or Medicine 1981–1990* (pp. 211–225). Singapore: World Scientific, 1993.

Jerne, N. K. (1984b). Idiotypic networks and other preconceived ideas. *Immunological Reviews, 79*, 5–25.

Jones, G. D. (2002). Detection of long-lived isomers in super-heavy elements. *Nuclear Instruments and Methods in Physics Research A, 488*(1–2), 471–472.

Juhl, C. (2012). On the indispensability of the distinctively mathematical. *Philosophia Mathematica*. https://doi.org/10.1093/phimat/nkr043.

Kafka, F. (1961). *Parables and paradoxes*. New York: Schocken Books.

Kant, I. (1992–2012). *The Cambridge edition of the works of Immanuel Kant* (P. Gayer and A. W. Wood, General Editors). Cambridge: Cambridge University Press.

Kant, I. (1998/1787). *Critique of pure reason* (P. Guyer and A. W. Wood, Trans.). Cambridge: Cambridge University Press.

Kastner, R. E. (2010). The quantum liar experiment in Cramer's transcendental interpretation. *Studies in History and Philosophy of Modern Physics, 41*, 82–92.

Kastner, R. E. (2013). *The transactional interpretation of quantum mechanics: The reality of possibility*. Cambridge: Cambridge University Press.

Kennedy, D. (2014). Neoteny, dialogic education and an emergent Psychoculture: Notes on theory and practice. *Journal of Philosophy of Education, 48*, 100–117.

Kerr-Lawson, A. (1997). Peirce's pre-logistic account of mathematics. In N. Houser et al. (Eds.), *Studies in the logic of Charles Sanders Peirce* (pp. 77–84). Indianapolis: Indiana University Press.

Kleiner, I. (1991). Rigor and proof in mathematics: A historical perspective. *Mathematics Magazine, 64*, 291–314.

Knox, E. (2016). Abstraction and its limits: Finding space for novel explanation. *Noûs, 50*, 41–60.

Kolb, D., & Marinov, A. (2004). *The Chemical separation of Eka-Hg from CERN W targets in view of recent relativistic calculations*, arXiv.org./pdf/nuclex/0412010pdf

Kostyghin, V. A. (2012). Superheavey elements: Existence, classification, and experiment. *UDC, 541*, 2. http://arxiv.org/ftp/arxiv/papers/1212/1212.1016.pdf

Kragh, H. S. (1990). *Dirac: A scientific biography*. Cambridge: Cambridge University Press.

Krause, D., & Bueno, O. (2010). Ontological issues in quantum theory. *Man, 33*(1), 269–283.

Kripke, S. (1980). *Naming and necessity*. Cambridge, MA: Harvard University Press.

Landini, G., & Foster, R. T. (1991). The persistence of counterexample: Re-examining the debate over Leibniz Law. *Noûs, 25*, 43–61.

Lanouette, W., & Silard, B. A. (1992). *Genius in the shadows: A biography of Leo Szilard: The man behind the bomb*. New York: Scribner.

Lapidus, M. L. (2008). *In search of the Riemann Zeros: Strings, fractal membranes and noncommutative Spacetimes*. Providence: American Mathematical Society.

Laudisa, F., & Rovelli, C. (2013). Relational quantum mechanics. In E. N. Zalta (Ed.), *The Stanford Encyclopedia of Philosophy* (Summer 2013 Edition). http://plato.stanford.edu/archives/sum2013/entries/qm-relational/

Laugwitz, D. (2008). *Bernhard Riemann 1826–1866: Turning points in the conception of mathematics* (A. Schenitzer, Trans.). Boston: Birkhaüser.

Lawrence, S., & Margolis, E. (2012). Abstracting and the origin of general ideas. *Philosopher's Imprint, 12*, 1–20.

Le Poidevin, R. (2005). Missing elements and missing premises: A combinatorial argument for ontological reduction of chemistry. *The British Journal for the Philosophy of Science, 56*, 117–134.

Leng, M. (2010). *Mathematics and reality*. Oxford: Oxford University Press.

Leng, M. (2011). Creation and discovery in mathematics. In J. Polkinghorne (Ed.), *Meaning in mathematics* (pp. 61–69). Oxford: Oxford University Press.

Levine, D., & Steinhardt, P. (1984). Quasicrystals: A new class of ordered structures. *Physical Review Letters, 53*, 2477–2480.

Levy-Tzedek, S., Riemer, D., & Amedi, A. (2014). Color improves 'visual' acuity via sound. *Frontiers of Neuroscience, 8*, 1–7.

Lidin, S. (2011). *Scientific background on the Nobel prize in chemistry 2011*. In http://www.nobelprize.org/nobel_prizes/chemistry/laureates/2011/advanced.html

Linsky, B., & Zalta, E. N. (1995). Naturalized Platonism versus Platonized naturalism. *The Journal of Philosophy, 92*, 525–555.

Linsky, B., & Zalta, E. N. (2006). What is Neologicism? *The Bulletin of Symbolic Logic, 12*, 60–99.

Littlefield, M. (May 2009). Constructing the organ of deceit: The rhetoric of fMRI and brain fingerprinting in Post-9/11 America. *Science, Technology, and Human Values, 34*(3), 365–392.

Locke, J. (1975) *An essay concerning human understanding* (The P. H. Nidditch edn) Oxford: Clarendon Press.

Lombardi, O., & Castagnino, M. (2008). A Modal-Hamiltonian interpretation of quantum mechanics. *Studies in History and Philosophy of Modern Physics, 39*, 380–443.

Lombardi, O., & Dieks, D. (2017). Modal interpretations of quantum mechanics. In E. N. Zalta (Ed.), *The Stanford Encyclopedia of Philosophy* (Spring 2017 Edition). https://plato.stanford.edu/archieves/spr2017/entries/qm-modal

Lombardi, O., et al. (2011). Foundations of quantum mechanics: Decoherence and interpretation. *International Journal of Modern Physics D: Gravitation, Astrophysics & Cosmology, 20*(5), 861–875.

Lord, E. A., Mackay, A. L., & Ranganathan, S. (2006). *New geometries for new materials*. Cambridge: Cambridge University Press.

Loux, M. J. (Ed.). (1979). *The possible and the actual: Readings in the metaphysics of modality*. Ithaca: Cornell University Press.

Lowe, E. J. (1995). The metaphysics of abstract objects. *The Journal of Philosophy, 92*, 509–524.

Lowe, E. J. (2006). Metaphysics as the science of essence. Presented at the conference *The Metaphysics of E. J. Lowe*. http://ontology.buffalo.edu/06/Lowe/Lowe.pdf

Lyon, P. (2006). The biogenetic approach to cognition. *Cognitive Processing, 7*, 11–29.

Lyon, P. (2007). From quorum to cooperation: Lessons from bacterial sociality for evolutionary theory. *Studies in History and Philosophy of Biological and Biomedical Sciences, 38*, 820–833.

Lyre, H. (2008). Does the Higgs mechanism exist? *International Studies in the Philosophy of Science, 22*, 119–133.

Mackay, A. L. (1976). Crystal symmetry. *Physics Bulletin, 27*, 495–498.

Mackay, A. L. (1981). De Nive Quinquangula: On the pentagonal snowflake. *Soviet Physics, Crystallography, 26*, 517–522.

Mackay, A. L. (1982). Crystallography and the Penrose pattern. *Physica A: Theoretical and Statistical Physics, 114*, 609–613.

Mackay, A. L. (1990). Crystals and fivefold symmetry. In I. Hargittai (Ed.), *Quasicrystals, networks, and molecules of fivefold symmetry* (pp. 1–18). New York: VCH Publishers.

Mackay, A. L. (1997). Lucretius or the philosophy of chemistry. *Colloids and Surfaces A: Physicochemical and Engineering Aspects, 129–130*, 305–310.

Mackay, A. L. (1997/1998). *Bending the rules*. lecture 1997/98, HLSI 24/02. http://met.iisc.ernet.in/~lord/webfiles/Alan/CVU12.pdf

Mackay, A. L. (2002). Generalized crystallography. *Structural Chemistry, 13*, 215–220.

MacKenzie, D. (1999). Slaying the kraken: The Sociohistory of a mathematical proof. *Social Studies of Science, 29*, 7–60.

MacKenzie, D. (2005). Computing and the culture of proving. *Philosophical Transactions of the Royal Society A, 363*, 2335–2350.

Maddy, P. (1980). Perception and mathematical intuition. *The Philosophical Review, 89*, 163–196.

Maddy, P. (2003). *Realism in mathematics*. Oxford: Oxford University Press.

Maddy, P. (2007). *Second philosophy: A naturalistic method*. Oxford: Oxford University Press.

Madell, G. (1988). *Mind and materialism*. Edinburgh: Edinburgh University Press.

Madell, G. (2003). Materialism and the first person. *Royal Institute of Philosophy Supplement, 53*, 123–139.

Maldacena, J., & Susskind, L. (2013). Cool horizons for entangled Black holes. *Fortshritte der Physik/Progress of Physics, 61*, 781–811.

Marinov, A., et al. (1971). Evidence for the possible existence of a superheavy element with atomic number 112. *Nature, 229*(1971), 464–467.

Marinov, A., Eshhar, S., & Kolb, B. (1987). Evidence for long-lived isomeric states in neutron-deficient $^{236}$Am and $^{236}$Bk nuclei. *Physics Letters B, 191*, 36–40.

Marinov, A., et al. (2003). New outlook on the possible existence of superheavy elements in nature. *Physics of Atomic Nuclei, 66*(6), 1137–1145.

Marinov, A., et al. (2010). Evidence for the possible existence of a long-lived superheavy nucleus with atomic mass number A=292 and atomic number Z=~122 in natural Th. *International Journal of Modern Physics E, 19*(01), 131–140.

Marmodoro, A., & Mayr, E. (2019). *Metaphysics: An introduction to Contmeporary debates and their history*. New York: Oxford University Press.

Marraffini, L. A., & Sontheimer, E. J. (2010). CRISPR interference: RNA-directed adaptive immunity in Bacteria and archaea. *Nature Reviews Genetics, 11*, 181–190.

Martinon, F., et al. (2009). The inflammasomes: Guardians of the body. *Annual Review of Immunology, 26*, 231–246.

Matthen, M., & Levy, E. (1984). Teleology, error, and the human immune system. *Journal of Philosophy, 81*, 351–372.

Matzinger, P. (1994). Tolerance, danger, and the extended family. *Annual Review of Immunology, 12*, 991–1045.

Matzinger, P. (2002a). The danger model: A renewed sense of self. *Science, 12*, 301–305.

Matzinger, P. (2002b). An innate sense of danger. *Annals of the New York Academy of Science, 961*, 341–342.

Matzinger, P. (2007). Friendly and dangerous signals: Is the tissue in control? *Nature Immunology, 8*, 11–13.

Matzinger, P. (2011). Tissue-based class control: The other side of tolerance. *Nature Reviews/ Immunology, 11*, 221–230.

Matzinger, P. (2012). The evolution of the danger theory. *Expert Review of Clinical Immunology, 8*, 311–317.

McEvoy, M. (2008). The epistemological status of computer-assisted proofs. *Philosophia Mathematica, 16*, 374–387.

McGinn, C. (1983). *The subjective view: Secondary qualities and indexical thoughts*. Oxford: Clarendon.

McKenna, M., & Coates, D. J. (2016) Compatibilism. In E. N. Zalta (Ed.), *The Stanford Encyclopedia of Philosophy* (Winter 2016 Edition). https://plato.stanford.edu/archives/win2016/entries/compatibilism/

McKinney, M. L., & McNamara, K. J. (1991). *Heterochrony: The evolution of ontogeny*. New York: Plenum Press.

McMichael, A., & Zalta, E. (1980). An alternative theory of nonexistent objects. *Journal of Philosophical Logic, 9*, 297–313.

McNulty, K. P. (2012). Evolutionary development in Australopithecus Africanus. *Evolutionary Biology, 39*, 488–498.

Melander, P. (1993). How not to explain the errors of the immune system. *Philosophy of Science, 60*(2), 223–241.

Menzel, C. (2011/2016). Actualism. In E. N. Zalta (Ed.), *The Stanford Encyclopedia of Philosophy* (Winter 2016 Edition). https://plato.stanford.edu/archives/win2016/entries/actualism/

Menzel, C. (2014). Actualism. In E. N. Zalta (Ed.), *The Stanford Encyclopedia of Philosophy* (Summer 2016 Edition).

Milgram, S. (1963). Behavioral study of obedience. *Journal of Abnormal and Social Psychology, 67*(4), 371–378.

Milgram, S. (1965). Some conditions of obedience and disobedience to authority. *Human Relations, 18*, 57–76.

Milgram, S. (1974a). *Obedience to authority: An experimental view*. New York: Harper and Row.

Milgram, S. (1974b, October 31). We are all obedient. *The Listener, 99*, 567–568.

Milgram, S. (1983). Reflections on Morelli's 'dilemma of obedience. *Metaphilosophy, 14*(3), 190–194.

Millgram, E. (2015). Chapter 7, "Lewis's Epicycles, possible worlds, and the mysteries of modality". In *The Great Endarkenment: Philosophy for an age of hyperspecialization* (pp. 155–187). Oxford: Oxford University Press.

Mills, C. D. (2012). M1 and M2 macrophages: Oracles of health and disease. *Critical Reviews in Immunology, 32*, 463–488.

Mislow, K., & Rickart. (1976). An epistemological note on chirality. *Israel Journal of Chemistry, 15*, 1–6.

Mixon, D. (1972). Instead of deception. *Journal for the Theory of Social Behavior, 2*(2), 145–177.

Mixon, D. (1989). *Obedience and civilization: Authorized crime and the normality of evil*. London/ Winchester: Pluto Press.

Molyneux, W. (1688). *A Letter to John Locke*. Bodleian Library, University of Oxford: MS Locke c. 16, fol. 92r.

Montagu, A. (1955). Time, morphology, and Neoteny in the evolution of man. *American Anthropologist, 57*, 13–27.

Montagu, A. (1956). Neoteny and the evolution of the human mind. *Explorations, 6*, 85–90.

Montagu, A. (1989). *Growing young*. New York: Greenwood Press.

Montagu, A. (2000). *The natural superiority of women*. Lanham: Rowman and Littlefield.

Moody, K. J. (2014). Synthesis of superheavy elements. In M. Schädel & D. Shaughnessey (Eds.), *The chemistry of superheavy elements* (2nd ed., pp. 1–81). Berlin: Springer.

Morelli, M. F. (1983). Milgram's dilemma of obedience. *Metaphilosophy, 14*(3–4), 183–189.

Nagel, T. (1979). *Mortal Questions*. Cambridge: Cambridge University Press.

Nagel, T. (1986). *A view from nowhere*. New York/Oxford: Oxford University Press.

Nelson, M., & Zalta, E. N. (2009). Bennett and 'Proxy Actualism'. *Philosophical Studies, 142*, 277–291.

Newstead, A., & Franklin, J. (2012). Indispensability without Platonism. In A. Bird et al. (Eds.), *Properties, powers, and structures: Issues in the metaphysics of realism* (pp. 81–97). New York/London: Routledge.

Norman, J. F., et al. (2004). The visual and haptic perception of natural objects shape. *Perception & Psychophysics, 66*, 342–351.

Norman, J. F., et al. (2006). Aging and the visual, haptic, and Cross-modal perception of natural object shape. *Perception, 35*, 1383–1351.

O'Leary-Hawthorne, J. (1995). The bundle theory of substance and the identity of Indiscernibles. *Analysis, 55*, 191–196.

O'Shea, D. (2007). *The Poincaré conjecture: In search of the shape of the Universe*. New York: Walker Publishing.

Orozco, L. (2007). Wave Goodbye. *Nature, 448*(August 2007), 872–878.

Panksepp, J. (1998). *Affective neuroscience: The foundations of human and animal emotions*. New York: Oxford University Press.

Parsons, C. (1982). Objects and logic. *The Monist, 65*, 491–516.

Parsons, C. (1990). The Structuralist view of mathematical objects. *Synthese, 81*, 303–346.

Parsons, C. (2008). *Mathematical thought and its objects*. Cambridge: Cambridge University Press.

Pascual-Leone, A., & Hamilton, R. (2001). The Metamodal Organization of the Brain. *Progress in Brain Research, 134*, 427–445.

Patten, S. C. (1977). Milgram's shocking experiments. *Philosophy, 52*(202), 425–440.

Pauli, W. (1994). *Writings on physics and philosophy* (C. P. Enz and K. von Meyenn, Ed., R. Shlapp, Trans.). Berlin: Springer.

Penrose, R. (1974). Role of aesthetics in pure and applied research. *Bulletin of the Institute of Mathematics and Its Applications, 10*, 266.

Perelman, Grisha (Grigory). (2002). *The entropy formula for the Ricci flow and its geometric applications*. http://arxiv.org/pdf/math/0211159v1.pdf

Pincock, C., Baker, A., Paseau, A., & Leng, M. (2012). Science and mathematics: The scope and limits of mathematical Fictionalism. *Metascience, 21*, 269–294.

Pittman, K., & Kubes, P. (2013). Damage-associated molecular patterns control neutrophil recruitment. *Journal of Innate Immunity, 5*, 315–323.

Poletaev, A. B. (2012). "Maternal immunity, pregnancy, and child's health." Chapter 3. In S. Sifakis (Ed.), *From preconception to postpartum* (pp. 42–56). Rijeka/Shanghai: InTech. http://www.intechopen.com/books/from-preconception-to-postpartum/maternal-immunity-pregnancy-and-childs-health

Poletaev, A. B., & Osipenko, L. (2003). General network of natural autoantibodies as immunological homunculus (immunculus). *Autoimmunity Rev, 2*, 264–271.

Poletaev, A. B., et al. (2008). Integrating immunity: The immunculus and self-reactivity. *Journal of Autoimmunity, 30*, 68–73.

Poletaev, A. B., et al. (2012). Immunophysiology versus immunopathology: Natural autoimmunity in human health. *Pathophysiology*. https://doi.org/10.1016/j.pathophys.2012.07.003.

Polkinghorne, J. (2011). Mathematical reality. In J. Polkinghorne (Ed.), *Meaning in mathematics* (pp. 27–34). Oxford: Oxford University Press.

Poole, C. P. (Ed.). (2004). *Encyclopedic dictionary of condensed matter physics, Vol 1*. Amsterdam: Elsevier.

Popper, K. R. (1968). *The logic of scientific discovery*. London: Hutchinson.

Pradeu, T. (2012). *The limits of the self: Immunology and biological identity*. Oxford: Oxford University Press.

Pradeu, T. (2013). Immunity and the Emergence of Individuality. In F. Buchard & P. Huneman (Eds.), *From groups to individuals: Evolution and emerging individuality* (pp. 77–96). Cambridge: MIT Press.

Priest, G., Tanaka, K., & Weber, Z. (2013). Paraconsistent logic. In E. N. Zalta (Ed.), *The Stanford Encyclopedia of Philosophy* (Fall 2013 Edition). http://plato.stanford.edu/archives/fall2013/entries/logic-pracosnsistent/

Prince, E. (Ed.). (2004). *International tables for crystallography, vol. C: Mathematical, physical, and chemical tables*. Dordrecht: Kluwer.

Putnam, H. (1972). *Philosophy of logic*. London: George Allen and Unwin.

Putnam, H. (1979). *Mathematics, Matter, and Method: Philosophical Papers, vol. I* (pp. 43–59). Cambridge: Cambridge University Press.

Putnam, H. (1991). Reichenbach's metaphysical picture. *Erkenntnis, 35*, 61–75.

Pyykkö, P. (2011). A suggested periodic table up to $Z\leq 172$, based on Dirac–Fock calculations on Atoms and Ions. *Physical Chemistry Chemical Physics, 13*, 161. https://doi.org/10.1039/c0cp01575j.

Pyykkö, P. (2012). Predicting new, simple inorganic species by quantum chemical calculations: Some successes. *Physical Chemistry Chemical Physics, 14*(43), 14734–14742. https://doi.org/10.1039/c2cp24003c. Epub 2012 Feb 15.

Quine, W. v. O. (1960). *Words and Object*. Cambridge, MA: M.I.T. Press.

Quine, W. v. O. (1969). *Ontological relativity and other essays*. New York: Columbia University Press.

Rabin, M. O. (1980). Probabilistic algorithm for testing primality. *Journal of Number Theory, 12*, 128–138.

Räikkä, J. (February 2010). Brain imaging and privacy. *Neuroethics, 3*(1), 5–12.

Räikkä, J., & Smilansky, S. (July 2012). The ethics of alien attitudes. *The Monist, 95*(3), 511–532.

Reichenbach, H. (1944). *Philosophic foundations of quantum mechanics*. Berkeley and Los Angeles: University of California Press.

Reichenbach, H. (1962 [1951]). *The rise of scientific philosophy*. Berkeley/Los Angeles: University of California Press.

Reid, T. (1997 [1764]). *An inquiry into the human mind: On the principles of common sense*. University Park: Pennsylvania State University Press.

Reines, F. (1995). The Neutrino: From Poltergeist to particle. *The Nobel Lecture*, pp. 202–221. http://www.nobelprize.org/nobel_prizes/physics/laureates/1995/reines-lecture.pdf

Rescher, N. (1999). How many possible worlds are there? *Philosophy and Phenomenological Research, 59*, 403–421.

Rescher, N. (2003). Nonexistents then and now. *The Review of Metaphysics, 57*, 359–381.

Restall, G. (2003). Just what is full-blooded Platonism? *Philosophia Mathematica, 11*, 82–91.

Rock, K. L., et al. (2005). Natural endogenous adjuvants. *Springer Seminars in Immunopathology, 26*, 231–246.

Rosen, G. (2011a). Comment on Timothy Gowers' 'Is mathematics discovered or invented'. In J. Polkinghorne (Ed.), *Meaning in mathematics* (pp. 13–15). Oxford: Oxford University Press.

Rosen, G. (2011b). *The reality of mathematical objects*. In op. cit., pp. 113–131.

Roskies, A. L. (2007). Are Neuroimages like photographs of the brain? *Philosophy of Science, 74*, 860–872.

Roskies, A. L. (2009). Brain-mind and structure-function relationships: A methodological response to Coltheart. *Philosophy of Science, 76*, 927–939.

Roskies, A. L. (2010). Saving subtraction: A reply to Van Orden and Paap. *British Journal for the Philosophy of Science, 61*, 635–665.

Rota, G.-C. (1997). The phenomenology of mathematical proof. *Synthese, 111*, 183–196.

Rota, G.-C. (2003). Nonexistents then and now. *The Review of Metaphysics, 57*, 359–381.

Sacks, O. (1993). A Neurologist's notebook: To see and not see. *The New Yorker*, May 10.

Sassen, B. (2004). Kant's on Molyneux's Problem. *The British Journal for the History of Philosophy, 12*, 471–485.

Sture, A. (Ed.). (1989). *Possible worlds in humanities, arts, and science*. Berlin: De Gruyter.

Saunders, N. T. (2000). Does god cheat at dice? Divine action and quantum possibilities. *Zygon, 35*(3), 517–544.

Saunders, S. (2013). Indistinguishability. In R. Batterman (Ed.), *Oxford handbook of philosophy of physics* (pp. 340–380). Oxford: Oxford University Press.

Scerri, E. (2013a). *A tale of seven elements*. New York: Oxford University Press.

Scerri, E. (2013b). Cracks in the periodic table. *Scientific American, 69*, 70–73.

Schädel, M., & Shaughnessey, D. (Eds.). (2014). *The chemistry of superheavy elements* (2nd ed.). Berlin: Springer.

Schäfer, L. (2006a). Quantum reality and the consciousness of the universe: Quantum reality, the emergence of complex order from virtual states, and the importance of consciousness in the universe. *Zygon, 41*, 505–532.

Schäfer, L. (2006b). A response to Ervin Laszlo: Quantum and consciousness. *Zygon, 41*, 573–582.

Schäfer, L. (2008). Nonempirical reality: Transcending the physical and spiritual in the order of the one. *Zygon, 43*, 329–352.

Schäfer, L., Ponte, D. V., & Roy, S. (2009). Quantum reality and ethos: A thought experiment regarding the Foundation of Ethics in cosmic order. *Zygon, 44*, 265–287.

Schwenkler, J. (2013). Do things look the way they feel? *Analysis, 73*, 86–96.

Seaborg, G. T. (1968). Elements beyond 100, present status and future prospects. *Annual Review of Nuclear Science, 18*, 53–152.

Searle, J. (1994). *The Rediscovery of the Mind*. Cambridge, MA: MIT Press.

Segrè, E. (1980 [2007]). *From X-rays to quarks: Modern physicists and their discoveries*. New York: Dover.

Shaner, D. E. (1989). Science and comparative philosophy. In D. E. Shaner, S. Nagatomo, & Y. Yasuo (Eds.), *Science and comparative philosophy* (pp. 13–98). Leiden: Brill.

Shea, B. T. (1989). Heterochrony in human evolution: The case for Neoteny reconsidered. *Yearbook of Physical Anthropology, 32*, 69–101.

Shears, T. G., Heinemann, B., & Waters, D. (2006, December 15). In Search of the Origin of Mass. *Philosophical Transactions: Mathematical, Physical and Engineering Sciences, 364*(1849), 3389–3405. http://www.jstor.org/stable/25190411

Shechtman, D., Blech, I., Gratias, D., & Cahn, J. W. (1984). Metallic phase with long range orientational order and no translation symmetry. *Physical Review Letters, 53*, 1951–1954.

Shimony, A. (1978). Metaphysical problems in the foundations of quantum mechanics. *International Philosophical Quarterly, 18*, 3–18.

Shimony, A. (1999). Philosophical and experimental perspectives on quantum physics (6th Vienna circle lecture). In D. Greenberger et al. (Eds.), *Epistemological and experimental perspectives on quantum physics* (pp. 1–17). Dordrecht: Kluwer.

Simchen, O. (2006). Actualist essentialism and general possibilities. *The Journal of Philosophy, 101*, 5–26.

Solzhenitsyn, A. (2001 [1968]). *Cancer Ward* (N. Bethell & D. Burg, Trans.). New York: Farrar, Straus, and Giroux.

Somel, M. et al. (2009, April 2009). Transcriptional Neoteny in the Human Brain, *PANS (Proceedings of the National Academy of Sciences of the United States of America)* 106: 14, 5743–5748.

Spector, F., & Maurer, D. (2009). Synesthesia: A new approach to understanding the development of perception. *Developmental Psychology, 45*, 175–189.

Spinoza, B. (1985). Ethics. In *The collected works of Spinoza* (E. Curley, Ed. & Trans.). Princeton: Princeton University Press.

Stalnaker, R. (2004). Assertion revisited: On the interpretation of two-dimensional modal semantic. *Philosophical Studies, 118*, 299–322.

Stalnaker, R. (2012). *Mere possibilities: Metaphysical foundations of modal semantics*. Princeton: Princeton University Press.

Stein, H. (1988). Logos, logic, and Logistiké. *Minnesota Studies in the Philosophy of Science, 11*, 238–259.

Steiner, M. (2002). *The applicability of mathematics as a philosophical problem*. Cambridge, MA: Harvard University Press.

Strange, C. (1995). Rethinking immunity. *Bioscience, 45*, 663–668.

Strawson, G. (1994). *Mental Reality*. Cambridge, MA: MIT Press/Bradford.

Strawson, G. (1996). *Freedom and Belief*. Oxford: Oxford University Press.

Strawson, G. et al. (2006). *Consciousness and its place in nature: Does Physicalism Entail Panpsychism?* (A. Freeman, Ed.). Exeter: Imprint Academic.

Stromberg, K. (1979). The Banach-Tarski Paradox. *The American Mathematical Monthly, 86*, 151–161.

Sture, A. (1989). Possible worlds in humanities, arts and sciences: *Proceedings of Nobel symposium 65*. Berlin: De Gruyter.

Suárez, M. (2004). Quantum selections, propensities and the problem of measurement. *The British Journal for the Philosophy of Science, 55*, 219–255.

Suárez, M. (2007). Quantum Propensities. *Studies in the History and Philosophy of Modern Physics, 38*, 418–438.

Susskind, L. (2016). Copenhagen vs. Everett, teleportation, and ER=EPR. *Fortshritte der Physik/ Progress of Physics, 64*(6–7), 551–564.

Tauber, A. I. (1997). *The immune self: Theory or metaphor?* Cambridge: Cambridge University Press.

Tauber, A. I. (1998). Conceptual shifts in immunology: Comments on the 'two-way paradigm'. *Theoretical Medicine and Bioethics, 19*, 457–473.

Tauber, A. I. (1999). The elusive immune self: A case of category errors. *Perspectives in Biology and Medicine, 42*: 459 ff.

Tauber, A. I. (2008). The immune system and its ecology. *Philosophy of Science, 75*, 224–245.

Tauber, A. I. (2012) The biological notion of self and non-self. In E. N. Zalta (Ed.), *The Stanford Encyclopedia of Philosophy* (Summer 2012 Edition). http://plato.stanford.edu/archives/ sum2012/entries/biology-self/

Tauber, A. I. (2013). Immunology's theories of cognition. *History and Philosophy of the Life Sciences, 35*, 239–264.

Taz, H. (2018). *A post* in https://hortenscleferrand.wordpress.com/2018/10/03/

Teller, P. (1983). Quantum physics, the identity of Indiscernibles, and some unanswered questions. *Philosophy of Science, 50*, 309–319.

Tolstoy, L. N. (1969) *Anna Karenin* (R. Edmonds, Trans.). Harmondsworth: Penguin.

Tomberlin, J. E. (Ed.). (1998). *Language, Mind, and Ontology: Philosophical Perspectives 12* (A Supplement to *Noûs*) (pp. 489–509). Boston/Oxford: Blackwell.

Tong, F., & Pratte, M. S. (2012). Decoding patterns of human brain activity. *Annual Review of Psychology, 63*, 483–509.

Tressoldi, P. E., Sella, F., Coltheart, M., & Umilta, C. (2012). Using functional neuroimaging to test theories of cognition: A selective survey of studies from 2007 to 2011 as a contribution to the decade of the mind initiative. *Cortex, 48*, 1247–1250.

Tugby, M. (2015). The alien paradox. *Analysis, 75*, 28–37.

Türler, A., & Pershina, V. (2013). Advances in the production of chemistry of the heaviest elements. *Chemical Reviews, 113*(2), 1237–1312.

Ugelvig, L. V., & Cremer, S. (2007). Social prophylaxis: Group interaction promotes collective immunity in ant colonies. *Current Biology, 17*, 1967–1971.

Unger, P. (1990). *Identity, consciousness, and value*. New York: Oxford University Press.

Vallicella, W. F. (1997). Bundles and indiscernibility: A reply to O'Leary-Hawthorne. *Analysis, 57*, 91–94.

Van Cleve, J. (2007). Reid's answer to Molyneux's question. *The Monist, 90*, 251–270.

Van Frassen, B. (1989). *Laws and Symmetry*. Oxford: Oxford University Press.

Van Inwagen, P. (1993). *Metaphysics*. Oxford: Oxford University Press.

Vance, R. E. (2000). Cutting edge commentary: A Copernican evolution? Doubts about the danger theory. *The Journal of Immunology, 165*, 1725–1728.

Vermaas, P. E. (1999). *A Philosopher's understanding of quantum mechanics: Possibilities and impossibilities of a modal interpretation*. Cambridge: Cambridge University Press.

Vetter, B. (2011). Recent work: Modality without possible worlds. *Analysis, 71*, 742–754.

Viereck, G. S. (1929, October 26). What Life Means to Einstein. *The Saturday Evening Post*, pp. 109–117.

Ward, J. (2008). *The frog who croaked blue: Synesthesia and the mixing of senses*. London: Routledge.

Westfall, R. (1980). *Never at rest: A biography of Isaac Newton*. Cambridge: Cambridge University Press.

Weyl, H. (1929). Consistency in mathematics. *The Rice Institute Pamphlet, 16*, 245–265.

Williamson, T. (1998). Bare Possibilia. *Erkenntnis, 48*, 257–273.

Williamson, T. (1999). Existence and contingency. *Proceedings of the Aristotelian Society Supplement, 73*, 181–203.

Williamson, T. (2000). The necessary framework of objects. *Topoi, 19*, 201–208.

Wilson, R. A. (2005). *Genes and the agents of life: The individual in the fragile sciences*. New York: Cambridge University Press.

Winnicott, D. W. (1971). *Playing and reality*. London: Tavistock Publications.

Winnicott, D. W. (1975). *Through paediatrics to psycho-analysis*. London: The Hogarth Press and the Institute of Psycho-Analysis.

Wolf Foundation. (2004). http://www.wolffund.org.il/full.asp?id=17

Wolpe, P. R., Foster, K. R., & Langleben, D. D. (2010). Emerging Neurotechnologies for lie-detection: Promises and perils. *The American Journal of Bioethics, 10*, 40–48.

Woodward, R. (2011). The things that Aren't actually there. *Philosophical Studies, 152*, 155–166.

Yagisawa, T. (2010). *Worlds and individuals, possible and otherwise*. Oxford: Oxford University Press.

Yost, D. M., & Kaye, A. L. (1933). An attempt to prepare a chloride or fluoride of xenon. *Journal of the American Chemical Society, 55*, 3890–3892.

Zalta, E. (1983). *Abstract objects: An introduction to axiomatic metaphysics*. Dordrecht: Reidel.

Zeh, H. D. (2010). Quantum discreteness is an illusion. *Foundations of Physics, 40*, 1476–1493.

Zimmerman, D. W. (1998). Distinct discernibles and the bundle theory. In P. van Inwagen & D. W. Zimmerman (Eds.), *Metaphysics: The big questions* (pp. 58–67). Malden/Oxford: Blackwell.

Zingrone, N. L., et al. (2010). Out-of-body experiences and physical body activity and posture: Responses from a survey conducted in Scotland. *The Journal of Nervous and Mental Disease, 198*, 163–165.

Zollikofer, C. P. E., & De León, M. (2013). Pandora's growing box: Inferring evolution and development of hominin brain from Endocasts. *Evolutionary Anthropology, 22*, 20–33.

Zuboff, A. (1981). The story of a brain. In D. Hofstadter & D. Dennett (Eds.), *The Mind's I: Fantasies and reflections on self and soul*. New York: Basic Books.

# Author Index

**A**

Abboud, S., 105
Adams, R.M., 27, 30, 34
Aerts, D., 192
Aharonov, Y., 204–216
Amedi, A., 110, 111
Anderson, C.D., 38, 39, 48
Arendt, H., 251, 252
Aristotle, 18, 191
Armstrong, D.M., 53, 61, 63, 82
Arnold, F.H., 134–138, 218
Atiyah, M., 116, 164
Atlan, H., 281
Ayer, A.J., 215, 218, 225
Azzouni, J., 85, 86, 89

**B**

Balaguer, M., 93
Bandel-Ruth, G. (Ruthie), viii
Bangu, S., 54, 56
Barbieri, R., 50
Bartlett, N., 126–130
Bell, J.S., 189, 193
Bemporad, J.R., 239
Bennett, K., 52, 114
Bernal, J.D., 122, 137, 141, 156, 157, 168
Bigaj, T., 193
Bigelow, J., 61, 106
Bindi, L., 170
Bjerring, J.C., 76
Bjorklund, D.F., 240
Black, M., 27–34
Blass, T., 251
Bles, M., 234
Blood, C., 191

Bloom, J.D., 136
Bracha, G. (Krasny), vii
Brading, K., 56
Bramlett, D.C., 72
Brandt, R., 176, 256
Braun, E., 279
Breznitz, S., 267, 268, 273–275, 280–284
Brout, R., 37, 38, 40–42, 140
Brumfiel, G., 179
Brustad, E.M., 136
Bub, J., 192
Bueno, O., 191, 192
Bufill, E., 239
Burgess, J.P., 89

**C**

Campbell, K., 102
Cartwright, J.H.E., 91, 158, 159
Cartwright, N., 91
Casullo, A., 27, 35
Changeux, J.-P., 90, 94
Chen, G., 269
Cheng, T., 110
Cheselden, W., 105, 109
Van Cleve, J., 105
Cocchiarella, N.B., 53, 115
Cohen, I.R., 204, 239–241, 243, 248, 270–276, 281–286
Coltheart, M., 218
Colyvan, M., 60, 81
Connes, A., 90, 94
Contessa, G., 114
Copenhaver, R., 105
Costello, E.K., 270
Cowan, C., 39, 48

© Springer Nature Switzerland AG 2020
A. Gilead, *The Panenmentalist Philosophy of Science*, Synthese Library 424,
https://doi.org/10.1007/978-3-030-41124-4

Cramer, J.G., 191
Crick, F., 129–131
Cross, C.B., 27

**D**
Dasgupta, D., 283
Davidson, D., 218, 266
Davis, P.J., 87
Day, M.V., 64
Degenaar, M., 109
Dellinger, F., 175
Denkel, A., 27, 34
Dennett, D.C., 276–279
Descartes, R., 135, 237, 267
Deutsch, D., 159
Dieks, D., 192, 194, 197
Dirac, P., 38–42, 44–46, 48–51, 54, 183
Dixon, D., 261
Dorato, M., 186
Dretske, F., 220
Du Sautoy, M., 65, 85, 90, 92
Dürer, A., 143, 151, 155, 169
Dyson F.J., 157, 158

**E**
Earman, J., 43
Eddington A.S., 156
Einstein, A., 145, 147, 200, 242, 246
Elitzur, A.C., 204
Emsley, J., 175, 176
Englert, F., 7, 37, 38, 40–42, 140
Euclid, 47, 117, 141, 147, 164, 168
Evans, G., 108

**F**
Farah, M.J., 218, 224, 228
Feldman, F., 74
Fermi, E., 39–42, 44–46, 48, 50, 51
Field, H., 82
Fine, K., 1, 63
Fitch, G.W., 12
Folescu, M., 102
Forrest, P., 106
Fossum, E.R., 196
Foster, J., 27, 34, 35
Fraser, A., 109
Van Frassen, B., 213

French, S., 21, 27, 34, 200
Freud, S., 223, 224, 243, 253, 256
Freund, M.A., 53, 115
Furka, Á., 122–123

**G**
Gamow, G., 133–134
Giaquinto, M., 46, 94
Gilead, A., 5, 6, 12, 21, 23, 28, 29, 31, 33, 43,
       49, 51, 53, 55, 66, 79, 89, 93, 95, 107,
       108, 114, 115, 118, 136, 137, 141, 151,
       154, 177, 181, 193, 200, 207, 209, 211,
       213, 218, 225, 228, 230, 247, 253, 257,
       259, 265–267, 272, 273, 278, 286, 288
Gillespie, R.J., 129
Godfrey, L.R., 239
Gonzales, G., 275
Gould, S.J., 239, 240
Gregoire, T.D.R., 50
Gross, E., 37
Gruss, E., 205, 207, 209

**H**
Hacking, I., 27, 34
Hamilton, J., 84, 109, 174, 175
Hargittai, I., 118–122, 124–134, 141, 142,
       144, 145, 149, 150, 152, 155, 156,
       162, 168–171
Hargittai, M., 124, 150, 167
Heisenberg, W., 186, 189–191
Held, R., 109, 110
Hellman, G., 89, 117, 166
Hersh, R., 72–76, 80, 87
Hershberg, U., 272, 285
Herzberg, R.D., 178
Hoffmann, R., 179
Hopkins, R., 105

**I**
Ingram, D., 106
Isaacson, W., 242

**J**
Jacobs, J.D., 1
Jacquette, D., 64
Jerne, N.K., 268, 275, 281

Jones, G.D., 178
Juhl, C., 60

**K**

Kafka, F., 5, 22, 23
Kant, I., 32, 34, 46, 61, 66, 77, 94, 96, 105,
106, 116, 117, 141, 165, 166, 168, 177,
204–210, 215, 241
Kastner, R.E., 189–191
Kennedy, D., 241
Kepler, J., 142, 143, 151, 153–155, 159, 169
Kerr-Lawson, A., 46
Kleiner, I., 72
Knox, E., 102
Kostyghin, V.A., 178
Kragh, H.S., 54
Krause, D., 191, 192
Kripke, S., 19, 44, 116, 220, 266

**L**

Landini, G., 27, 34, 35
Lapidus M.L., 157
Laudisa, F., 197
Laugwitz, D., 94
Lawrence, S., 108
Le Poidevin, R., 174
Leng, M., 84, 93
Levine, D., 140, 143, 145, 169–172, 177
Levy-Tzedek, S., 110
Lewontin, R., 276, 277
Lidin, S., 162, 163
Linsky, B., 106, 107
Littlefield, M., 234
Locke, J., 32, 75, 99–103, 105, 109–111
Lombardi, O., 186, 190, 192, 197
Lord, E.A., 150, 168
Lovelock, J., 124, 125, 279
Lowe, E.J., 1, 102
Lyon, P., 269
Lyre, H., 43, 44

**M**

Mackay, A.L., 4, 7, 108, 118, 120, 137–160,
163, 168–172, 177
MacKenzie, D., 72, 73
Maddy, P., 65, 66
Madell, G., 219, 220

Maldacena, J., 199
Marinov, A., 175–178, 180
Marinov-Cohen, R., 176
Marmodoro, A., 1
Marraffini, L.A., 269
Matthen, M., 270
Matzinger, P., 273, 274, 280–284, 286
Mayr, E., 1
McEvoy, M., 121, 166
McGinn, C., 220
McKenna, M., 205
McKinney, M.L., 239
McMichael, A., 53, 115
McNulty, K.P., 239
Melander, P., 270
Mendeleev, D.I., 7, 12, 13, 15, 127, 130, 147,
148, 178, 180
Menzel, C., 12, 52, 114, 211
Milgram, S., 249–262
Millgram, E., 1
Mills, C.D., 283
Mislow, K., 152
Mixon, D., 250, 251, 255
Molina, M.J., 43, 123–125
Molyneux, W., 99–111
Montagu, A., 240
Moody, K.J., 178
Mordechai, G. (Siegler), vii
Morelli, M.F., 252

**N**

Nagel, T., 219, 220
Ne'eman, Y., 37, 140
Nelson, M., 52, 114
Newstead, A., 82
Newton, I., 78, 208, 240, 242
Norman, J.F., 109

**O**

O'Leary-Hawthorne, J., 27
O'Shea, D., 66, 84, 87, 95
Orozco, L., 196

**P**

Panksepp, J., 240
Parsons, C., 53, 95–97, 114
Pascual-Leone, A., 109

Patten, S.C., 251, 252
Pauli, W., 39, 40, 45, 50
Penrose, R., 7, 142, 143, 145, 151, 152, 155,
      157, 163, 164, 166, 168, 169, 177
Perelman, G. (Gregory), 84
Pittman, K., 283
Plato, 16, 97, 145
Podolsky, B., 199
Poletaev, A.B., 275
Polkinghorne, J., 91
Poole, Ch.P., 156
Popper, K.R., 42
Pradeu, T., 266, 268, 272, 280, 281, 285, 287
Priest, G., 53, 78, 79, 114
Prince, E., 166
Putnam, H., 46, 60, 66, 67, 80, 92–94, 166,
      215, 278
Pyykkö, P., 181–183


Q
Quine, W.v.O., 3, 19, 24, 80, 106, 108, 166,
      196, 266, 287


R
Rabin, M.O., 72
Räikkä, J., 232
Redhead, M., 192, 200
Reichenbach, H., 215, 216
Reid, T., 103–105
Reines, F., 39, 40, 48, 50, 51
Rescher, N., 11–25, 28, 49
Restall, G., 89
Rosen, G., 91
Rosen, N., 199
Roskies, A.L., 218, 226, 228, 235
Rota, G.-C., 47, 84
Rowland, S.F., 43, 123–125
Rutherford, E., 49, 132, 181


S
Sacks, O., 109
Sassen, B., 105
Saunders, N.T., 193, 194, 202
Saunders, S., 196, 197, 202
Scerri, E., 174, 177, 178, 180–182
Schäfer, L., 191
Schwenkler, J., 110
Seaborg, G.T., 175, 176
Searle, J., 219, 220
Segrè, E., 39, 49
Shaner, D.E., 240

Shea, B.T., 239
Shears, T.G., 37, 42, 45, 50
Shechtman, D., 4, 7, 86, 118, 122, 124,
      127–129, 131, 132, 138, 140–147, 149,
      152, 154, 155, 160–172, 176, 177, 259
Shimony, A., 189
Siderer, Y., 176
Silard, B.A., 131
Simchen, O., 3–5
Solzhenitsyn, A., 123
Somel, M., 239
Spector, F., 109
Spinoza, B., 6, 90, 194, 253, 267, 279
Stalnaker, R., 52, 114
Stein, H., 88, 117, 166
Steiner, M., 39, 51, 54, 55, 183
Steinhardt, P., 140, 143, 145, 169–172, 177
Strange, C., 281
Strawson, G., 237, 238
Stromberg, K., 103
Suárez, M., 186, 192
Susskind, L., 199
Szilard, L., 43, 131–132, 181


T
Tauber, A.I., 268, 273–277, 279, 281, 286
Taz, H., 138
Teller, P., 195
Tolstoy, L.N., 21
Tong, F., 218
Tressoldi, P.E., 218
Tugby, M., 106
Türler, A., 175, 176, 178


U
Ugelvig, L.V., 269
Unger, P., 218, 225


V
Vallicella, W.F., 27
Vance, R.E., 280, 281
Vermaas, P.E., 189
Vetter, B., 1, 52, 114
Viereck, G.S., 242


W
Ward, J., 109
Watson, J.D., 120, 126, 129–131
Westfall, R., 242
Weyl, H., 117, 166

Williamson, T., 28
Wilson, R.A., 127, 133, 134, 276–279
Winnicott, D.W., 243–245
Wolpe, P.R., 234
Woodward, R., 52, 114


Y
Yagisawa, T., 54

Yost, D.M., 128, 129


Z
Zalta, E., 52, 53, 106, 114, 115
Zimmerman, D.W., 27
Zingrone, N.L., 275
Zollikofer, C.P.E., 239
Zuboff, A., 218

# Subject Index

**A**

Abstract, 13, 16–18, 46–48, 51, 53, 67, 76, 94, 95, 97, 102, 106–108, 114, 115, 121, 157, 183, 184, 190, 191, 228, 245, 255, 256, 271, 272

Abstractions, 17, 46–48, 51, 53, 60, 61, 101, 102, 106–108, 114, 117, 147, 164, 165, 220, 281

Acquaintance, tactile, 104, 111

Actualism, 11–25, 28, 52, 66, 89, 114, 118, 132, 138, 191, 197, 211, 213, 214, 251

Actualist fallacy, 19, 117, 127–129, 131, 138

Actualities, 2, 12, 28, 38, 59, 101, 113, 164, 173, 186, 228, 245, 251, 276

Actualization, 8, 11, 13–16, 22, 24, 28, 29, 31, 43, 48, 51, 52, 54, 60, 68, 69, 76, 82, 88, 103, 104, 115, 116, 121, 125, 126, 129, 130, 132, 146, 157, 173, 174, 182–184, 187, 189, 191, 195, 196, 198–203, 216, 228–230, 234, 236, 277, 287, 288

Alpha-helix, 129

*Anna Karenina* (Tolstoy), 21

*a posteriori*, 46, 54, 61, 94, 108, 115, 116, 121, 147, 166

*a priori*, 46, 47, 50–55, 61, 83, 94, 96, 101, 103, 105, 108, 114–117, 120, 121, 123, 128, 129, 133, 145–150, 165, 166, 172, 200–202, 214, 241

Art, 4, 18, 22, 23, 129, 164, 241–245, 247, 273

Artifact, 243

Artist, 243

Authority, 249, 250, 252–259, 261

Authority, malevolent, 253, 256, 258, 260

**B**

Bacteria, 125, 137, 156, 269–271, 274

Behavior, 41, 42, 78, 208, 209, 218, 221–223, 226, 231–234, 249, 251, 254, 256, 257, 260, 262, 264, 265, 268, 269, 271, 273, 277, 283, 284

Belief, 278

Big Bang, 48, 133–134

Blindness/blind, 33, 49, 77, 99, 101, 104, 107–111, 126, 131, 138, 163, 171, 222, 236, 261, 277, 279

Body, 4, 9, 45, 60, 68, 101, 162, 190, 218–221, 228–233, 235–238, 240, 253, 256, 264–267, 270–274, 276–283, 286, 287

Brain, 278

Brain imaging, 110, 217–238, 265

**C**

Calculation, 45–46, 63, 78, 94, 116, 117, 164, 165

Causal connection, 143, 170, 203

Causality, 17, 45, 80, 91, 200, 203–210

Causality, causal, 2, 3, 5, 6, 8, 9, 16, 17, 29, 30, 48, 49, 59, 60, 65, 68, 69, 75, 81, 83, 86, 88, 89, 95, 97, 102, 103, 115, 139, 143, 147, 149, 151, 153, 186, 187, 190, 197, 198, 200, 202–204, 206–209, 213, 236, 247, 282, 287

Causation, 203

Cause, 131, 190, 208, 209, 236, 237, 251, 252, 259, 260

CERN, 7, 37, 38, 44, 46, 50, 140, 192

Chain reactions, chemical, 181
Chain reactions, nuclear, 131–132
Chemistry, 13, 14, 43, 56, 113, 122–124, 126,
    128–130, 132, 134–137, 140, 145,
    147–150, 156, 159, 161, 162, 171,
    179–184, 188, 264, 280, 282, 288
Chemistry, combinatorial, 122–123
Chlorofluorocarbons (CFCs), 124, 125
Cinematic experience, 244
Cognition, cognitive terms, 246, 268–273,
    275, 276, 278
Complexity, 155, 159, 179, 268, 279
Conjectures, 40, 42, 44, 66, 84, 259
Consciousness, 53, 115, 218, 220, 229, 238,
    246, 269, 270, 272, 278, 284, 285
Contingency/contingent, 5, 18, 22, 23, 46–48,
    51, 54, 63, 79, 80, 82, 84, 88–90, 92,
    94, 100, 115, 116, 127, 129, 133–135,
    163, 174, 182, 188, 200–202, 204, 246,
    247, 253, 277
Control, 244
Creativity, 1, 91–93, 214, 240–245, 248, 264
Cross-modal sensory, 108, 109, 111
Crystallography, 4, 7, 56, 108, 113, 118, 121,
    137, 141–143, 146, 147, 149, 152,
    155–158, 160, 162, 166–171, 177
Crystallography, generalized, 141, 149,
    150, 156–159


D
Description, 21
Determinism, 23, 200, 202, 204–210, 213,
    215, 216, 241, 251, 253, 279
Discoverer, lonely, 127, 128
Discovery, 4, 14, 28, 37, 61, 108, 113, 139,
    161, 174, 188, 242, 259, 279
Disillusionment, 244
Distinctness, 32, 34, 76
DNA, 120, 122, 125, 126, 130–131, 137, 156,
    157, 159
Double helix, 120, 130
Dualism, 219, 237
Duality, 206–210


E
Elements, chemical, 7, 12, 14, 151, 173, 183
Elements, eka-, 5, 7, 12–15, 17, 19, 24, 29, 33,
    55, 126, 147, 151, 173–184
Empiricism, 108, 135, 192, 215
Entanglement, quantum, 197, 198,
    200, 202–204

Epistemic access, 2, 59, 83, 97, 101, 104–107,
    111, 177, 220, 223, 224, 226, 229,
    231–233, 248, 265, 267
Essence, 76, 142, 156, 168, 252
Evil, 252, 254
Evolution, 24, 53, 72, 115, 134–137, 148, 149,
    202, 204, 239–244, 246, 248, 269, 274,
    280, 286
Existent, 5, 6, 8, 41, 45, 49, 50, 60, 70, 82, 85,
    95, 184, 187, 197, 287
Experience, 243
External reality, 243, 244


F
Fantasies, 243
Fantasy, 244
Fiction, 1, 20–22, 44, 92, 95, 181, 219,
    230, 243
Freedom, 243


G
Geometry, 29, 45, 46, 60, 64, 77–79, 82, 87,
    92–94, 101, 103–105, 111, 116–118,
    129, 141, 146–148, 152, 155, 157, 164,
    165, 168


H
Higgs boson, 7, 8, 37, 38, 40–46, 48, 50, 51,
    57, 114, 139, 140, 188, 192
Hilbert space, 191
Hypothesis, 39, 40, 42, 44, 53, 61, 65, 87, 94,
    128, 157, 158, 279


I
Idealism, 66
Identity, 13, 27, 49, 62, 104, 115, 154, 179,
    186, 225, 247, 266
Illusion, 244, 278
Imagination, 2, 3, 8, 9, 19, 20, 28–30, 43, 51,
    52, 60, 83–86, 91, 92, 94, 97, 101, 102,
    104, 105, 107, 108, 116, 118, 121, 124,
    126, 134, 138, 151, 152, 154, 164, 165,
    177, 181, 183, 187, 188, 211, 212, 214,
    231, 241–248, 252, 278
Immune system, 225, 263–288
Immunity, 125, 225, 266, 268, 270, 274,
    281, 286
Indiscernible, the identity of indiscernible,
    27–35, 62, 179, 180, 193, 195, 200, 211

Indispensability argument, 80–82
Individual/individuality, 1, 11, 28, 41, 61, 101, 139, 171, 179, 186, 218, 243, 253, 264
Information, 100, 110, 111, 119, 130, 132, 133, 137, 149, 150, 157–159, 203–205, 218, 226, 227, 232, 235, 257, 270, 271, 276, 278, 281, 283–285
Innateness, innate, 100, 105, 107, 109–111
Instantiation, 43, 53, 61, 64, 82, 218, 219, 228
Intellect, 2, 3, 8, 9, 51, 52, 60, 83, 84, 86, 92, 97, 101, 102, 104, 105, 107, 108, 116, 118, 121, 124, 134, 145, 151, 152, 154, 165, 183, 187–189, 208, 241, 245, 246, 248
Interference, quantum, 202–204
Intermediate area, 243
Intermediate region, 243
Interpretation, 21
Intersubjective reality, 18, 19, 219, 220, 222, 223, 261, 267
Invention, 1, 15, 25, 41, 48, 145, 188, 217, 240
Irrational, 278

**L**
Language, 5, 14, 19, 22, 51, 88–89, 95, 111, 143, 146, 158, 166, 169, 179, 222, 241, 242, 246, 247, 252, 261, 268, 276, 281
Law, 22
Literature, literary, 17–19, 21–23, 95, 123, 190, 241, 244, 276–278, 283

**M**
Madness, 244
Magic, 278
Materialism, 266
Mathematics/mathematical, 1, 14, 28, 40, 59, 100, 114, 139, 163, 174, 187, 242
Measurement, 45–46, 63, 86, 190, 193, 194, 199, 201–206
Mental, 6, 21, 41, 125, 190, 191, 198, 201, 217, 218, 243, 259–262, 272
Mental reality, 201
Mind, 4, 20, 28, 40, 61, 101, 114, 145, 174, 186, 218, 246, 264
Mind-body inseparability, 237
Mind-independence/mind-independent, 1, 5–8, 53, 75, 86, 88, 92–94, 102, 105, 106, 114, 115, 118, 151, 187, 191, 211, 212, 220
Mind-reading, 229–233

Modality/modal, 1–3, 5, 6, 18, 19, 33, 38, 53, 54, 62, 63, 67, 72, 96, 100–111, 114, 150, 163, 166, 186, 187, 189, 192, 194, 197, 213
Modal metaphysics, 8, 19, 105
Model, 37, 45, 48, 50, 87, 108, 119–122, 129–134, 143, 145, 147, 149, 153–155, 170, 171, 181, 183, 274, 277, 280, 284, 286
Moral, 206–210, 232, 233, 241, 249, 251, 253, 256, 258, 260, 262
Morality, 206, 207, 210, 215, 232, 241, 254, 255, 260, 262

**N**
Naming, 22
Narrative, 21
Necessity/necessary, 2, 15, 29, 41, 100, 113, 158, 163, 173, 187, 227, 244, 253, 264
Neoteny, 239–248
Neurosis, 243
Nobel Prize, 7, 37, 126, 140, 161, 176
Noble gas compound, 126, 128–130
Numerically distinctness, 32, 34

**O**
Obedience and disobedience, 250, 251, 256
Object, 244
Objective, 244
Objectivity, 244
Omega-minus, 17, 56, 140, 188
Ontology, 1, 23, 25, 76, 191, 220
Order, 244
Organism, 129, 135, 155, 159, 239, 264–267, 269, 270, 272, 274, 276–281, 283, 285–288
Ostension, 21, 22
Ozone layer, 123–125

**P**
Panenmentalism, 193
Particular(ity), 21, 22
Periodic table, 7, 12–15, 23, 29, 55, 126, 134, 147, 148, 151, 173–184
Person, 5, 13, 14, 18, 22, 30, 99–101, 105, 109, 111, 137, 175, 194, 212, 218–230, 234–236, 240, 249, 250, 252–258, 262, 264–267, 272, 273, 275, 280, 285–288
Personality, 253–256, 272

Phenomenology, 6
Philosophy, 147
Physicalism, 225, 237–238, 266
Physics, classical, 187, 189, 192, 193, 195,
    197–201, 203, 206, 216
Play, 278
Playing, 23, 239–248, 261
Positron, 38, 39, 44, 45, 48, 50, 55, 57
Possibilism, 6, 12, 13, 15, 16, 19, 24, 25, 28,
    30, 52, 53, 66, 89, 106, 114, 115, 191,
    193, 211, 213, 215, 251, 266
Possibilities, mental, 6, 21, 190, 198
Possibilities, physical, 17, 20, 65, 82, 141,
    160, 178, 182, 261
Possibilities, psychical, 9, 21, 228–230, 236,
    262, 287
Possibilities, pure, 1, 11, 28, 48, 61, 101, 114,
    139, 163, 173, 186, 228, 242, 251, 287
Potential space, 243
Prediction, 39, 42, 44, 113, 133, 134, 137,
    143, 169, 173, 178, 181
Privacy, brain, 217–219, 232, 237, 238
Privacy, psychical, 217–219, 232, 237, 238
Projection, 244
Proof, 38, 47, 51, 60, 64, 72–75, 78–80, 84,
    85, 87, 90, 92, 94, 129, 140, 158, 162,
    229, 241
Protein, 119, 121, 122, 129, 135–137, 156,
    157, 283
Psychical, 6, 18, 41, 102, 198, 217, 243,
    253, 276
Psychical reality, 23, 223, 230, 243, 253,
    261, 288
Psychoanalysis, 243
Psychology, psychological terms, 263,
    267–268, 270, 273, 277, 282,
    284–286, 288
Psychophysical assumptions, 263–267,
    269–273, 280–288
Psychophysical identity, 229, 231, 237
Psychophysical problem, 219, 266, 287
Psychophysical reduction, 237, 266–267, 277
Psychophysical unity/inseparability, 9, 66, 68,
    226–231, 233, 236–238, 265, 267, 279,
    281, 286–288


Q
Quantum mechanics, 33, 56, 148, 181, 189,
    190, 192–194, 197–202, 204, 205, 208,
    210, 213–215
Quantum physics, 56, 192–195, 198, 199,
    201, 216

Quantum possibilities, 134, 159, 187, 189–201
Quantum reality, 186, 192–195, 197–202, 216
Quasicrystal, 4, 86, 118, 127, 140, 142–145,
    152, 154, 155, 157, 158, 160, 162–164,
    169–171, 177


R
Rationalism, 135
Realism, 11, 12, 16, 19, 24, 53, 54, 61, 62, 65,
    82, 91, 93, 115, 191, 204
Realism about individual pure possibilities, 93
Reality-testing, 243
Reductionism, 174, 199, 220, 278, 279
Reference, 21, 22
    direct r., 21
Relationality, 2, 14, 31, 48, 61, 115, 146, 165,
    177, 188, 222, 247, 261


S
Sanity, 244
Scientific discoveries, 42, 43, 113–138, 141,
    163, 168, 242
Singularity, 217, 226, 227, 236, 253, 256,
    258, 263–288
Social organisms, 269–270
Space, 6, 7, 32, 34, 47, 61, 65, 67, 74, 82, 93,
    95, 97, 105, 117, 118, 141, 142,
    147–149, 153–155, 157, 159, 162, 164,
    165, 168, 172, 179, 191, 197, 215, 243,
    246, 252
Spatiotemporal, 2, 3, 5, 6, 8, 9, 16, 17, 22,
    28–30, 32–34, 45, 48, 49, 53, 59, 61,
    62, 66, 68, 69, 75, 76, 80, 81, 83, 86,
    88, 89, 95, 97, 102, 103, 106, 115, 116,
    139, 146, 151, 153, 154, 165, 179, 186,
    187, 190, 191, 197–201, 203, 212, 225,
    229, 236, 247, 253, 282
Square circle, 72–80
Standard Model, 37, 38, 40–42, 45, 46, 48,
    50, 55–57
Stipulation, 4, 44, 86, 281
Subjectivity, 6, 84, 217, 219–222, 225, 226,
    231, 235, 236, 264
Subjects, 257
Substances, 198


T
Teleology, 204–207
Temporality/temporal, 34, 80, 89, 91, 136,
    149, 172, 204, 213, 247

Thinking machine, 278
Thought-experiment, 28–30, 33, 43, 44, 60,
        82, 100, 101, 103, 118, 121, 124–126,
        132, 242, 246
Time, 3, 12, 30, 41, 61, 101, 113, 142, 161,
        174, 196, 218, 241, 254, 269
Transference, 243
Transitional phenomena, 243
Truth, 2, 3, 16, 55, 60, 63–65, 71, 73, 78, 79,
        82, 90–93, 113, 116, 152, 171, 221,
        242, 247, 251, 260
Two-slit experiment, 202–204


U
Uniqueness, 14, 52–55, 225, 263–288

Unthinkable, 243


V
Valence shell electron pair repulsion (VSEPR)
        model, 129
Veridicality, 278
Visual-to-auditory sensory substitute devices
        (SSDs), 110, 111


W
Weaning, 244
Wolf Foundation, 37, 38
Worlds, 12

Printed in the United States
by Baker & Taylor Publisher Services